IRRIGATION SYSTEMS

DESIGN, PLANNING AND CONSTRUCTION

Adrian Laycock

www.cabi.org

Irrigation Systems: Design, Planning and Construction

Adrian Laycock, CEng, BSc(Eng), MSc(Agr Eng), FICE

CABI is a trading name of CAB International

CABI Head Office	CABI North American Office
Nosworthy Way	875 Massachusetts Avenue
Wallingford	7th Floor
Oxfordshire OX10 8DE	Cambridge, MA 02139
UK	USA
Tel: +44 (0)1491 832111	Tel: +1 617 395 4056
Fax: +44 (0)1491 833508	Fax: +1 617 354 6875
E-mail: cabi@cabi.org	E-mail: cabi-nao@cabi.org
Website: www.cabi.org	

A catalogue record for this book is available from the British Library, London, UK.

A catalogue record for this book is available from the Library of Congress, Washington, DC.

ISBN-13: 978 1 84593 263 3

Printed and bound in the UK by Cromwell Press, from copy supplied by the author

For Ghalia, my favourite Irrigation Engineer !

The Author

Adrian Laycock is a Civil Engineer with a Masters degree in Agricultural Engineering. He first studied hydraulics at Aberdeen University under Professor 'hydraulic Jack' Allen. A yearning for overseas travel, a lifelong interest in water, and a secret desire to be a farmer sent him first to work in Central Africa then back to study irrigation at the National College of Agricultural Engineering at Silsoe, England. For three and a half decades since then he has lived and worked in some 27 countries from West Africa to South East Asia, and has visited many others out of professional interest.

He lives on a small island on the West Coast of Scotland, which is about as far removed as you can get from the average irrigation scheme in a waterless desert. From there he runs a small consulting company and thanks to computers and modern telecommunications keeps in contact with the rest of the world.

PREFACE

Of all the confrontations man has engineered with nature, Irrigation schemes have had the most widespread and far-reaching impact on the natural environment. Over a quarter of a billion hectares of the planet are irrigated. Entire countries depend on irrigation for their survival and indeed for their very existence. Until the massive oil and infrastructure projects of the 20th century, the oldest and largest civil engineering projects of the world were always irrigation schemes. And until recently the technology and principles of design applied to their design and construction had changed hardly at all in 4,000 years.

But now Irrigation is changing fast. Or rather, it should be, if the author's provocatively worded exhortations to all those involved in the developing world to move with the times are heeded. Engineers used to rule the roost in planning and designing irrigation projects. They often did so without much thought for the consequences, or without much imagination or inclination to change the way things were done. Several projects executed in the latter half of the 20th century rapidly turned into ecological and social disasters, and countless others never worked properly. Because of these, engineers are frequently (and inadvisably) asked to take a back seat in the planning of new schemes. This should not be so, for irrigation is dependent more than anything else on getting the engineering right to begin with. But engineers have only themselves to blame for this widespread erosion of confidence in their capability. Most engineers still cling unthinkingly to ancient beliefs and customs for no better reason than "it was done this way before, so it must be right". Modern thinking on irrigation engineering has benefited from a cross-fertilisation of ideas from many other fields including social sciences, control theory, political economics, management and of course agriculture. Many engineers involved in irrigation design will be unfamiliar with some of the ideas presented in this book.

This is a book for anyone involved with irrigation schemes or canals in any walk of life, but especially for engineers and others working in developing countries where irrigation schemes are commonplace and frequently not working very well or difficult to manage. It is a book that draws on the author's 38 years' practical experience of irrigation in the developing world, and puts some of the more esoteric mysteries of hydraulic engineering into simple terms that may be understood by those on the periphery of engineering but in the thick of management and operation and policy-making that remain integral parts of an irrigation scheme long after the engineers have gone away.

For the engineers there is always something new to learn, and this book introduces some new ideas on the design of canals and pipeline systems as well as pulling together some important issues from further afield in the realms of social conflict, management, and political thinking. It goes into detail on parabolic canals and computerised design.

For students of development studies and engineering there is a wealth of ideas to be turned over now as a preparation for the real world where things don't always go according to the rules. So this is not purely a text book, but more of a practical guide on how to get by and avoid making some of the blunders that have bedevilled many irrigation schemes in the recent past.

The author is a great believer in learning from mistakes, especially his own. Throughout the book, candid anecdotes illustrate how not to do things as well as how to do them properly. It is not the mistake that matters; it's what is done about recognising it and rectifying it afterward that really counts.

Acknowledgements

This is my chance to say a big thank you to all those third-world farmers who've taught me so much over the past 40 years. And to all you irrigation people that I've had the good fortune to work with in the following countries:-

Albania, Greece, Syria, Italy, Kazakstan, Kirgystan, Uzbekistan, Indonesia, India, Pakistan, Afghanistan, Sri Lanka, Burma, Brunei, Malaysia, Sudan, Nigeria, Zambia, Zimbabwe, Botswana, Ethiopia, Tanzania, Belgium, Germany, Switzerland, Denmark, England and Scotland.

And even if we didn't quite see eye-to-eye all the time, at least maybe we all came away a little bit wiser, didn't we?

Quite a few people have helped me with timely criticism of the text and computer program:- Ian Smout, Robert van Bentum, Dave Meldrum, Tim Shreeve, Alison Shreeve, the late John Merriam, Mike Chegwin, Mike McKenzie, Grant Davids, Linden Vincent, Chris Cronin, to name a few. And thanks to John Hendry of Pict Design for the website and Stephen Hall for upgrading the software.

Website and CD

All the figures and photographs used in this book, and many more related by subject, are posted in full colour on the website www.adrianlaycock.com from where they may be downloaded and used free of charge. A CD is also available, which includes these and many other pictures at high resolution, plus drawings and computer routines useful in irrigation design. The software for canal design described in Chapter 10 is available on the CD, as a windows-based program that takes the pain out of designing canal sections and hopefully will encourage designers to be a bit more adventurous in choosing curved profiles. A demonstration program can be downloaded from the website.

Figures referred to in the text and numbered with the prefix *A* are to be found on the website and CD, but are left out of the book for reasons of space.

The CD may be ordered from the website and costs £100.

CONTENTS

LIST OF TABLES

LIST OF FIGURES

LIST OF SUPPLEMENTARY FIGURES AVAILABLE ON THE WEBSITE AND CD

(Supplementary figures are prefixed with the letter A and numbered with relevance to chapters).

CHAPTER 1   EVOLUTION AND A PRELUDE TO CHANGE

Irrigation changed little in the ten millennia leading up to the end of the 19ᵗʰ century. It played a part in the rise and fall of many civilisations and societies which depended on it for food, then as now. But the techniques of getting water onto crops hardly changed in all that time. The scenario of present-day irrigation is different, having endured a dramatic period of evolution through twentieth century upheavals in geopolitics and economic conditions. It is a scene on which much of the planet depends for its food, but which is marred by devastation, failed schemes, corruption, and an inability of many involved in design and planning to appreciate modern concepts of operation, management and construction. It is a scene which demands for its full appreciation not only knowledge from a wide range of professional disciplines but above all the ability to think laterally around its manifold problems. There are clear indicators for the future direction of irrigation, which lies in the hands of engineers, planners and decision makers in governments and financing agencies around the world. Whether irrigation in the 21st century succeeds or fails will depend on their adaptability and readiness to consider these ideas.

1.1   A World of Canals

There are at the beginning of the 21st century about 280 million hectares of irrigated land in the world, and another 170 million hectares of agricultural drainage projects. This half-billion hectares of irrigation and drainage mostly uses canals for water conveyance. A rough calculation based on an average 20 metres of canal per hectare gives 10 million kilometres of man-made canals, channels and ditches of one kind or another world-wide. Most of them are small, and if past performance is anything to go by, most of them will need upgrading, reconstruction or replacement in the next 20 years.

Canals have been used since irrigation first began to convey water from one place to another. The ancient irrigation civilisations of Sri Lanka, Egypt, China, Persia, India and the Roman Empire all used remarkably complex networks of canals, many of which still work today and the engineering of which makes some of our modern efforts look feeble in comparison. We may never know how the people of Parakrama Bahu's kingdom over one thousand years ago in Sri Lanka built the Yoda Ela Canal which ran half the length of the country, linking dozens of irrigation tanks along the way. In order to construct these canal systems through such rugged terrain they would have needed a means of accurate surveying as well as a profound understanding of engineering principles governing water flow. Two thousand years ago in what is now Syria, the Romans built canals in the air. They lifted water from the River Orontes with *nourias*, giant water wheels 15 metres in diameter, and ran it over great distances through intricately constructed stone aqueducts to irrigate crops and to serve the domestic needs of their townspeople. Nowadays we would do it slightly differently with pumps and pipelines, but the underlying engineering strategies remain unchanged. But the most astounding thing about these ancient machines is that they are still operating today, after two millennia.

In more recent times the ever-increasing need for irrigation water has steadily extended the ingenuity of engineers. The great canal systems of India and Pakistan developed under the British Administration in the 19th century were not only masterpieces of planning and administrative organisation, but extended the frontiers of civil engineering beyond anything that had gone before. The 20ᵗʰ century has seen the face of the planet changed forever by the construction of canals of one kind or another in every country of the world. It seems that no desert is too remote, no mountain too high, no rainforest too dense that a canal cannot be thrust across it or through it.

In some cases we have gone too far. The arrogance of the planners of the Karakum Canal, which takes a third of the flow of the Amu Darya and runs it 1100 kilometres into the Turkmenistan desert, has been a major factor in the demise of the Aral Sea. The Jonglei Canal in Upper Sudan would have had a similar effect on the Sudd swamps but for violent intervention by national political groups. The Great Man-Made River (in reality a pipeline, the largest ever built) brings fossil water from beneath the Sahara to the coastal plains of Libya at phenomenal cost, measured not only in petro-dollars but in terms of the exhaustion of its underground reservoir sometime in this century. Like the water in the Aral Sea, the fossil reservoir beneath the desert was seen as a resource to be used up.

Mega-projects like these are not all bad. Three hundred years ago the Rhine valley was little more than a continual swamp, and only massive amounts of river engineering works and canalisation have made it into the international lifeline its waterway is today. Pakistan is dependent for its very existence on its 16 million hectares of irrigation, which is the largest contiguous block of irrigated land in the world. Let us hope that centuries from now the engineers of the future will derive inspiration from at least some of the canals we build today.

The recent history of irrigation is none too inspiring. The Romans and the ancient Lankans would have laughed at some of our modern efforts at canal building. Over the past 25 years or so many billions of dollars have been spent on new and rehabilitated irrigation canals. The saddest thing of all is that much of the rehabilitation has had to be carried out on canals that are themselves almost new. The fact is that the majority of irrigation works built today do not last very long, sometimes not even two years, let alone two millennia.

figure 1.1   *Arch aqueduct on the Upper Swat Canal, Pakistan*

1.2   The Importance of Small Canals

Although big projects usually have some big canals, the vast majority of their canals are small, carrying less than 1 cumec[1]. When big schemes go wrong it is usually to its small canals that we should turn. These are the canals that do most of the work; they distribute the water to farmers, they are used by animals and people for drinking and bathing, they take up most of the effort and money spent on water management, and they lose most of the water that is lost from the whole scheme. The way in which these canals are conceived, designed and built will influence the success or failure of the entire scheme.

It is easy enough to build a small canal well enough to last a very long time, but there are many reasons why it may not survive its first decade. These include shortage of money, technical naiveté on the part of designers and decision makers, and deliberately built-in obsolescence. And small canals, because they are small, are often thought of as being cheap and unimportant, and not worthy of detailed attention from experienced engineers.

Suffice it to say that small canals are not getting the attention they deserve, and this book aims to rectify the situation. Chapter 11 presents a few horror stories from some modern irrigation schemes,

---

1 Cubic metre per second.

and throws more light on the thorny subjects of deliberate damage and deliberately bad design and construction. The underlying reasons are often of a sociological nature rather than a purely engineering one, although it is usually the engineers who get the blame, and rightly so, because engineering should be concerned with very much more than just cement and concrete.

figure 1.2   ***The parabolic Pehur High Level Canal, Pakistan***

1.3   All- purpose Canals

A canal is defined as an artificial watercourse, a duct or passage that conveys fluids. That means it can also be a pipe. In this book we shall be stretching the definition to include low-pressure and medium-pressure pipelines. Irrigation planning strategies are now starting to turn away from the supply-scheduled, government-run schemes almost universally applied in the 19th and early 20th centuries, toward schemes in which the farmers have more control over their water, and flexibility and choice in how much water to use and when. Pipelines offer advantages in irrigation distribution that canals alone cannot give. But their design entails a radical change in thinking on the part of irrigation planners. This book seeks to consolidate these new ideas into a format that can be easily applied to practical design.

Most irrigation schemes distribute their water through an open canal network. The concept is simple enough. Water flows at a rate which is governed by the canal size, roughness and longitudinal slope. Usually for irrigation we are trying to keep the water delivery point at as high an elevation as possible by minimising the canal longitudinal slope. But this has to be balanced against the cost involved in making the canal deeper and wider as the slope gets flatter. A wide variety of structures can be inserted in the canal to control water levels, discharges, turbulence, sediment content and velocity, to convey water around or across obstacles, and to measure the flow. All this may seem obvious and trivial, but there are a surprising number of canals around the developing world that actually try to flow uphill! This book aims to present straightforward concepts of water control in a format readily understood by non-engineers and those engineers not routinely engaged in irrigation design.

Although irrigation distribution is by far the commonest use of canals, there are many other applications including drainage, power generation, navigation, flood relief, industrial and domestic water supply and sewerage. Normally at least a part of the agricultural drainage system will be in the form of open channels. Its function is to remove flood water from excess rainfall and over-irrigation, and to control groundwater levels to prevent waterlogging in the soil. Flood relief schemes may enlarge or supplement natural arterial drainage channels with new canals in order to convey intermittent flood flows. Power canals may be used to convey water to a point where it can be dropped steeply and run through turbines. Canals may be used at various stages in industrial and urban drainage and sewerage, in which water is used to transport effluent in a system which is predominantly piped. And water supply for domestic and industrial use also employs canals to supplement a ubiquitous system of pipes.

In this book, although we are concentrating on irrigation canals, many of the principles apply equally well to channels which are used for other purposes. Irrigation canals, by virtue of their location and their symbiotic association with the people they serve, tend to generate a few special problems of their own which require their engineers to look beyond the normally-accepted rules of engineering in order for them to operate successfully.

1.4   Pipelines- why and when

Pipelines are slowly gaining acceptance as a viable alternative to open canals as a means of distributing irrigation water. Low-pressure pipeline systems can lead to easier distribution and management. Land tenure problems can be lessened to the point of elimination with a pipeline, especially when a distribution system has to be routed through existing farmland having small, irregular and fragmented holdings. A pipe underground occupies no land that can be used for crops, nor does it interfere with land boundaries. Management losses, the biggest single contributor to canal water loss and low efficiencies, are potentially close to zero with a closed or semi-closed pipeline. Flexible delivery systems, in which the farmer is encouraged to take water as and when he requires it, are achievable with a pipeline but far more difficult with open canals.

Pipelines have their place in history too. In the Sri Lanka jungle can still be found the remains of ancient tanks[2] which feed into clay pipelines a thousand years old. At Palmyra in the Syrian desert there are well-preserved Romano-Greek stone pipelines for domestic water and garden irrigation, their spigot and sockets tightly jointed with hot wax.

The main factor in opting for a pipeline is the availability of head. For a given discharge capacity, a pipeline needs more head or level difference to operate than does an open channel. In flat terrain, a canal may be the only possibility for conveying water. In steep terrain, there may be excess head, which an open canal system would need to dissipate by means of drop structures. A pipeline in this case could utilise this excess head and might prove cheaper than a canal if the land is steep. The more head available, the smaller the diameter and cheaper the cost of the pipe.

However, for a multi-user flexible system it is important to maintain a stable operating head, which is most easily achieved at low-pressures and large diameters, usually with the inclusion of pressure-reducing valves. A pipeline also offers some built-in intermediate storage and a zero response time, which is a prerequisite for most demand-scheduled water management.

Their detractors usually cite cost of construction as the main reason why pipelines should not be used. It is true that pipelines may be more expensive to construct than canal systems. However, their economics are a different matter. If the potential benefits of a pipeline are higher than a canal (as they usually are) then the higher construction cost becomes insignificant when discounted over the lifetime of the project. Moreover, pipelines have so much to offer in terms of ease of management, flexibility of supply, economy of labour and efficiency of water use that it is short-sighted to dismiss the idea merely because they cost more to construct.

Most canal systems also utilise pipelines for certain structures such as aqueducts, culverts and inverted siphons, and the two types of distribution system may also be combined in the same scheme, with pipelines serving the lower end and canals feeding the points of intermediate storage or themselves providing storage.

On the Swabi SCARP irrigation rehabilitation scheme in Pakistan, the designers were initially requested to consider a pipeline-based demand system as a follow-up to a small pilot project. It soon

---

2 *'Tank' is the South Asian term for a reservoir behind an earth dam. Some of them are huge.*

became clear though that the drastic management upheavals involved in changing from the existing proportional-distribution canal system would lead to unacceptable social strife, and so the idea of a pipeline network with demand scheduling was dropped. A compromise was reached on Swabi by increasing the minor canal capacities to allow for some future conversion to a demand-scheduled system.

However, the adjacent Pehur High Level Canal project, which commenced just 3 years later[3], incorporated some pipeline distribution systems and automatic main canals. This milestone in the history of irrigation development in Pakistan was due to some forward thinking and a readiness to break with tradition on the part of the engineers concerned. Likewise the nearby Palai Dam project was also designed for pipeline distribution.

Attitudes of engineers usually veer toward the tried-and-tested rather than the search-for-a-better-way approach. The great temptation for most design engineers is to design something standard that is uncontroversial and has been done before and consequently that nobody will call into question. Low-pressure pipelines remain a mystery to most irrigation designers. Designing a pipeline is not merely a question of sizing a pipe for the same discharge that the equivalent canal would carry. The entire philosophy of moving water around the system needs to be rethought. The engineer needs to think like a farmer. This means considering more than just the physical problem of moving a certain volume of water from A to B. It has to be delivered to the field in such a way that it can be used to the greatest effect.

Low-pressure pipelines are possible when there is available head, and when there is intermediate storage in the system. Pipelines are desirable whenever flexibility in water use is required. Flexible supplies to the farmers can lead to more social harmony, better crop yields and less wastage of water. These ideas are developed in chapter 7.

1.5 Evolution of Irrigation Systems

Precepts of irrigated development are changing in several ways. Technical equipment is evolving in response to reducing availability of water and the need to use less water more efficiently than before. Geopolitics, as ever, has a large but vacillating role in financing and promoting many new developments. In many of the old irrigation societies the strategic functions of irrigated development are themselves changing, from merely preventing famine to efficiently producing surplus food and fibre.

We can trace a progression of development strategies that have affected the way that irrigation schemes are designed and expected to operate. First came ***run-of-river*** schemes, with diversion weirs directing water into open canals. Some of the biggest schemes in the world were conceived as run-of-river. They can irrigate only when water is available. Management methods range from the laissez-faire, do-nothing principle to sophisticated Warabandi scheduling when maximum command area is reached for the available water supply. As area increases, these schemes get ***congested*** and run out of *flexibility*[4].

Then came dams, ***impoundment and regulation*** schemes which store some of the water and release it when needed. Storage gives more *flexibility* for the farmers to irrigate whenever they want to. Storage at intermediate levels in the system makes it practicable to maintain stable canal flows even though the end users take water at irregular times. However, at main system level the use of storage is normally to enable the cropped area to be increased if water is a limiting factor.

Conjunctive use can be applied where an aquifer exists beneath a run-of-river scheme. Groundwater can often be a useful form of intermediate storage reservoir.

---

3 *Construction started in mid 1998 and commissioning finished in 2003.*
4 *See section 5.4 for an in-depth explanation.*

Flexibility and consequent production efficiency can be increased again by ***automation***, ***demand scheduling***, and low-pressure pipelines.

The adoption of advanced technology in the form of first sprinklers, then drip irrigation, then microjets, is starting to happen in the private sector in many developing countries such as Brazil, India, Sudan and Zambia (figures A1.7 - A1.9). Whenever water or the labour to distribute it costs enough money these systems will come into use. These devices originated in the developed world primarily to save the farmer money and as an indirect consequence to save water. The adoption of centre-pivot sprinklers in Brazil was paradoxically hastened by the hyper-inflation of the 1980s; farmers would take out a government loan, convert it into dollars, order the equipment from a local manufacturer, and when the time came to pay they would convert some of the dollars back into local currency which by this time had reduced in value by a factor of 100 or so.

Another paradox is the increasing sophistication of surface irrigation methods, notably in conjunction with flexible delivery schedules and semi-closed pipeline distribution networks. Furrow and border strip irrigation can with flexible scheduling cost less to install and operate, and be more efficient, than many mechanical methods. A recent example from the USA is the Pima-Maricopa project (Ref. 3), and several more are frequently cited by John Merriam (Ref. 4). With the adoption of these methods it is slowly being realised that the route to better efficiency need not necessarily follow the conventional wisdom of sprinkler and drip.

In many instances a single scheme will in its lifetime evolve through several stages of development strategy. An example is given in Chapter 2 of a major scheme in Pakistan, which started as an unregulated run-of-river scheme a century ago and is now gradually progressing toward automation.

1.6 Aid, Finance and Politics

By their nature, size and cost, irrigation schemes in developing countries tend to assume significant political and economic importance. They are, more often than not, receptors for large amounts of overseas aid and funding from international lending agencies. When entire societies and countries are totally dependent on irrigation, a scheme can be a ready tool for political and geopolitical manipulation at all levels.

When it comes to building new irrigation schemes, many third-world countries continue to waste billions of dollars of their own and other peoples' money. Many well-meaning aid agencies have discovered the hard way that constructing an irrigation project can be an effective and high-profiled way of wasting taxpayers money. Many now avoid the irrigation sector altogether, after burning their fingers first on large scale projects, then on small scale schemes and latterly on software[5] and management.

It is interesting to trace how things have changed in irrigation development over the last few centuries. It is also instructive because it offers some explanation for the anomalies and misdirected efforts on design and execution which are highlighted in this book. Because so many of the large schemes of Asia and Africa were instigated under colonial rule we can broadly divide the history of irrigated development into several chronological and economic groups as follows:

- Historical.
- Pre-colonial.
- Colonial.
- Post-colonial.
- Socialist economic decree.
- The European Union.
- Oil wealth.
- Developed countries.

---

5 *'Software' here meaning the people, management and institutional side of things rather than concrete and steel, or for that matter computer programs.*

- Private development and self-help.
- Commercial.

Historical

In historical and pre-colonial times, including the present time in countries such as Afghanistan which never had an effective colonial period, irrigation was practised usually at village level in a non-formal way with the aid of small run-of-river canals. Notable exceptions in which ancient schemes were more extensive include: the ancient schemes of Sri Lanka, in which the enigmatic remains of long-abandoned village tanks occupy every available small dam site in the country; the Mediterranean Basin schemes of the Roman Empire in countries such as Syria and Spain; and the Ma'arib Dam in Yemen, the world's oldest.

Colonial

In Colonial times, roughly speaking from the mid-19th to the mid-20th century, many African and Asian countries especially under British, Dutch and French rule embarked on massive engineering projects to provide irrigation over vast tracts of land. Ostensibly the aims were to prevent famine and enhance the standard of living of the local people. The geopolitical reasons were to consolidate political stability, to promote internal economic development, and in some cases such as Sudan's Gezira Project and numerous sugar estates, to supply cheap raw material such as cotton and sugar to the domestic markets of the colonial power. Finance was assured by the colonial power or its commercial companies, and often a declared objective of the schemes would be to generate revenue either indirectly through increased agricultural output or directly through taxes and water charges.

There were a few spectacular failures, such as the infamous Tanzanian groundnut scheme of the 1950s. (This was not in fact an irrigation scheme, but nonetheless one which demanded far more preparatory investigation into soils, land and climate than it actually received.) But lessons were learnt and vast experience was gleaned from the development of plantations and large schemes.

figure 1.3   A minor headsluice in the Gezira, Sudan. This scheme was built for commercial cotton production in the 1920s. It was well-engineered but has had its own severe problems, particularly of siltation

During this time many technological advances were made both in terms of engineering construction practice and engineering design theory. Reputations were made and occasionally lost on the design of huge dams, barrages and canals. Massive resources of labour and materials were mobilised to construct large projects in a very short time. The legal and managerial frameworks for operation and maintenance were drawn up, in the form of irrigation ordinances and acts of Parliament. In most ex-colonies these documents remain unchanged to this day, after a century and a half, as the basis for operation and maintenance and collection of water charges. Ironically the legacy of the old irrigation laws is sometimes a stumbling block toward progressing with more modern concepts of water management. So is the institutional structure of government, with its line departments which hinder communication throughout the system. Although these institutions and laws were adequate for their time, they can today provide a hiding place for do-nothing bureaucrats and an excuse for operational staff not to address field problems in ways which are appropriate to the problems of today.

Post-colonial

In post-colonial times, i.e. the second half of the 20th century, many more large schemes were built or extended, and latterly rehabilitated, often with financial assistance in the form of loans from international lending agencies such as the World Bank, Asian Development Bank, and African Development Bank. Multinational agencies such as FAO and the European Union tended to provide mainly 'soft' assistance in the form of visiting experts and consultants.

More money was provided by rich countries as overseas aid, in the form of grants or soft loans and the provision of services in the form of experts and goods such as tractors. Much of this aid was 'tied' to the country of origin, ensuring that only the goods and services from the donor were used. So a million dollars of tied aid would in fact be used to commission a feasibility study or buy earth-moving machinery from the donor country, and perhaps only a few per cent of this would actually filter down to the poor farmers on the ground. But it was aid nevertheless, and served several purposes, it appeased the consciences of rich people in Europe distressed by television images of famine in India, Bangladesh, the Sahel, Ethiopia; it consolidated political relations between donor countries (often the ex-colonial power) and the recipient; it gradually helped to advance the knowledge and capabilities of professional people in the recipient countries; and some of it actually resulted in farmers becoming better off. Irrigation was of course not the only sector to benefit from overseas aid, but it was initially a popular one because of its high profile and the promise of visible results in turning the desert green. Irrigation was a recurrent theme of Popular Causes which were mooted in the form of grandiose declarations such as the Green Revolution, the International Water Decade, the International Decade for Disaster Relief, most of which went largely unnoticed by both the developed world and the developing world alike.

The engineering lobbies were initially strong in the lending agencies, within donor government organisations, and especially in recipient governments. To begin with irrigation development was typified by big projects, some of them designed for national prestige rather than practical need, and in particular it meant engineering. So the engineers got their way and designed big schemes like the South Chad Project in Nigeria, Rahad, Managil and Rosieres Dam in Sudan, Tana River and Boura in Kenya, Narmada and Tungabhadra in India, the Mahaweli in Sri Lanka, and many more. No sooner were these schemes under way than realisation dawned of the difficulties of their operation, bilharzia and malaria demonstrated their side effects on health, water logging and salinity showed up their inefficient management and irrigation methods. Social and environmental problems appeared that had never been properly addressed because that was not the engineers' job. Reservoirs such as Bakolori in Nigeria displaced almost more people than their new irrigation area could support. Many projects were driven by engineering principles (*'we can divert this much water over here, what shall we grow with it?'*), rather than by social need (*'these people need an assured supply of water, how can we get it?'*). As evidence of engineering myopia mounted, so did resistance to big projects that spent billions and did damage. Engineers in the corridors of financial power were replaced by social scientists and economists. Large projects became unfashionable. It could also be said that opportunities for large-scale developments were forever diminishing in any case, as the best sites and more and more water resources were used up. But by the late 1970s irrigation engineers in general had a thoroughly bad name and upcoming aid projects were targeted in a different way to avoid the embarrassment of more high-profile failures.

That is, they concentrated on water management, farmer participation, integrated rural development, water users associations, institutional support. The emphasis shifted to small scale irrigation - village tanks in Upper Burma, hill torrents in Nepal, village irrigation schemes in the Western Ghats of India, vegetable plots in Zimbabwe. No massive engineering structures were needed here. Local materials, local skills, local manpower could do the job, with guidance from social anthropologists from Europe of course. Funding small projects was less fraught with the risk of spectacular failure and at the same time satisfied the increasingly socialistic and liberal attitudes of people in rich countries toward those less fortunate across the world.

But many of these failed too, often because not enough emphasis was placed on the need for good engineering. However important the social factors may be, engineers always have a vital role to play, despite their past sins.

Rehabilitation of necessity became increasingly prominent on the shopping lists of aid-seeking countries. This is not surprising since by this time many of the big irrigation schemes were nearing the end of their first century, and were falling apart. Maintenance had too often been neglected for want of money or good management. Funding on rehabilitation was a safe way of avoiding criticism since existing schemes could not be labelled as white elephants.

So over the past half century, the emphasis on irrigated development has swung away from big projects, first to medium and small scale schemes, then to rehabilitation of older schemes, and to socially-oriented concepts based on institutional strengthening, farmer participation, social engineering. The status of the irrigation engineer waned as a result of his early failures to take a sufficiently wide view of the irrigated world. Rightly or wrongly, engineers were branded as uncaring narrow-minded technocrats who were unconcerned with management of their schemes, with social or economic side effects, with agriculture, or with the host of other disciplines which all contribute in some way to a successful irrigation project. But all this is changing fast, and it is an aim of this book to help irrigation engineers prove to the rest of the world that they do care, and are aware, and are able to see across wider horizons than before.

Socialist economic decree

Meanwhile, half the globe was labouring under a different form of colonialism. Where schemes were centrally planned and executed by Socialist economic decree, disasters on a scale hitherto unimaginable were bound to happen. In the USSR there was no safety valve of public opinion or investigative journalism to highlight potential disasters. The results are only now becoming apparent. The Aral Sea is almost no more. The Karakum Canal pours water into an empty desert. A million hectares of pumped schemes in Rumania are uneconomic and unsustainable, as are others in the Uzbekistan steppes which lift Amu Darya water through hundreds of metres in order to grow wheat, in a place where rain-fed wheat is obtainable from neighbouring Kazakstan at a fraction of the price.

Central planning can have bad side effects on even apparently unimportant issues. The Ghab project in Syria was rehabilitated with a view to maximising - as opposed to optimising - the area under irrigation. So the villages that were strung out through the centre of the scheme conveniently close to the fields were removed to the edge of the valley in order to gain a few hectares of extra land for irrigation. Henceforth the farmer's day was disrupted by having to travel 10 kilometres or more to his fields that used to be literally at his back door. Then came the inevitable social strife caused by needlessly moving all those people.

A similar thing happened in 1970s Tanzania, in the days of the *Ujamaa* villagisation programme. Under Ujamaa, entire villages were moved from their age-old scattered locations on rich soils of the floodplains, and re-established alongside newly-built roads which gave ready access to schools and newly-developed piped domestic water supplies. The new sites were invariably situated on poor upland soils, which resulted in an immediate loss of income for many farmers. The problems were compounded by the social experiment of communal farming, in which the entire village was expected to line up every morning, on land owned by everyone and therefore by nobody, and wield their hoes in unison to the throbbing beat of the drum. To entice the farmers into this unnatural existence, their ancestral lands in the fertile Rufiji floodplains were decreed out of bounds. The hippos and monkeys moved in. The admirable thing about Julius Nyrere was that after it all went pear-shaped, he had the good grace and courage to admit that it had been a mistake.

An inevitable side effect of central control is that when official crop production targets are not met, the records have to be falsified in order to prove that the local bureaucrats are doing their job. It was an almost impossible task to comprehend the real state of agriculture in the Syria of Hafez Assad. The official crop production figures showed a steadily increasing rate of production year by year. This was patently not the case on the ground. In the Ghab scheme, cotton fields had been converted into fish farms, yet the reams of official figures showed actual production of cotton to be right on target.

The beautiful but beleaguered country of Burma has had for the past two generations a particularly nasty version of centralised control. Once upon a time Burma was a net exporter of rice to much of south-east Asia. But then the socialist government of the 1970s onwards decreed that every village should deliver a certain quota of rice and other crops to central government warehouses. Some bureaucrat in Rangoon, given a target tonnage for export (in turn no doubt worked out on the basis of the foreign exchange needed to pay for the latest shipment of weapons), would then work out the tonnage required from every village throughout the country. When I was working in the dry zone of Upper Burma in the mid 1980s, I was horrified to walk though field after failed field of desiccated paddy rice, planted on rolling sandy soils in a drought. When I asked why they had not planted sesame or pulses, either of which would have stood some chance of survival, the reply came that the village was merely obeying orders from central government, who told it what crops and what hectareage to plant, without regard for agronomic reality.

The upshot of all this of course was that crop return records had to be falsified in order to satisfy the official target figures; people had to live on whatever food they could hide away from the government officials, and many starved, in a country potentially able to feed not only itself but half of Asia too.

Ethiopia under the brutal 1970s and 1980s regime of Mengistu Haile Meriam was another example of central control gone mad. The horrors of famine and starvation of those years were beamed in televised technicolour into western living rooms, but the real reasons for it were obscured. Teff, the tiny grain that is the staple diet over much of the Ethiopian highlands, was all this time being exported by the shipload to the Soviet Union to make vodka, in payment for armaments, the services of Cuban soldiers to fight the Eritreans, and economic advice that the liberal *Economist* news magazine once described as 'spectacularly incompetent'. It was this same economic advice that led to the establishment of state farms carved out of the tsetse-ridden jungle to grow vast quantities of crops such as Kenaf, for which there was no market.

Albania is a small country but has nearly half a million hectares of irrigation schemes that vary in size frrom 1000 to 7000 hectares. Most of these no longer work. Most were originally built under the autocratic regime of Enver Hoxha, in which the rules of conventional economic theory did not apply. As a result the country is an irrigation engineer's paradise, with every conceivable scenario of scheme imaginable. There are schemes which pump water into contour canals high up a mountainside, to feed into piped sprinkler networks under gravity. There are schemes which drain the coastal marshes and reclaim saline-alkali soils. Another scheme in Korce (started by Mussolini in the short period of Italian rule) drains a peat swamp; with every 5 metres of ground subsidence, another bank of control gates has to be built, so that there are now three separate structures with a fourth at the planning stage, each with a lower invert level than the last. Another collects water high in the hills above Elbasan and pipes it along roller-coaster mountain ridges to irrigate olive groves. One of the schemes around Fier shares its drainage canals with an oilfield; no weed maintenance problems there, but the fish tastes peculiar. The Kurjanit Dam was built too big for its catchment yield, so it was filled up every winter by pumping from the Semanit River, until the harsh realities of pumping costs became apparent. Every scheme has its own idiosyncrasies that may be anathema to the economist but are a joy to the engineer.

Hand-in-hand with the legacy of grandiose and economically unjustifiable schemes go previously unfamiliar problems of land tenure. Under communist rule there was no land tenure problem since all land belonged to the state. It was farmed under the all-encompassing administration of collectives, communes or villages. But now in every ex-communist state the land is being redistributed, with consequent difficulties in management and operation which never arose before.

Albania's farming was everywhere based on Brazda, the layout of ploughed plots each with a ridge down the centre that was ideal for tightly-regimented surface irrigation with no flexibility. Land has now been redistributed to give everyone two hectares, although not usually in a single plot. The canal irrigation network that was ideal for 24-hour continuous irrigation is hopelessly inadequate for the flexible scheduling now required by private farmers. Fortunately Albania's economy no longer depends on agriculture - most Albanians live abroad and send money home, and there are flourishing twilight industries such as smuggling refugees, drugs and women into Western Europe, that for the moment ease the pains of transition. But other ex-communist countries such as Moldova, which is totally dependent on its collapsed agriculture, are less fortunate.

In Kyrgystan the most successful post-privatised farmers seem to be those who have remained in their original collectives and continued irrigation and farm operations much as before, except that there is a fairer distribution of profit and produce which is no longer sent off to some state-owned warehouse but sold in a fledgling local market.

Adjustments in land tenure from the state-owned collectives to private holdings will take a couple of generations but has already begun over parts of the FSU[6]. Time will tell whether this will lead to more efficient and better irrigation, or merely compound the ongoing disasters. My guess is that things will improve dramatically, given the right kind of macro-economic adjustments (such as developing freer markets, an efficient infrastructure for manufacturing equipment and inputs and for processing produce) that these countries will need. And these improvements will be hastened as engineers come to accept concepts such as flexible scheduling, begin to think laterally, and learn to think like farmers.

The early years of the 21st century will undoubtedly see a readjustment of schemes like these. Furthermore it would be unfair to suggest that all is doom and gloom with the schemes of the FSU. Much good development work was done, often in imaginative schemes which were developed unfettered by the need to minimise cost. Furthermore the natural resilience of farmers everywhere to physical adversity, and the inescapable fact that half of humanity needs irrigation in order to survive, will ensure that irrigation goes on wherever there is water.

figure 1.4   ***An irrigation channel flows through the Fier oilfield, Albania***

---

6 Former Soviet Union.

The European Union

Central planning fiascos such as these are not limited to less-developed countries. The European Central Agricultural Policy is an extreme case of centralised policies leading to wasteful excess. Few people in the world can possibly understand every clause of the CAP. Although the guiding principles behind it are without doubt honourable, the practicalities of stimulating agricultural wealth in the most needy and underdeveloped parts of Europe are fraught with political hurdles which are almost insurmountable.

Greece has benefited hugely since joining the Union in 1980. Its farmers were subsidised for everything, from buying tractors and machinery to fertilisers and chemicals, to planting wheat. Many of them planted the wheat, collected the planting subsidy and then promptly forget about the crop. Many bought tractors far too big for the needs of their tiny farms, as the row of 200 horsepower four-wheel-drive monsters parked outside any village bar would testify. Irrigation multiplied, with hose-reel rainguns becoming a popular way of saving labour. Nowhere were the effects of this explosion in agricultural activity manifested more dramatically than in the demise of Lake Koronia, north of Salonica. This beautiful lake, famous as a tourist spot and a site of special scientific interest, supported several fishing villages. But as the irrigation developed in the surrounding fields, dairies and piggeries and dyeing factories opened up in the surrounding villages, so the water level subsided and the nitrates and BOD built up, until by the mid 1990s the fish had all died and the life of the lake itself was estimated at only four more years, before it dried up completely. Abnormally high rainfall in the winter of 1997/98 granted the lake some reprieve, but the only real answer was unpalatable and clear. The demise of the lake could be correlated directly to the date of Greece joining the EU, and the consequent subsidies that came with it.

The agricultural gravy train was due to be wound down from the year 2000 onwards. As soon as this was announced, the well-organised Greek farmers blocked the main road from Athens north with tractors and the first of many such demonstrations began.

In the same vein, the streets of Paris are a regular target for cartloads of manure, courtesy of French farmers. In Spain the Genil-Cabra canal winds through mile after mile of olive groves, funded by subsidies which achieve nothing other than irrigating the burgeoning olive oil lake of the EU.

Wherever the farm lobby is powerful, there is a political reluctance to reduce the subsidies which guarantee inefficiency and waste.

Oil wealth

Not all developing countries are poor. Some had great wealth in the form of oil. Nigeria, Angola, Libya, Indonesia, Iraq, Iran and the rest of the Middle East, all perceived a need for irrigation but all had different population densities and hence vastly different wealth per capita. The oil 'crisis' of the early 1970s resulted in a sudden increase in disposable national income in all these countries. Some of their new-found wealth was spent on weapons, some on infrastructure and irrigation projects.

Nigeria had money to burn in the 1970s. Consultants arrived from everywhere to help burn it. With a population then[7] of 80 million dependent on the agricultural sector it had a clear need for irrigation especially in the north where annual rainfall averages less than 400 mm. Irrigation schemes were dreamed up all over the country, some of them eminently sensible and others on the flimsiest of pretexts. There was a scheme proposed for 10,000 hectares of sprinkler irrigation in hilly jungle country. It could only be justified economically if it produced a high value crop. Ergo, 10,000 hectares of chillies and tomatoes, probably enough to satisfy the urban markets of the entire continent. Fortunately this project and others like it never saw the light of day because sagacity prevailed in the government, and anyway the money ran out.

Other oil countries fared differently. Indonesia had a population approaching 250 million, and an already diverse economy with a well-developed irrigation sector, into which extra oil revenues were

---

7 *In 1975 the population was about 80 million; 25 years later it is over 120 million and the oil wealth has been squandered.*

absorbed without fuss. Libya had a small population of only 5 million, a healthy income from oil and gas, and large reserves of fossil water beneath the Sahara. Resolving to exploit this hidden resource, a large scheme of centre-pivot irrigation was set up at Kufra Oasis, fed by deep tubewells and growing initially alfalfa as fodder for sheep. The scheme worked - in technical terms - but being 1,000 kilometres from the nearest big population centres on the coast it was difficult to persuade anyone to go and live there. Transport of live sheep was also a problem over this distance in a hostile environment. And so was begat the Great Man-made River. It is the world's biggest pipeline, with concrete pipes 6 metres in diameter. So far it has cost in the region of USD 5 billion and it is still not finished. Coastal lands are now under irrigation using fossil desert water, although most farm labour is imported from Egypt and Sudan, because the local people don't want to work in the fields even on the coast.

Similar projects were established in other rich desert countries such as Saudi Arabia, albeit on a more modest scale. Even desalinated sea water has been used, at phenomenal expense, for irrigating wheat and fodder. These schemes are interesting aberrations of world-wide irrigation development, and serve only to show the perceived value of water if you happen to live in a place which has none.

figure 1.5   ***Centre pivots in Nebraska. Each circle fills a half-mile square***

Developed countries

Technical advancement in irrigation hardware design has in the past half century been driven largely by developed countries with important irrigation sectors such as the USA and France. Israel has, through geographical necessity, contributed hugely to development of specialised drip and micro irrigation. Automation of canal systems was made practical for the first time in developing countries with the introduction of passive structures and self-regulating gates from France. It is taking a long time however for these exciting concepts of automation to break through the barriers of engineering prejudice that persist in the very countries that have the most need of them. Recent developments in surface irrigation mainly in the USA using surge and pulse methods have yet to become applicable over a wider area.[8] Centre-pivots were developed in the USA to run on cheap power, but in recent years increases in power cost have caused manufacturers to look for ways of reducing operating pressures. Concepts of flexibility and demand scheduling have been developed and tested there too, with important implications for large irrigation schemes world-wide. But these ideas are still treated with scepticism almost everywhere, and it is an aim of this book to bring them to a wider audience.

---

8 *These are mechanical methods dependent on both a flexible supply and correct flow rate. John Merriam says that if a large enough flexible supply is available then manual operation of furrows and border strips is just as efficient and requires little labour. I believe he is right, at least in the context of the large fields to be found in the USA.*

These developments came about partly because of the large irrigation private sector in the USA with large privately-owned farms that could support the expenditure on new equipment. Government-controlled main canal systems in both the USA and France benefited from developments in active automation, much of which has been centrally funded.

But there are major differences between irrigation schemes in the USA and developing countries. The most obvious is the preponderance of small farms, often no bigger than a hectare each, that make up the average large scheme in Asia and Africa. The implication of this is that no matter how large the scheme, the amount of water delivered to an individual farm is minuscule by comparison, is impossible to measure accurately, and moreover impossible to control by anyone other than the farmer. Therefore any attempts to make farmers pay for volume of water used invariably founder through the impracticality of external control at this level in the system. Special problems arise in operation and maintenance due to the sheer number of individual users. And few of these users have financial resources to purchase more than the basic of farm inputs, let alone irrigation equipment.

Private development and self-help

In every country there has always been some private irrigation development but usually on a small scale. Self-help schemes have taken place at village level through necessity or through prompting by overseas aid agencies. Villages in Afghanistan have for many centuries relied on co-ordinated self-help to construct the *karezes* - tunnels that bring undergroundwater from the mountains to the village, in a similar way to the *qanats* of Iran and the *aflaj* of Oman. Many small dams in Africa were built through self-help schemes but failures were common because the engineering was not done properly (usually compounded by a basic error such as inadequate soil compaction or omission of a spillway). Figure A2.2 shows a typical community effort in Burma where villagers have long ceased to rely on officialdom to keep the water flowing. Here the whole village turns out at the start of each irrigation season to unblock the lead canal from the river.

figure 1.6   *A linear-move sprinkler on a commercial farm in Sudan*

Commercial schemes

In Africa the perennial problem of large schemes has been poor management. This has generally been compounded by difficulties in obtaining agricultural inputs on time, inadequate communications and transport facilities, and poor financial services. It is gratifying to find that in some hitherto economic basket-case countries like Sudan and Zambia, commercial enterprises are successfully establishing large-scale irrigated farms using state-of-the-art technology. Many such companies have foreign origins, but are encouraged to invest in the country with the promise of cheap water, land and labour.

Northern Sudan is completely desert, but has under-utilised water resources in the Nile and in underground aquifers. The soils are sandy and the terrain is undulating, conditions for which the ideal method of irrigation is by large-scale centre-pivot sprinklers. Neither small farmers nor government ministries are willing or able to establish this type of scheme, but commercial farming companies are. By 2005 there were around 150 pivot and linear-move machines in Sudan, and future developments such as the Merowe Irrigation Project will encourage many more.

The immediate benefit goes to the commercial enterprise, but spin-off benefits will filter through the entire agricultural sector. Once a base of expertise and spare parts is established in the country, there is no reason why smallholder farmers also should not irrigate under centrally-managed pivots. They already do so in Mauritius and Brazil. Under these systems in-field water management, the source of most irrigation inefficiency and wasted water, is simplified to the point of waiting for the rain to fall in a predetermined quantity and at a predetermined time each day.

Virtual water and self-sufficiency

It is common amongst developing countries for the stated aims of national policy to include self-sufficiency in food production, or at least import substitution of staple foods. Most of Nigeria's post-colonial irrigation adventures were justified at least partially by this argument, which aims at increased political security through a reduction in free trade. Its efforts were initially hampered by macro-economic policies which artificially over-valued the local currency and made it cheaper to buy packets of rice imported from the USA rather than locally-produced rice in the village market.

The thought-provoking concept of virtual water has recently been postulated (Ref. 1). This really considers cross-border trade in agricultural produce as effectively a trade in water. On the presumption that it takes roughly 1,000 tonnes of water to produce 1 tonne of wheat, for example, the argument runs that if a country is short of water it would be far cheaper to import food from water-rich countries rather than spend money on expensive irrigation schemes or, worse, steal water from trans-national river systems. This would appear to be no more than common sense. Unfortunately the world is not yet in this state of liberal free-trade utopia, but we can live in hope that simple economics will eventually rule over national pride. The threat of global water wars predicted by some, using such examples as Turkey's abstractions from the Euphrates or Israel's from the Jordan, has not receded yet, despite the fact that the cost of running even a small war for a week would if put to better use fund the irrigation needs of an entire country for a lifetime.

1.7   We have an Attitude Problem

Some changes are overdue, in the ways that canals are habitually designed and constructed, and in the jobs they are expected to do. Traditional methods of construction and lining are being superseded by new materials and machinery. The gradual acceptance of new concepts of water management will mean canals performing as more than just conduits to pass water. In many cases they will be replaced or supplemented by pipelines. In most countries the automation of canals will become increasingly more sophisticated. We already have the means to update or improve obsolete canal systems, but usually we can't or won't recognise the need for change. We have an attitude problem.

The need for change is apparent, with broken down schemes, land going out of production through salinity and waterlogging, social strife and poverty on the very schemes that were built to alleviate it. The technology and the management concepts for change are in place, but often the will to change is not. Old traditions die hard. Whether it entails merely using high strength precast concrete in place of the old bricks and mortar, or automating a canal with passive gates, or learning a radically new method of designing a pipeline distribution system, engineers will grasp at all manner of excuses to design things in the old way because they don't want to understand the new way. Most design projects are well endowed with engineers who know about concrete, steel and even water requirements and canal design, but usually the principles of water management and control, and especially farmers' aspirations and capabilities, are lost on them. They might know ***how*** to design, but not ***why***.

1.8   Prelude to Change

So the scene is set for change. Irrigation engineers have been accused, not entirely without justification, of being too insensitive to the wider issues of people, agriculture, and the social and environmental consequences of their actions. The planet is strewn with projects that do not work properly, that have been built in the wrong place or with the wrong aims, that have been badly designed and carelessly constructed, that are collapsing for want of maintenance. Worse, precepts of design and planning that were relevant a century ago are now obsolete after the most tumultuous period of change in human history. The massive advances in technology and science, in education and social development, in the way people think and live, and the things people aspire to, ought to be reflected in the way that engineers go about planning and designing and constructing and operating their irrigation schemes. Too often we are stuck in our own conservatism, with the reassuring illusion of knowing that if we did it this way before, then we need not change it.

figure 1.7   *Abandoned land due to salinisation in the Aral Sea basin, Kazakstan*

References and further reading for chapter 1

(1) Allan, J. A., 'Water Security Policies and Global Systems for Water Scarce Regions', ICID 16th Congress, Cairo, 1996.

(2) Plusquellec, H., Burt, C., Wolter, H.W., 'Modern Water Control in Irrigation', World Bank Technical Paper 246, 1994.

(3) Lindstrom, S., 'Pima - Maricopa Flexible Irrigation Supply Project', Proc. XXVII Congress IAHR/ASCE, San Francisco, 1997.

(4) Merriam, J. L., Styles, S. W., 'Flexible Irrigation Systems Concept Design and Application', ASCE Journal Irrigation and Drainage Eng, 1998.

(5) Easterly, W., 'The White Man's Burden – why the West's efforts to aid the Rest have done so much ill and so little good.', Penguin, 2006.

(6) Reisner, M., 'Cadillac Desert – The American West and its Disappearing Water', Penquin, 1986.

(7) Postel, S., 'Pillar of Sand – can the Irrigation Miracle Last?', Worldwatch, 1999.

PART 1 - PLANNING

All new or remodelled or rehabilitated schemes undergo a gestation period in which the conceived scheme is assessed from many viewpoints. In the case of a smallholder diverting a stream the planning phase could be merely 5 minutes' thought followed by an instant decision to start digging, but on all formal irrigation schemes the planning phase is normally measured in years and even decades.

During planning of a big project there may be several tiers of study, including appraisal, pre-feasibility, feasibility, pre-design, environmental impact, and if private sector money is involved, corporate planning and finance. The viability of the scheme and alternative outline designs will be tested from technical, social, financial, economic, and environmental viewpoints.

It is at this stage that the aims and direction of the scheme are established and fixed. Bad decisions here can eventually break the project even if it goes ahead. If imagination and forward-thinking are lacking then the final scheme may suffer from obsolescence as soon as it is built. All too often this is the case, with engineers particularly guilty of following the conservative path of repeating what went before without giving thought to the exciting possibilities that await around the corner.

This section aims to present all involved in the planning phase with a background wide enough to view a scheme from many angles. So I am asking the engineers to consider not only hydraulics and concrete, but water management and the intrigues of automation, as well as the social, agricultural and economic side of things. I am asking the trainers and extension workers and agronomists to understand the reasons for particular canal structures and engineering strategies. I am asking the economists and financial planners to consider more than just the black figures on the bottom line. I am setting out the stall for political decision makers to sleep easier in the knowledge that the future need not be impoverished by the decisions we make today.

CHAPTER 2   ELEMENTS OF IRRIGATION

The need for irrigation may be obvious, but here we take a deeper look into the underlying reasons for irrigation, some of which may not be so obvious to either the farmer or the engineer. Nor, for that matter, to the planner and financier.

2.1   What can irrigation do?

In some places, plant growth is impossible without it. Irrigation can guarantee a wet season crop if there is a danger of dry periods, and can increase the yield by eliminating periods of water stress. Where water but not rainfall is available throughout the year it may be possible to grow two or three crops under irrigation, in comparison to only one crop without it. Irrigation may be full, to give the plant all the water it needs for maximum yield, or partial or protective only, in order to let the crop merely survive. Even in high rainfall areas such as south-western India and Sri Lanka irrigation is vital in order to redistribute water over time so that it can be better utilised during the cropping period. The ancient tank systems of Sri Lanka not only store water for the dry season but even out the water supply during the wet seasons, when it arrives as short and violent storms. Irrigation can improve the quality and hence the value of crops such as potatoes and tomatoes. It can be employed in cultivation practices such as land levelling for paddy rice or basin-grown wheat, and as a pre-irrigation to germinate weed seeds which are then allowed to die off before planting the main crop.

2.2   Productive, Partial and Protective Irrigation

The strategy employed in planning a scheme will have far-reaching consequences. Even for old schemes needing rehabilitation it is important to understand the strategies applied to their initial planning, which in many cases took place in the 19th century and are no longer relevant to the needs of today's farmers.

There are several strategies which may have conflicting aims. They are aimed respectively at crop **production** and crop **protection**, and at **economic** benefits versus **financial** benefits. Consider the following questions that ought to be posed by an irrigation planner:

- Do we have enough water to feed the plant everything it needs to get maximum yield (*productive*)? Or are we just trying to keep the crop alive (*protective*)?
- Are we trying to maximise the agricultural output given a limited amount of water (*productive*)? Or are we trying to maximise the irrigated area and number of farmers that can subsist on it (*protective*)?
- Are we looking for an optimum *economic* yield, as is usual in modern irrigation planning, balancing the costs of irrigation and intensive agriculture against the crop's economic return to the country as a whole? Or shall we let the farmer grow whatever he wants, which will normally mean maximising his own *financial* returns?
- Consequently, are we going to control the amount and frequency of irrigations, or are we going to let the farmer irrigate whenever he wants?

The reason why many old schemes in India, Pakistan and elsewhere are now undergoing serious social and organisational conflict is that they were originally designed for *protective* irrigation, but are now required to *produce* far more than before. Many of these large schemes were instigated under colonial rule with the object of stabilising (cynics might say *pacifying*) the local population. They were therefore designed for protective irrigation only, insuring against widespread total crop failure and consequent famine. So irrigation duties were high and water allowances were low. Cropping intensities were also low, typically no more than 40 per cent which meant irrigating a crop of wheat, say, during Rabi (winter) only, covering less than half the total area. Any other crops which survived either the winter or summer were a bonus. Farmers were not allowed to irrigate all their land; only up to 40 per cent of it, hence the term *partial* irrigation. No cross regulators were provided on the main canals, since there was no attempt at varying the flow. The story of the Upper Swat Canal system in north-western Pakistan is typical:

The Upper Swat Canal, evolution from protective to productive

For over 2500 years the Swat Valley has been a focal point of civilisation. The armies of Alexander the Great and Ghengis Khan both used it as a route from central Asia into India, and it was a major feeder route of the Great Silk Road. The Ghandara civilisation flourished there for a thousand years, leaving Bhuddist ruins and relics of great sophistication. In 1897 the Malakand Pass was the scene of a battle between local tribesmen and the British Army, intent on extending the boundary of the British Empire. The Empire was duly extended, but only on paper, and its practical limit was never to be more than a few miles from Malakand.

In the decade of uneasy and sporadic peace that followed, the Benton Tunnel and Upper Swat Canal Project was conceived to bring Swat River water to the thousand square miles of fertile but parched land of the Peshawar Vale. It was justified in a variety of ways. For the imperial government the immediate need was to mollify the local tribes and establish a wider perception of its honourable intent, and so the scheme was born with its water as a political tool. Secondly, since it was from the outset intended to levy a water charge on every farmer, the scheme was also seen as a financial instrument for raising taxes. Thirdly, in humanitarian terms a scheme so bold had to be of benefit to the local populace, who had for generations been subject to the vagaries of a near-desert climate. Fourthly, a longer-term view would have the scheme helping to boost the economic development of the country. And lastly we might be forgiven in conjecturing that its main aim was simply to glorify the status of the British engineer. If these were indeed the original objectives, the first was a qualified success, the second a failure, the third and fourth were a lasting success, and the fifth in passing left behind for the generations of engineers that followed a wealth of lessons and experience that we ignore at our cost even today.

Tunnelling beneath the Malakand Ridge commenced in 1911 and was completed 3 years and more than 2 miles later. By the outbreak of the First World War a massive canal network was in place, remarkable for its complexity and boldness. Distributaries were constructed across dauntingly rough terrain to bring the irrigation command area to around a quarter of a million acres. Although eclipsed in size by many of its predecessors throughout northern India, the Upper Swat Canal remains to this day one of the most challenging and technically exciting irrigation schemes ever built. Being routed across the foothills at the extreme southern edge of the Hindu Kush, numerous streams necessitated elaborate cross drainage structures, and the steep slopes not only required a large number of drop structures but also offered opportunities for hydro power development on a small and medium scale. These were duly exploited with commendable foresight, by providing flour mills at many of the canal falls. Some of the structures, like the towering arch aqueducts, used an ancient but finely-honed technology of masonry construction that reached its zenith at that time and which we find impossible to reproduce today. Others, like the massive 10 feet diameter steel tube inverted siphons, were then at the forefront of technological development.

With the termination of the third Afghan War in 1920, the region became more settled and farmers began to look for more than just subsistence and a hedge against famine. All the while the population was increasing apace with its production of wheat, sugar, tobacco, maize, fruit and vegetables. One-horse villages swiftly grew into prosperous little towns. Little houses expanded outwards around hidden courtyards, as is the Pathan way. Holdings were split up between sons and grandsons. Land became more and more valuable, and spare land scarce. They started growing poplar trees for match-wood alongside every watercourse; the roots wrecked the brick canal lining, but it gave an income from land that could be used for little else. They started growing two crops a year instead of the single crop originally planned, and then many grew three, with manifold combinations of inter-cropping amongst the fruit orchards.

So, over its first 50 years the irrigation scheme brought undoubted prosperity. Against this, the burgeoning population placed ever more onerous demands on water and land. Ironically, the success of the scheme and others like it has no doubt contributed to the excessive population growth rate that now contrives to keep Pakistan poor. Tail-end problems appeared, as the carefully-set irrigation outlets were interfered with and inequity set in to the simple proportional distribution system that had sufficed admirably at the outset. There were gunfights over water, and people died.

The original design duty was 3 cusecs[1] per 1000 acres, with an intended cropping intensity of 40 per cent. This meant that only part of any farm could be irrigated in any one season; partial protective irrigation. The warabandi system was introduced from the outset, with continuous flow and proportional distribution along the primary and secondary canal system. There was no regulation of the river, but the dry season flows were sufficient to meet the irrigation requirements for winter wheat.

By the time the British left in 1947 the scheme duty had almost doubled[2] to 5 cusecs per 1000 acres, the crop intensity had increased to over 100 per cent, and main system rotation had been introduced in winter to eke out the available water supply which was in some years inadequate for the increased area under winter crops. Now, half a century later, the scheme duty has doubled again to 10 cusecs per 1000 acres, cropping intensity is targeted at over 180 per cent, and major rehabilitation works have doubled the canal capacity. The problem of water shortage in winter is being solved by tapping into the adjacent water resources of the River Indus, through construction of the Pehur High Level Canal. This was completed in 2002, and supplements supplies to the tail end of the Upper Swat system and permits new irrigated development in the middle reaches of the Upper Swat Canal. It is an automated downstream-controlled canal, the first in Pakistan, conveying up to 30 cumecs (1000 cusecs) from Tarbela Reservoir.[3]

figure 2.1   Machai Branch, Upper Swat system, Pakistan.
One of the world's most beautiful canals

---

1 *Cubic feet per second.*
2 *Duty is used here as suggested in section 3.3, and not as often referred to in India and Pakistan as an area served per unit of water.*
3 *Tarbela Dam was built on the Indus in the mid-1970s, and is still one of the world's largest. It is a multipurpose dam, primarily for regulating irrigation supplies to the Punjab, and also for generating 3,000 MW of hydro power, which is a significant proportion of the needs of the entire country. In 1991 the four provinces of Pakistan signed the Indus Waters Apportionment Accord, which allocated a guaranteed portion of Indus water for the use of North West Frontier Province. As a direct result of this agreement, the go-ahead was given for the Pehur High Level Canal as a means of transferring Indus water into the command area of the Upper Swat Canal.*

More than a million people now depend for their livelihood directly and indirectly on the Upper Swat Canal.

Critics of the ongoing rehabilitation works[4] question the need for more water, avidly promoted by the powerful engineering lobby which exists both in the national government and, until recently at any rate, in several major international financing agencies. The critics point to the Kharif (summer) rainfall figures which theoretically are quite adequate for a reasonable crop, especially if supplemented by the existing level of protective irrigation. They point to the rampant corruption which is prevalent throughout the government agencies charged with administering the engineering works. They cite the old schemes in the Sind, which still operate quite well on a duty of 3 cusecs per 1000 acres. They point to the sugarcane fields which use a great deal of water and are unnecessary as a staple food crop. They see the high crop yields emanating from the head of the canal system and ignore the parched lands at the tail. They read the high illiteracy figures and decide that these people cannot cope with any further sophistication of the distribution system.

But ignored most of all is the fact that in the century since the scheme was started the population has not only increased sixfold, but the people aspire to far more than their great grandfathers ever did. They have the vote. They need money to send their children to school. They are largely emancipated from the slave labour which is still imposed by powerful landlords in other parts of the country. They want *production*. They need maximum output from every field. Protective irrigation here is a century out of date.

Deficit irrigation – the strange case of Albania

These days we are starting to pay more attention to water use efficiency, and increasingly realise that on many schemes where water resources are being stretched to the limit, some compromise is necessary between theoretical water demands and those lesser amounts which it is possible or practicable to give. Enter the concept of Controlled Deficit Irrigation, and Deficit Management.

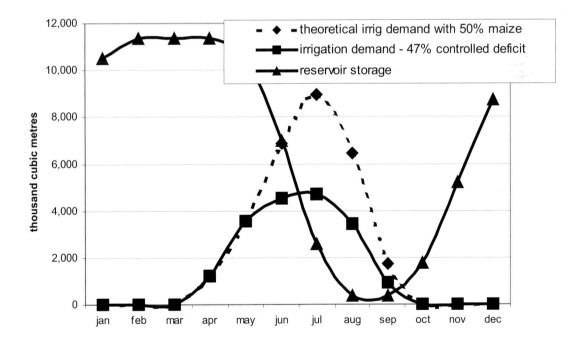

figure 2.2   ***Irrigation demands and reservoir storage curve – Durres, Albania***

---

4 *Costing around USD 300 million. The Pehur High Level canal and associated works are of a similar magnitude.*

If we look at a graph of typical crop water requirements, very often we find the peak demand occurring for only a short time. Figure 2.2 elates to maize in Durres, Albania, where it is planted in May and harvested in September. The peak irrigation requirement in July is 40 per cent more than in June, and 45 per cent more than in August. So, do we design the canal system for the peak demand in July, or something less than that? The answer depends on the type of system, and how much flexibility it has.

Most of the 400,000 ha of irrigation in Albania is in schemes of 3,000 to 8,000 ha each. They are ostensibly designed to meet this theoretical peak demand. However, most of them were planned, in the bad old days of centralised communist control, on wildly optimistic estimates of water availability, river flow and catchment yield. We don't know if they ever worked properly, because the political upheavals in 1990s Albania ensured not only that most of the irrigation hardware was destroyed, but that all records and all collective memory of how things worked before the revolution have been wiped out. We are now faced with rehabilitating the schemes under radically different social and economic conditions. In many cases we find that the reservoirs are destined to dry up half way through the season, or that the pump stations require twice the pumping capacity that they were originally built with. Given that the country is broke and cannot afford millions of dollars on each project, we are faced with two options:

a) Reduce the irrigated area. Invoke the wrath of a thousand nouveau-capitaliste farmers. Start another revolution.

b) Reduce the design capacity to something below the theoretical peak demand. Stress the crop. Accept a lower than maximum crop yield. Enter into managed-deficit irrigation.

figure 2.3   Tarinit Dam, Albania

The sensible option is of course to reduce the irrigation deliveries at times of peak demand and either:

a) Set a limit on mm /month given to the crop e.g. 100 mm/month, or
b) Reduce the irrigation release by a certain percentage - the managed deficit.

The first is the more practical, since it implies a cap on the canal discharges and gives a constant discharge for the peak demand months.

Financial and Economic Benefits

It is useful to bear in mind the difference between financial and economic benefits, in the context of irrigated farming. Financial analysis applies only to the farmer, the farm gate prices, the monetary value and returns to the farm or family unit. It ignores the distorting effects of government-imposed subsidies or taxes. At an elementary level, the Greek farmer might decide that he can afford the four-wheel-drive tractor that is 80 per cent subsidised, especially when he receives a wheat-planting subsidy and can then work 6 months of the year on a building site. These are financial decisions and are made by the farmer.

Economic analysis considers the wider picture of the project in the context of cost and benefit to the nation, state or region. World prices are used for produce and inputs, stripped of the taxes or subsidies that the government pays to itself. The cost and availability of labour is considered on a national scale, with labour that would otherwise be unemployed having a lower opportunity cost than the cost to the farmer. At the economic level, the Government might decide that there are indirect benefits to be gained in paying for 80 per cent of the cost of a tractor, if that allows more land to cultivated, more wheat to be produced, and less imported wheat to be paid for.

Both financial and economic analyses are relevant, and the outcome of either may affect decisions on whether to proceed with or modify a project. International financing agencies have long required that these separate analyses be included in feasibility studies. Decisions to proceed are more often than not judged on the basis of the internal rates of return thus derived. For smallholder based irrigation projects, an economic internal rate of return higher than 10 per cent might be considered very attractive. Normally the corresponding financial internal rate of return for the same project would be less.

2.3   Equity and Equality

Equity means justice, or fairness. One dictionary definition is: *a system of law supplementing the ordinary rules of law where the application of these would operate harshly in a particular case; sometimes it is regarded as an attempt to achieve natural justice.* **Equality** is not the same thing. *Equity* of water distribution does not necessarily mean that every farmer gets the same, equal, amount. It means that every farmer gets an amount that is *fair*. Thus in some Warabandi schemes there are deficiencies in the watercourse distribution because the allocation times do not take into account canal absorption losses. Each farmer is given an equal time-share of the flow but does not end up with an equal amount of water. In many places this lack of equity is recognised, and measures such as those described in section 3.4 have been introduced as standard water management procedure.

Similarly, a farm on permeable soils might require a lot more water than one on heavy soils. In the dry zone of Sri Lanka there frequently occur two soils side by side: the red-brown earths on the upper slopes, and the low-humic gleys below them. The RBEs are permeable and take a lot of water, which percolates through and reappears further down the slope on the LHGs. Some farmers on the lower soils do very well on seepage water and take very little from the field channels. If they took an equal amount of water to their upper neighbours, they would be waterlogged in no time. The water distribution is not *equal*, but it is *equitable*.

The corollary to this is that if those same farmers were charged for water on the basis of the volume that each took from the canal, the up-slope farmers would have to pay more, since they require more, due to the prevailing physical conditions of the soil. The system of water charging would then be neither *equal* nor *equitable*. If every farmer paid for water on the basis of the area of his farm, this would be more *equitable*.

Clearly the same arguments do not apply for a 5000 acre farm in California's Imperial Valley. Equity here must mean measurement and payment for water on a volumetric basis, according to an agreed schedule of charges.

Equity of ***income distribution*** is another idea sometimes mooted by social engineers. Whether this ought to be an aim of irrigation management is questionable. It is a misplaced ideal based on the premise that everyone is equal. This is not so. Some people work harder than others and deserve to get more income than others, and nowhere is this more obvious than on an average third-world smallholder irrigation scheme.

2.4 Sustainability

This word is beloved of all political strategists who believe that in order to make an irrigation scheme[5] successful it is imperative to declare its objectives to be ***sustainable***. The word is rarely defined precisely, and is therefore open to a wide variety of interpretations, but its normally-intended meaning is that once the project is established it will keep itself going without external help either from financial sources, overseas aid, external management or outside supervision. ***Institutional strengthening*** is invariably included in the terms of reference for all new projects to be studied by an international consultancy. That implies that with a little foreign help, an inexperienced, inadequately-staffed and dubiously-motivated government department will soon be able to handle its own affairs competently and efficiently. Much money is thrown in the direction of inexperienced and incompetent local constructors in the belief that since they are the ones whose land it is and who will therefore benefit most in the long term as the prosperity of the country increases about them, they will be keen to build the project well and efficiently. Noble though these concepts may be, it is naive in the extreme to believe that their guiding principles will be shared by all involved.

Much more money is usually hived off in corruption, which in one of its more virulent forms ensures that no canal is properly constructed and everything collapses and has to be rebuilt every two or three years. This is a disease which feeds upon itself and becomes a vicious circle which many countries find it hard to break out of. Far from being sustainable, many projects are contributing to their countries' rapid economic demise.

The one thing that always seems to be overlooked is that in order to make any project sustainable it has to be built properly. Most small irrigation canals are not built properly, either through ignorance, malevolent intent, or plain engineering incompetence. And the only answer is good engineering.[6]

The following catch phrases were proposed at a recent ICID conference (Ref. 2). To be sustainable a scheme should be:
- naturally adaptable;
- technologically appropriate;
- financially affordable;
- socially viable;
- economically viable;
- institutionally manageable;
- politically attractive;
- environmentally sound.

Desirable as these conditions may be, rare indeed is the project that fulfils them all. Too often there are conflicting interests at stake. Furthermore, they are open to interpretation and may take on quite different meanings in different countries. And how often do the environmentalists step in and declare war on a scheme for the flimsiest of reasons? Take the case of the Spynie Canal, a pumped drainage scheme in northern Scotland:

Farmer Mike Geddes had a prize-winning herd of Holstein cattle that thrived on the lush low-lying grasslands north of Elgin. The land was drained by the Spynie Canal, into which water from the feeder

---

5 *And not only does this apply to irrigation projects, nor even to any project in developing countries; recently in Scotland we encountered identical terminology in applying for funding the local community hall.*
6 *'Engineering' means planning, design, and supervision of construction. See chapter 11 for a more thorough explanation.*

ditches was continuously pumped. Then several fields adjacent to his own were officially declared 'Wetlands of Special Scientific Interest', and the water table was allowed to rise to the surface. This was to encourage no-one to go there and to allow the natural plants comprising mainly the common rush, and the natural fauna comprising mainly the common frog, to thrive unchecked. Consequently a large proportion of Geddes' lands were lost to production as the cattle could not walk in the fields without sinking in to their knees. Sustainability for the common frog spelled disaster for the dairy farmer.

I find some of these terms a little subjective and propose the following. To be sustainable a scheme should be:
- Properly constructed
- Able to be maintained
- Perceived as necessary by the people it is designed to serve

Adequate maintenance of the scheme is crucial to its continuing operation, that is, to its sustainability. If the scheme is properly constructed it will not need much maintenance, other than routine cleaning. Responsibility for maintenance is usually split between the farmers and the Government Irrigation Department or local councils. There are two kinds of maintenance, routine maintenance caused either by normal wear and tear, or repairs caused by unexpected but usually avoidable damage. Some schemes get badly afflicted with deliberate damage. There is always a reason for it however, and by looking at the signs of damage on an existing scheme, we can often conclude that there are social problems behind it. Two very different but sustainable schemes are shown in figures A2.1 and A2.2. The first is a large formal project, the Mahaprabha in Karnataka, India, which is kept operating through routine maintenance such as desilting by these paid labourers. The second is a village canal at Thijauk in Upper Burma, where the canal headreach gets destroyed each year by river floods, and the entire village turns up to dig it out.

2.5   Guaranteed Flow

The principle of guaranteed flow, or reliability of flow, is one upon which the success of all water management effort hinges. It is especially important on a demand schedule. It is essential on a supply schedule. It means not so much that water must be there all the time, but that water must be there when the management *says* it will be there.

One of the most difficult management vicious circles to break out of occurs when a supply schedule is not going according to plan. Having established a water delivery schedule, the initial deliveries may be a few days late because of an unforeseen delay in receiving bulk issues from the main storage reservoir. Irate farmers downstream complain that their rice nurseries are drying out and demand water from their block manager. He gives it to them, in the process upsetting a farmer at the head end who wasn't quite ready with his field when the water came. He breaches a distributary canal to get water. More time is wasted as the whole schedule is delayed further. Farmers near the tail-end realise that if they want water they'll have to do something about it themselves. They threaten the gate operators. They break the gate. Soon everything is out of control. This sort of thing is common in countries like Sri Lanka, where the big schemes are still run on a supply schedule.

2.6   The Downside – tragic environmental side effects

It is well-known, and all too often ignored, that irrigation schemes bring with them attendant risks of health, especially in the tropics. Canals are used for recreation, washing clothes, animals and people, and for drinking water. But in hot countries especially, disease vectors abound in open canals. They are also used as open sewers and convenient garbage dumps. Malaria, rift valley fever and dengue amongst others are killer diseases spread by mosquitoes. Diarrhoeal diseases are major killers of children and spread mainly through contaminated drinking water. Bilharzia or schistosomiasis is prevalent and spreading, through its water-snail vectors, in Africa and South Asia.

Chemical control of vectors and their habitats is well-documented as short-lived at best and foolhardy at worst, as it generates resistance in target species and frequently destroys an ecological balance including biological predators of the target species. Biological control including fish that feed on snails and weed, and the release of sterile male insect vectors, have an instant attraction for the ecologically-sympathetic mind, but success stories are rare. Engineering control is possible to a certain degree, but rarely practised. Some ideas are presented in chapter 7. Sociological control, in the form of education and provision of basic sanitation and drinking water, is of fundamental importance but too often ignored in the planning of a large irrigation scheme. It is too difficult to 'justify' these things in terms of economic internal rates of return.

There are also increased physical safety hazards. The Pehur High Level Canal drowned 14 people and countless other animals in its first year, despite designing-in exit steps every kilometre and trash racks at every major structure. Fencing off the entire canal was never a sensible option, especially as many people and animals quickly came to depend on it for washing and bathing as well as drinking. Provision of more frequent washing places might have helped. An educational programme certainly would have, but neither funding agency nor government saw these as sufficiently interesting to expend money on.

References and further reading for chapter 2

(1) Withers, B. and Vipond, S., 'Irrigation Design & Practice', Batsford, 1974.
(2) Viljoen, M.F., 'Strategies for promoting Sustainable Irrigation', ICID Workshop on Sustainable Irrigation in Areas of Water Scarcity and Drought, Oxford, September 1997.
(3) Jobin, W., 'Dams and Disease – Ecological Design and Health Impacts of Large Dams, Canals and Irrigation Systems', Spon, 1999.
(4) Stern, P., 'Small Scale Irrigation', ITDG, 1979.
(5) Kay, M., 'Surface Irrigation', Cranfield, 1986.

CHAPTER 3   WATER MANAGEMENT

The way in which water is to be distributed will affect the whole design of the scheme. There are several concepts of water distribution, which need to be understood by anyone involved in irrigation design and management. They are supply scheduling, demand scheduling, arranged scheduling, and continuous flow. One or any combination of these might be relevant to a particular scheme. One of the first stages of design should be a consideration of how water is going to flow around the scheme, and how much flexibility needs to be designed into it. All too often in the past these concepts have been anathema to design engineers, but this needs to change.

Institutional management is discussed here in some detail, because often there are serious flaws in traditional organisations which can be ironed out at minimal cost, provided they are recognised in time. A modular system of management is presented as an improvement over the more traditional ideas of line management.

3.1   Levels of Water Management

Different levels of the system may have greatly different priorities and objectives of management. Refer to table 6.1 for a definition of these levels. Consequently there may be several different modes of water delivery at various levels along the system. The bulk issues will probably be on a supply schedule, perhaps dependent on hydroelectric generation rather than irrigation demand. The main canals may also operate on a supply schedule but with continuous flow. The branch canals may be supply-scheduled with rotational flow. And the distributary canals using minor tanks for intermediate storage may be on a demand schedule.

Level 0 - Bulk issues

On a large irrigation network such as the Mahaweli system in Sri Lanka, the water may come from any one of several sources. To take the Mahaweli as an example, it has several major hydroelectric dams having different priorities for water release other than just irrigation. There is a complex network of tunnels and feeder canals (some of them not much different from the ancient schemes built more than a thousand years ago). These transmit water to large holding reservoirs, which in turn feed several separate irrigation canal systems. The entire system covers a quarter-million hectares of irrigation, together with half a dozen major dams, 50 km of tunnels, a score of large storage reservoirs, a thousand kilometres or so of main and branch canals, hundreds of minor tanks, and a great many minor canals. Clearly this has the makings of a management nightmare.

The management of all this has to be split up according to the physical levels of the water distribution system. Bulk issues refer to the sending of water through the primary distribution system, tunnels and hydroelectric works, before it gets to the holding reservoirs for the separate irrigation systems. The management organisation in charge of bulk issues might be completely separate from the one controlling the main irrigation canals, and certainly not the same as the ones controlling water flowing into the fields. Indeed it has to be separate, because there are conflicting demands between the end users of water for hydro power and irrigation. In fact in the case of the Mahaweli there are two separate authorities each managing major reservoirs, the Central Electricity Board for power generation, and the Mahaweli Authority for irrigation. Hence there are conflicting requirements even within the same Level 0. A third factor is brought into play when major reservoirs are used for flood control, in which the operating rules may be at variance with those for either irrigation or power generation.[1]

---

1 *Most large multi-purpose dams encounter such problems of conflicting interests. Cabora Bassa should have reduced flooding in the lower Zambezi but in fact exacerbated it as the reservoir was kept full to maximise power generation. Grand Coulee was originally built for irrigation of the Columbia Basin but for decades power production has been its main purpose. Tarbela controls a large proportion of Pakistan's irrigation water without which the country would not exist, and each year not only has to contribute to an equitable balance of the irrigation needs of four provinces but also generate a large proportion of the country's power. Since the reservoir is rapidly silting up the conflicts will become acute in the first half of the 21st century.*

Sri Lanka, being an island, is at least spared some of the problems which beset other countries situated on international rivers. The political disintegration of the Soviet Union in the early 1990s suddenly created Level 0 management problems that never existed before. For example, the Syr Darya River runs through Kirgystan, Uzbekistan and Kazakstan on its way to the Aral Sea. The headwaters and almost all the hydro power potential are in Kirgystan, whilst most of the irrigated lands are in Uzbekistan and Kazakstan. One of Kirgystan's main exports is hydroelectric power, but in order to optimise its returns from power production it needs to generate and hence release water throughout the winter. Unfortunately this timing does not coincide with the downstream irrigation requirements which occur mostly in summer. Moreover the lower reaches of the Syr Darya have been partly canalised and cannot accommodate the additional flows imposed by winter releases from the upstream dams. The excess flows are then wasted by spilling from Chardara Reservoir into the vast Arnasai depression in the Uzbekistan Desert. Whilst the Aral Sea inexorably continues to shrink, a new lake in the desert is being created. In the central-control days of the USSR, power shortfalls in Kirgystan were met by transfers of power from further north, and consequently timely bulk irrigation supplies to the lower Syr Darya were guaranteed. Political reform has brought with it some unwanted problems of water management.

Levels 1 and 2 - Main system

The main system loosely includes main and branch canals whose prime function is to get water around the system rather than delivering it to the field. They may have some direct offtakes into watercourses or field channels, but in general the main system delivers water to the head of the distributary canals and may include intermediate storage reservoirs. Main system operation may be designed for intermediate storage in the form of operational spillage, or with automation for downstream control.

Level 3 - Distribution

The distributaries and minors typically feed groups of small farms. Since these canals have a shorter response time than higher levels, they may often be designed for rotation and hence they flow intermittently. It is at this level that government control usually ends, it being impractical to take maintenance and policing any further down the system.

Level 4 - Watercourses, blocks and farm groups

This level requires a quite different approach to management. The delivery canals are small, and each may need to share out the water in some way to 50 or 100 farmers. It is at this level that social pressures and problems come to the fore and are manifested in physical damage or abuse of the delivery system. This is where the effects of water shortages or uncontrollable inflows are felt first. This is where fights break out, where people get killed, and where local politicians stand or fall. This is the level for which Water User Associations, familiar catchwords of all international funding agencies, are usually targeted.

In the case of low-pressure pipeline distribution systems this is the farm group level, amongst which the main operating restrictions have to be enforced. Potentially, this is where the advantages of pipeline distribution systems will be manifested the most.

3.2 Delivery Scheduling

There are several different ways of sending water around a canal system. The simplest way is to let the water flow continuously without control down all the canals at once. But this laissez-faire state of affairs is far from perfect, as the flow cannot be regulated and the tail-end receives only a trickle when it really needs a flood.

So introduce some control, with a planned operating philosophy and the physical structures with which to serve it. Control can be imposed from the top down, or induced from the bottom up. In either event it can operate according to the service concept, of providing water for the farmers' convenience, or otherwise if supplies are restricted or farmers' needs indeterminate, according to old-fashioned ideas of equality[2], in which everyone gets an equal share of water no matter what crop he grows and when. On a large scheme there might be different types of control at different levels in the system.

Then modify the delivery by altering the rate of flow, the frequency of flow, and the duration of flow. The system becomes more flexible, more adaptable to varying crops and weather conditions, and the farmers save time and labour in the field. The scheduling system becomes crop-based[3].

For the most efficient management, ultimate control of irrigation water must be in the hands of the farmer. In particular he must have the capability of turning off the water when he has had enough. Above all, he must have an assured supply of water, which is always available when he needs it.

Irrigation water management can be visualised as four evolutionary stages:

1) Uncontrolled continuous flow
  - wild flooding, including spate irrigation
  - run-of-river
  - proportional distribution

2) Supply scheduling (control from the top down)
  - proportional distribution
  - rotation
  - indented schedules
  - upstream control of canals

3) Demand scheduling (control from the bottom up)
  - limited flow rate
  - limited duration
  - intermediate storage
  - in-built flexibility
  - downstream control of canals
  - low-pressure pipelines

4) Flexible arranged scheduling (control from the bottom up via a Water Users Association imposing some operating restrictions)
  - semi-demand
  - limited-rate arranged schedules

These evolutionary states may co-exist in the same scheme. Rehabilitation may for instance introduce intermediate storage to enable demand scheduling to be carried out at the tail-end of a scheme whilst delivering water on a supply schedule on the main canals. In general they are arranged in ascending order of cost of capital works, and also in ascending order of operational efficiency. As limited water resources have to be spread amongst an increasing number of people, so the potential benefits of increased efficiency make it worthwhile to spend extra money on capital works for more sophisticated schemes. In many countries this threshold of efficiency over cost has long been reached, even though engineers have been slow to recognise it.

Uncontrolled continuous flow is in many instances a misnomer, because in schemes which rely on wild flooding the flow is anything but continuous. The spate irrigation schemes of the Arabian peninsula and Balochistan receive a highly unreliable water supply which has to be utilised to the full whenever it arrives. The water is directed into intermittently flowing channels, and there is no intermediate storage nor any regulating structures on the canals. This type of scheme is to be found in many parts of the world where farmers have built their own diversion weirs, which often have to be re-

---

2 *Not to be confused with equity, as discussed in chapter 2.*
3 *'Crop-Based Scheduling' is a term loosely used in reference to Demand Scheduling, but in fact it can apply equally well to Supply Scheduling. It infers enough flexibility in the supply system to enable the irrigation applications to be varied according to the actual needs of any particular crop in any particular field. A better term would be 'farmer based', in which many more variables are taken into account.*

built each year after being destroyed by floods. In recent years many small schemes have received international aid to build permanent diversion structures, like the *bhandara* in figure A3.1. Inevitably these schemes are run-of-river, meaning without reservoir storage. The more sophisticated may have some means of ensuring proportional distribution, with a few division structures in the canal system. However, especially in regions of very uncertain rainfall, many have no means of control at all and distribute water on a nearest-the-head-takes-all basis.

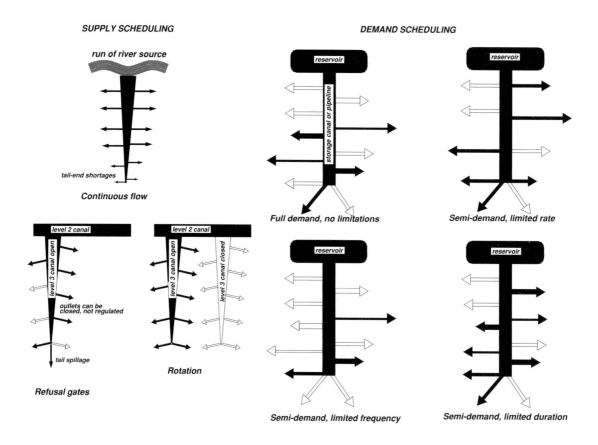

figure 3.1   Delivery scheduling

Supply scheduling has up to now been applicable to practically every large scheme in the world. Typically the overall scheme is operated by a government authority such as the Ministry of Irrigation, which has complete control over water distribution and maintenance down to a level beyond which control becomes impractical. Proportional distribution was of necessity applied to many of the large protective irrigation schemes. For productive schemes such as the Gezira, indenting was introduced at the outset. Rotation between secondary canals or parts of the main system is often applied during the dry season. The canals are designed for **upstream control**.

Demand scheduling is now in vogue amongst many financing agencies, although still not practised on a large scale in many places. Its basis is **downstream control** of the canals or pipelines, although downstream control does not necessarily mean demand scheduling. The discharge is controlled by the end user, from the downstream end of the system. The advantages of demand scheduling are that water can be supplied to the crop at the optimum time and when the farmer finds it most convenient. This offers the chance of increased crop yields, a saving in labour costs, a reduction in water wastage and a consequent reduction in problems of salinity and drainage. It means a free choice of crops as long as water is available, but also an increased capacity of the downstream end of the system.

Arranged scheduling, sometimes known as modified- or semi-demand, is a derivation of pure demand. The need for it arose when the concept of demand scheduling as developed on a pilot scale was applied to increasingly larger projects. The problem with a free-for-all demand situation is that if too many users on a section of the scheme want to take water at the same time, then unless the capacity of the system is very large the physical limitations of the system will be surpassed and some users will experience a failure of or reduction in the water supply. Arranged scheduling imposes some degree of self-management, usually through a water-users association. In its most elementary form a record is kept of the number of users on each lateral or tertiary unit wishing to take water on any particular day. If this exceeds a threshold amount established during design, then some users will be scheduled for the following or previous day. This concept leads on to the idea of ***congestion,*** which is discussed in Chapter 5.

Figure 3.1 gives a visualisation of modes of water delivery. The length of the turnout arrows indicates relative duration of flow, their width indicates their relative discharge. It is clear that demand scheduling implies a wide and disordered range of outflows.

3.3   *Uncontrolled continuous flow*

As the term implies, water flows continuously through the canal system with no intermediate regulation or control. If a large scheme is run entirely on a continuous basis it can be very wasteful of water, and invariably there will be a tail-end problem in which farmers close to the water source take most of the water, and farmers at the tail-end of the canal system get no water at all.

However, small schemes for paddy can nevertheless be run successfully with continuous flow, especially if the source of water is a run-of-river diversion weir and there can be no attendant risk of wastage of stored water as in a tank, nor of power as in a lift scheme. Larger schemes are often operated on a partially-continuous flow basis, with the headreach of the main canal flowing constantly and rotational scheduling taking over further down the canal system.

> Imagine how inconvenient it would be if your domestic water supply at home didn't have any valves or taps, yet you still used the same amount of water as you do now. There would be a steady trickle of water out of the end of the pipe, and you'd have to catch it in a bucket so as not to waste it, and empty the bucket every hour or so.

figure 3.2   ***Paddy fields in Java with a continuous throughflow, using bamboo spiles***

Basin flooding of paddy rice

The steep volcanic mountainsides of Java and the Philippines are covered in ancient paddy fields. Terrace upon terrace, stretching up into the clouds. They are irrigated by a continuous flow system which runs from field to field. Interceptor channels high up in the hillsides trap all the rainwater that runs off, and spread it around to the head fields. In this particular terrain there is no better system of soil conservation, nor for that matter of distributing water. It has to be said that in this region the cropping pattern is intimately linked to the climatic conditions, so there is not much opportunity to change the crop from paddy rice in the wet season, followed by a dryfoot[4] crop on residual moisture in the early dry season.

The Talli project - wild flooding

Here is an example of a continuous-flow scheme that at first sight seems desperately unfair, especially to the tail-enders. It comes from Balochistan, in Pakistan. The soil is sandy and dry. The rainfall is unreliable to say the least; on average 200 mm/year, but frequently nothing at all, not enough to grow anything. Summer temperatures regularly top 55°C. Yet people live there, and have done for centuries. They irrigate sorghum, millet and vegetables, and they survive in one of the most hostile environments on earth. They have a long-established tradition of water rights, which at first sight might appear iniquitous.

The Chakar River comes down from the dry mountains and flows intermittently for only 4 months a year, and sometimes hardly at all. Because there is no vegetation in the steep catchment area, the rainfall when it comes causes sudden and violent flash floods which can rise from 0 to 1000 cumecs overnight, and then diminish to nothing in a few days. The river in flood spills over the peneplain at the foot of the mountains, frequently changing its course, creating new channels and filling in others with sediment. The farmers' dilemma is how to use this water in the best way.

The people there, being excellent engineers, have evolved a system of canals fed from diversion weirs across the main river channel, which can be 500 m wide in places. The canals or *wahs* are unlined, up to 20 km long and the diversion weirs are temporary affairs of stone, earth and brushwood, which have to be rebuilt in most years. In the Talli area there are half a dozen diversion weirs feeding a total area of 6000 ha.

When a flood comes, all the water is diverted into the first *wah* until the diversion weir overtops or breaches. Any water spilling and not percolating into the deep gravelly soils beneath the river channel runs on to the next diversion point downstream, which takes all the water it can. And so on downstream until there's no water left in the river. The upstream *wahs* flow continuously at their maximum capacity until all the water in the river dries up. Water is flooded onto the fields continuously and any excess is diverted into the drainage system where it percolates into the ground or ends up in the river miles downstream. There is no attempt to divide the available water between all the potential users. Only in a very wet year, say once a decade, will the entire 6000 ha get water. In an average year only half the farmers get any water at all, and even then it's only those near the upstream offtakes that get enough for a reasonable crop. It is a tail-end problem with a vengeance.

This might appear to be unfair. It seems a long way removed from the sociological ideal of equity in water distribution. Would it not be better to spread the water evenly around, perhaps by rotating the use of the offtakes alternately for each flood, so that everyone has a fair chance of getting water?

At the start of any given year there is no guarantee of water for anyone. Even if two early floods come in quick succession, there is no certainty that there will be any more afterwards. If the first flood were taken by the upstream farmers, and the second by the next group, what happens if there are no more floods that year? No crops survive anywhere and everybody starves.

At least with the continuous-flow, first-in-line-takes-all system, there is a realistic chance of a good crop over at least a part of the area. And any attempt to impose equality over the whole area would result in certain disaster, with regular crop failures assured over the whole area. The Talli project is a *protective* scheme, but it does not protect every individual farmer, rather the community as a whole.

---

4 *The more accurate term 'aerated root zone' is sometimes used instead of 'dryfoot'.*

Controlled wild flooding on the Rufiji

The Rufiji in Tanzania is the biggest river on the east coast of Africa, and the lower floodplain covers an area the size of Holland and has some of the most fertile soils in the country. It gets flooded every year, not always to the same extent, nor at the same time. Because of the unpredictability of the floods the farmers have over the past century developed a sophisticated farming technique of inter-cropping several different crops which have differing resistance to flood and drought and different water requirements. So in a year of heavy floods the rice does well but the cotton gets drowned and the maize does well on residual moisture. And in a dry year the cotton does well even though the rice can only be grown in the wet depressions, and the sorghum is grown as a flood-recession crop. So every year, whatever the weather, irrigation and soil conditioning in the form of new silt deposits came with the floods and the farmers did all right.

This delicate system of cropping was threatened by two things. First, a socialism-inspired and misplaced government policy of villagisation ('Ujamaa') which removed people from the floodplain and made it illegal to live there. Secondly, a proposed hydroelectric dam at Steigler's Gorge upstream, which would have reduced the annual flooding to the point where the farmers couldn't depend on the floods for irrigation as they always had in the past.

figure 3.3   The Rufiji Delta, Tanzania

The passage of time and the good grace of Julius Nyerere eventually killed off the Ujamaa idea and for economic reasons Steigler's Gorge dam is no nearer to being built. But the dam was nevertheless designed, and incorporated in the original design was a facility to release water downstream in huge bursts in order to simulate the natural flood regime and irrigate the floodplain as before. Unfortunately the designers didn't think through the other side effects of the dam, principally that the sediment-trapping capacity of the reservoir would lead to degradation of the river bed downstream and make it physically impossible to irrigate by wild flooding. The best and most practical solution was to install a canalised system of irrigation in the floodplain. It was commendable of the dam designers to take into account the existing 'natural' method of farming. But it needed to be recognised that any change at one point of the system, could have deleterious effects on other parts of the system. After all, the inter-cropping system was not the ideal, it was only a sensible adaptation of available resources. With a dam and reservoir to guarantee the water supply, a more formal system of irrigation which will be far less wasteful of water and far more reliable in terms of water distribution, becomes possible. So here is a recommendation for canals, in the face of maintaining the status quo.

The Gezira project flows continuously against the rules

The Gezira in Sudan was one of the best engineered projects ever. It was built in the 1920s when labour was cheap and plentiful. It was designed as a supply-scheduled indenting system with night storage in the minor canals. Farmers don't use it that way any more. It's much easier and cheaper to let the water run in continuously for four days and nights, rather than the old method of daytime-only irrigation with constant field supervision. The critical factor is the shortage of labour. An average tenant's farm of 10 feddans[5] now uses for irrigation labour only 20 man-hours per crop as compared with over 400 man-hours previously. All the farmer has to do is open the field outlet pipe and leave it for 4 days, whereas in the old days he had to constantly supervise the water flowing from basin to basin within the farm.

The Gezira farmers have been forced to do it this way by economic pressures such as the high cost of labour and low price of crops, and by practical necessity because the storage capacity of the canals is reduced with silt. But they can get away with it for one simple reason. The soils are cracking montmorillonite clays (the infamous black cotton soils), which swell and close up when they get wet. They are 'self-irrigating'. Even with basin irrigation it is possible to evenly and automatically irrigate the fields because the soil surface rises when it is wet enough and induces the flow across the field onto dry soil.

figure 3.4   *Gezira cotton*

But don't think that this is the correct thing to do just because the Gezira people do it. Even after a full rehabilitation it may never be possible to revert back to the tightly-controlled indent system that they had before. And with this continuous flow system there is no guarantee of downstream farmers receiving water. Whereas before it was simple to ensure that each reach of a minor received its full complement of water, now the head reaches get an assured supply but the tail reaches don't. And the Gezira's soil and billiard-table topography are quite unusual in their adaptability to this kind of management mutation, so it might be inappropriate to attempt it on more permeable soils or rougher topography.

However, a further and highly damaging variation of this management method has afflicted similar soils at Sennar (just south of the Gezira and recently the subject of a privatisation initiative), when

---

5 *1 feddan=0.42 Ha, or1 .038 acre. This unit is used in Egypt and Sudan.*

farmers, not fully comprehending the physical nature of black cotton soils, irrigate too often. Many farmers irrigate as soon as the soil surface appears dry. The small surface cracks rapidly close, preventing water from filling the deeper cracks below. The subsoil remains dry and the root zone remains in the top 25 cm. Yields are disastrously low due to drought, despite the surface of the fields being kept almost constantly wet.

Proportional flow

The great irrigation schemes of the Indian subcontinent that were developed in the latter half of the 19th century were all designed for protective irrigation based on proportional distribution of water. That is, the canal system subdivided repeatedly along its length with each branch and sub-branch carrying an amount of water proportional to the area under its command. No adjustment of flow is possible downstream of the main system, which normally means downstream of the distributary headgates. There are no cross regulators on the main system. There are no gates on the distributary and minor canals. All outlets were fixed, ungated flumes or orifices.[6] Theoretically, this results in a fair distribution of all available water, in which any unauthorised abstraction is immediately noticed. In practice, unless due allowance is made for seepage and other losses which tend to be proportionally greater toward the tail of the system, the tail-end will often be short of water. And if the upstream farmers can get away with it they may abuse the system by stealing water[7] and leaving less for those downstream. Nevertheless many of these old schemes have worked tolerably well for a century or more. In north-western Pakistan this is conceivably due in no small measure to the use of the ubiquitous Kalashnikov in solving all disputes, water-related or otherwise. But the times are changing, and the need for more control at all levels in the system is becoming more pressing.

figure 3.5   ***Katlang Distributary of the Upper Swat Canal***

3.4 Supply scheduling

In a supply-scheduled system, water is controlled from the supply end, that is, from the top down. For supply scheduling to be in force there must be a management authority firmly in control, who delivers

---

6 *In Pakistan the flumes are called open flumes, based on a design by Crump, and the orifices are APMs (Adjustable Proportional Modules) or Stoddard-Harvey type rectangular orifices. All are designed to operate proportionally by delivering a constant proportion of the parent canal flow through the outlet, no matter what the water level in the parent canal. None of them are truly proportional, and all only operate well under a narrow range of flow.*

7 *A favourite method is to encourage one's water buffalo to wallow in the parent canal immediately downstream of one's outlet, thereby raising the water level and increasing the discharge through the outlet.*

water down the canal system according to a strict programme. The programme may be determined in response to indenting or advance requests from farmers, or simply by dividing the available water on a proportional basis.

For supply scheduling to work properly, there must be tight control, and the supply programme must be adhered to rigidly. It is an inflexible system, which presupposes that all farmers will want or be able to use the water when it arrives. There is little scope for farmers to deviate from agreed crops or planting dates. Shortages of labour or machinery may cause the farmers to delay cultivation, and there can then follow

> Imagine the water supply in your house to be turned on and off only by the water company. The water comes only twice a week and there are no taps in the house, just an open pipe leading into the bathroom. The whole family has to have a shower on Thursday morning 3 am and Sunday afternoon 3.40 when the water is on. If you want a cup of tea on Wednesday you'll have to store water in a bucket until then.

extreme pressure on the management to alter the water delivery schedules. This usually results in a complete breakdown of water management.

figure 3.6   *Cross regulator on Machai Branch*

Rotation

In supply-scheduled schemes, all but the smallest will use a rotational schedule at some point along the canal system. Water is fed down each canal for a set time period, and the canal is then closed until its turn comes around again. Rotations are often made on a regular weekly, 10-day or two-weekly basis, depending on the crop. Rotations are usually made from level 3 downwards, that is between distributaries, tertiaries or field channels.

If the geography of the scheme permits, then rotation can be effected on the level 1 main system canals. The Upper Swat Canal in Pakistan rotates in winter between two halves of the scheme. The pivot point - the centre of gravity of water use - happens to be near the boundary of two administrative divisions. It is convenient from a management viewpoint to alternate supplies between Malakand

Division and Swabi Division. The rotation period in this instance is 8 days. This apparently strange period is because the watercourse-level warabandi schedule is weekly, based on 7 days. Due to the long response time of the main canal system (between 5 and 7 days from head to tail) the flow rate at the end of a rotation period will reduce gradually to the detriment of farmers with a late turn. Furthermore, water will arrive at an inconvenient time for many farmers. By making the main system rotation non-synchronous with the warabandi, these inequities are spread more fairly between all farmers.

It follows that if a canal is only going to flow intermittently, its design will have to account for a larger capacity than if it were to flow continuously, because a given irrigation volume is being transmitted in a shorter time. So in designing a canal, its mode of operation must be known from the outset.

Pivot points

The point in the system downstream of which the water is periodically closed off during rotation can be termed the *pivot point*. In planning a rotational schedule the location of the pivot point has to be fixed with regard to irrigation requirements upstream and downstream, losses in the system, and the possibility of regulating the incoming flow.

Figure 3.7 demonstrates the pivot point in the Upper Swat Canal system in Pakistan. This is a run-of-river scheme with no regulation of inflow. During Rabi (winter) all available flow in the river is used in the canals to be rotated between two halves of the system. Recent rehabilitation work included the installation of 12 cross-regulator structures along the Machai Branch Canal (figure 3.6) in order to control water levels and outflow into the distributaries. The pivot point will be situated at one of these cross regulators which will be used to isolate half the scheme during rotation. However, since 2002 the new Pehur High Level Canal has added another dimension to the whole scheme, delivering more water to the tail of the existing system. In turn this has freed up more Swat water for expanding irrigated areas in the upper part of the system. The centre of gravity of the irrigated area under rotation will change, and the pivot point will move upstream to a different cross regulator.

Traditional supply schedules

The way irrigation is normally planned and executed in many countries is on the basis of a *supply schedule*. In designing the scheme, the capacities of the canals are sized according to the water allowance multiplied by the area under command at each point in the system. Hence the canal capacities reduce down the system as far as the watercourse head. The scheme is controlled from the top (supply end) down, by the Irrigation Department, and no control at all is given to farmers. On the level 4 canals the discharge is shared between each farm according to one of the procedures such as *warabandi*. In essence the system is completely inflexible, and cannot accommodate any deviations from the planned flow rate, duration or frequency.

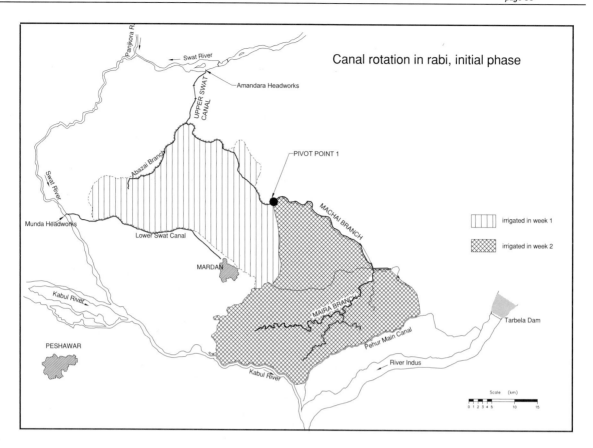

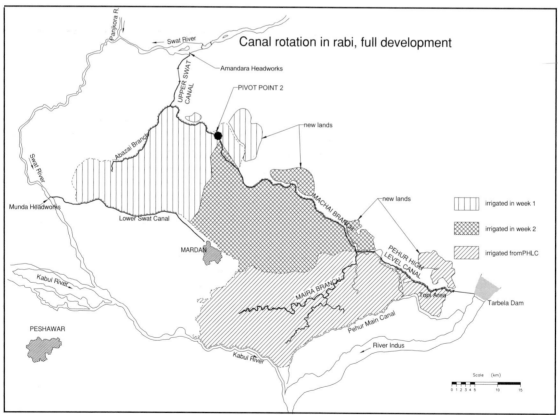

figure 3.7   ***Pivot points on the Upper Swat Canal system***

Warabandi

Warabandi is a famous example of rotational supply scheduling, which evolved in the 19th century on the large schemes of northern India as they developed run-of-river supplies ever more thinly throughout an expanding population. It means in Hindi 'Water for Land'. Its guiding principles are to ensure protective irrigation for all rather than intensive irrigation for a few, working on a pre-arranged roster for distributing water to the head of each watercourse in turn. The time interval between irrigations is constant and easily understood, usually a week or 8 days. The duration of irrigation at each turn is also constant. It is aimed at equitable distribution, in that the effects of water shortage are spread evenly amongst the farmers on the basis of irrigable land holding size thereby stimulating efficient water use in the field. Government control is down to level 3, and farmers manage their own watercourses. When water is flowing past their field, farmers are obliged to use it, or lose it. Parent canal outlet structures are ungated flumes which are pre-set to give a specific discharge for each watercourse, based on its command area. It is not perfect, and a major drawback is in the difficulty of making allowances for response time and conveyance losses in the minor canals. Neither has it been found to work with cropping patterns based on paddy rice, which are sensitive to the intermittent supply of water.

Adjustments are made on some schemes to account for response-time losses while the watercourse is filling the reach approaching a particular farm. The correction allowances are called *Bharai* and *Jharai* in the Punjab, and are described in detail in an excellent book by S.P. Malhotra (Ref. 5). The same author introduced another correction for absorption loss, called *Chusai*, which would vary depending on the length of lined watercourse serving each farmer. All these corrections are added to the allocated flow time for each farmer.

A drawback of this as of any supply-scheduled system is the lack of flexibility in water availability at the farm end. Thus a farmer growing a thirsty crop of sugarcane theoretically gets the same amount of water per hectare as one growing not-so-thirsty groundnuts. In practice it is also common for warabandi time slots to be traded between farmers, a self-regulating form of water management which although 'unofficial' can be quite efficient.

Shejpali

Shejpali is another Indian system which is practised in Maharashtra and Gujarat. Similar to the warabandi system, it was introduced to counter the tail-end problem, by giving priority to the tail-end farmers and serving the head-end farmers last in each rotation. In practice the system breaks down because it is too rigid, and unenforceable by available management resources. A typical case study is given in Ref. 15.

Indenting

Indenting is the process of requesting water deliveries on a supply-scheduled system, such as the million-hectare Gezira scheme in the Sudan. In American terminology indenting would be described as *fixed-rate arranged scheduling,*[8] implying little or zero flexibility in water usage other than the date of delivery.

On paper it is simple. The end-user requests water a certain time before he needs it, and the management releases it to him down the canal system. That way the farmer gets the water he needs, and the management retains full control. In the USA and Australia it is a common arrangement. When he needs water, the farmer merely picks up the telephone. Farms are large, 1000 ha or more. Discharges are comparatively easy to keep track of. Records are easy to keep and payments are made on time by farmers.

To succeed, indenting needs three essential things;

---

8 *The American Society of Civil Engineers categorises scheduling into three types being Demand, Arranged and Rotational, with modifications for frequency, rate and duration. They are relevant to the farm turnout level and may need extending or qualification to fit all situations met at main system level, especially in some less developed countries.*

- A canal system that is fully controllable by the management.
- A good communication system.
- Accurate records of water deliveries.

When the Gezira project first got under way it had all of these. A telephone network, at that time a major technological innovation, linked all the distribution points on the system, and impeccable records of all canal discharges were maintained. Even today the sub-divisional engineers still keep their flow record books filled in, but it doesn't mean so much any more thanks to the physical breakdown of the system. Block Inspectors are responsible for indenting water from the sub-divisional engineers. The amount of water required on each minor canal (which may be 20 km long and serve 2500 ha) is calculated daily during rainy periods or weekly at other times from Penman estimates of water use. It is then a straightforward matter for the engineers to calculate the water required at each division point in the canal system and operate the gates accordingly.

For half a century the Gezira project was a model indenting scheme. But by 1985 it had lost its internal communication system, its canals were silted to the brim and its farmers could not afford the cost of labour. As a result, its performance deteriorated to the point of catastrophe. It has since recovered somewhat, through the increased availability of government funding for maintenance, in turn deriving from recent oil revenues.

figure 3.8   *Imperial Valley, California. Indenting on a massive scale*

The *negative* points of rigid supply scheduling are:

- Water arrives on the farm at the scheduled time and for the scheduled duration, whether or not the farmer needs it and whether or not it is convenient for him.
- If the crop does not require irrigation at the time (e.g. during harvesting, or after rain) excess water can only be turned into the drainage system and wasted or creates waterlogging.
- If the crop needs more water than the average allowance (e.g. during a peak growing stage), it can't get it.
- If the crop needs more frequent irrigations than the average frequency (e.g. during early root development), it can't get it.
- The discharge on a small farm will be small, and each irrigation will cost the farmer more in time and labour than with a larger discharge. The farmer has no control over the flow rate.
- Since the farmer has no control over the flow rate, no improvements in field irrigation practices are possible.
- A communication network is required between the head of the system and the watercourses.
- The operating agency may have a difficult job to effectively control a large complex scheme.

The *positive* points are:

- The farmers each get a nominally equal share of the water per unit area of land. However, this may not be a fair (i.e. equitable) share, since it may not arrive at an optimum time or in an optimum quantity.

The *positive* points of flexible supply are:

- Water is taken on the farm only when the farmer needs it and when it is convenient for him.
- If the crop does not require irrigation (e.g. during harvesting, or after rain) the water is not taken and is therefore not wasted nor creates waterlogging.
- If the crop needs more water than the average allowance (e.g. during a peak growing stage), it can get it.
- If the crop needs more frequent irrigations than the average frequency (e.g. during early root development), it can get it.
- If several different crops are grown on one farm, each can be treated individually.
- The discharge at farm level will be controlled up to a set, limited rate, but is not dependent on farm size.
- The farmer has complete control over the flow rate and can take as much or as little water as required, within the physical and managerial constraints of the system.
- Irrigation durations can be reduced, thus saving labour costs to the farmer.
- Yields and economic returns will increase.
- Wastage of water will be reduced.
- Social friction amongst farm families can be reduced.
- The centralised operating agency does not have to operate gates below main system level.
- No communication network is required between head and tail of the system.

The *negative* points of flexible supply are:

- Intermediate storage and spare capacity is required in the distribution system, which will increase construction costs. However, the long term economic benefits usually far outweigh the extra construction cost.
- A Water Users Association is required to arrange irrigation schedules, and a certain amount of discipline amongst the users is a prerequisite for success.
- Initial training of engineers, irrigators and farmers in the design and operation of the scheme will be essential.

3.5 Flexibility

A *flexible delivery system* can dramatically increase crop yields and reduce water wastage. The essence of a flexible supply is that it is *downstream-controlled* and *demand scheduled*. The farmer must be allowed to make his own decisions relating to all farm functions and must not be restricted by the water supply. Frequency and duration of irrigation streams is decided by the farmer with as few imposed restrictions as possible.

Flexibility of operation can be built-in to the system with **automation**. Once the main system is fully or partially automated with intermediate reservoirs or level top canals, a logical development is to extend the automation further down the system to farm level. This can most easily be done with the use of pipelines. With a semi-closed pipeline system in place of open canals and watercourses, no regulation is required on the system which is controlled completely from the downstream end. When water is removed at the farm outlet, the pipeline automatically communicates to the source, which automatically replaces the removed water. With such automation in place down to farm level, dramatic improvements are possible in crop yields, social benefits, economic returns, reduction in water wastage and reduction in waterlogging damage.

The advantages of flexible supply are readily apparent. Even though it may not be applicable to many traditionally-designed existing schemes at the present time, it is likely that in the future it could be. **Low-pressure pipelines**, fed by service area reservoirs or level top canals are the easiest way to achieve flexibility.

3.6   Demand Scheduling

Water on demand

In a demand-scheduled system, water supply is controlled from the bottom up, that is, from the farmers, who take water as and when they need it. There must be some degree of built-in intermediate storage that allows them to take water at any time, or there may be an automatic canal system which allows the farmers to abstract water at will.

The advantages of demand scheduling are that it is flexible and can accommodate individual farmer's needs, thereby reducing problems in management, social conflict and shortages of labour or cultivation equipment. Consequently it can be very efficient in water use. Drainage problems that are not really drainage problems at all because they are created by over-irrigation, can be avoided without recourse to expensive tile drainage. And crop yields can be vastly increased. Farm economics can be enhanced with lower labour costs, and when water is taken at the farmer's convenience he can arrange his own time in a more productive way.

> The domestic water supply in your own home runs on demand. Whenever you want water, you just turn on the tap. But don't forget to turn off the tap afterward!

In order to work well, there must be storage of water either at source or at some intermediate point in the distribution system. And the response time of the system must be rapid, as is possible with the parabolic canals shown in figure 9.11 or, even better, with an underground pipe system of sufficiently large capacity. A demand system requires at least four times the farm outlet capacity of a continuous flow supply-scheduled system.

Methods to achieve farmer control all rely on some degree of intermediate storage and at least partial automation. Operational spillage on the canal system is the simplest and cheapest way of meeting both requirements, if water supplies are available and can be re-used downstream. Low-pressure pipelines offer an economical solution to the problems of storage and automation. Ponds and reservoirs, or in-canal storage, are other possibilities to be considered.

In practice however, with all smallholder schemes except the very smallest it is not possible to allow a full demand schedule to be operated without some restriction on flow rate, frequency or duration. Practical application of a demand schedule needs such limits to be built in. The general term '**semi-demand**' was coined by Merriam to apply in cases where full demand is not possible for practical reasons. It is a term which is really superseded by the more precise variations of '**arranged**' schedule: limited rate, limited duration or limited frequency.

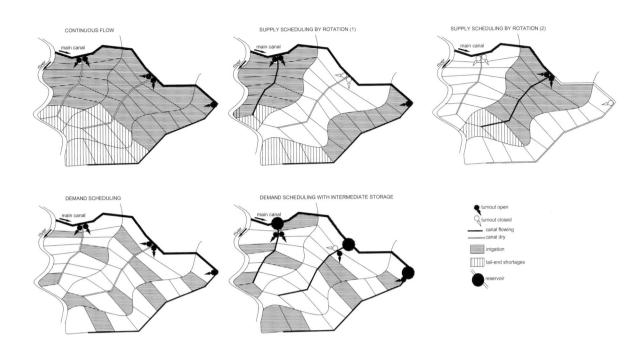

figure 3.9   **Continuous flow, rotation and demand scheduling, ways of mitigating the tail-end problem**

rrAnged scheduling

Usually the supply of water is limited, and a demand scheme must have some controls built in. Such controls can restrict either the rate of flow, the duration of flow, or the frequency of irrigations. These variations are demonstrated in figure 3.9.

John Merriam (Ref. 2) has been instrumental over the years in extolling the virtues of low-pressure pipeline systems. They usually comprise a level-top feeder canal or storage reservoir, feeding into a buried concrete pipeline which operates under gravity pressure. Field outlets are orchard or alfalfa valves having a limited capacity, hence limiting the flow rate, which is further controlled by pressure-reducing float operated Harris valves in the pipeline system. The trend in California is to replace the Harris valves with computer-controlled piped networks, usually with pumping.

In California, surface irrigation is becoming less unfashionable with the limited rate demand scheduling that is possible with buried pipeline distribution networks. In developing countries, there is almost infinite scope for improving irrigation operation through adjusting the water management philosophy. So far there have been a handful of pilot projects and one or two notable failures because nobody turned off the tap. It's a long way from the orchards of California to the paddy fields of Sri Lanka. One pipeline project failed after the farmers destroyed the outlet valves to get more water, perceiving the restriction of flow to be a retrograde step from the heavy flows they were used to. Another failed because the supervising engineers did not understand the demand concept and reverted to the old rotational schedule. Another failed because the design engineers did not appreciate the philosophy of sizing the pipes for low congestion and maximum stream size (see chapter 8), and the pipes were designed too small. But the 21st century is already upon us, and all this will change. It is an aim of this book to make such unfamiliar concepts accessible to more engineers.

Semi-demand, arranged scheduling

Unless there is an ocean-sized reservoir close at hand, everyone cannot take all the water they need all at the same time. Fortunately the probability of every farmer on the system wanting water at once is close to zero, but even so in practice there have to be limits on the number of farmers taking water, which means arranging irrigations in advance. (Probability and the concept of congestion is discussed further in section 5.4). This in turn means someone being in charge of localised sections of the distribution network, to keep records of water issues and an order-book of requested irrigations. And this in turn means that farmers need to be literate, and flexible in their attitudes toward their neighbours.

In many countries the necessary changes in educational standards and social attitudes can be a long time coming. Even in the USA it has taken 50 years to instigate this type of system control, and even now its acceptance is only gradually being realised.

Limited rate, arranged scheduling

The practical manifestation of demand-scheduled systems culminates in a combination of limiting the flow rate and arranging the day of irrigation in advance. A semi-closed pipeline system operating on a limited rate arranged schedule relies on physically limiting the farm discharge through alfalfa or orchard valves and by restricting the pipeline operating pressure, and on arranging the schedule by limiting the number of farmers in a distributary group that can irrigate on the same day (limiting the *congestion*). The design of a pipeline system is described in Chapter 8.

The limited rate arranged schedule is a semi-demand schedule which adequately meets the needs of flexibility. The operating agency, which should be a Water Users Association, retains a minimum degree of control by requiring that the farmer arranges the date on which he irrigates. Water is taken by the farmer under this schedule from his own turnout valve. This system requires no project personnel for operation, since the farmer controls the flow rate and duration on the arranged day and the rest is automated. The farmer can fix his own duration up to 24 hours and rate up to the design *unit stream*. He is not restricted to any frequency of irrigation.

3.7   Intermittent Flow

When a canal doesn't flow all the time, as in minor canals on a rotational supply system or on a demand schedule, there are several things to consider in its design and management.

Response time

If the canal is expected to flow for frequent short periods such as on a tubewell-fed demand system, a rapid response time is likely to be essential. This is especially so if the farmer is required to pay for water on the basis of pump running time or under the *warimetric* system (see 3.9 below). A fast response time is under these circumstances a major factor in opting for canal lining and especially a parabolic profile that is sensitive to fluctuating flow rates.

Filling time

Filling canals with gated outlets from the tail-end upwards is quicker than the normally-perceived procedure of filling from the head and can lead to savings in distribution times. This perverse-sounding statement means that outlets along the canal should be kept closed until all reaches in the entire canal are full. If any gates are open on the way, the advance wave slows down and levels in the canal take longer to stabilise, and consequently so do flows through any open outlets. The result of this is that outlet discharges are difficult to measure and a fair distribution of water is less easy to achieve.

The design of cross regulators can have a dramatic effect on filling time. Open undershot sluices allow the water to proceed rapidly, whilst overshot weirs require each reach to be completely filled before passing any discharge. Hence, where a weir is used as a cross regulator on a large canal, a combined gate for filling and draining is often a good idea.

Absorption

An earth canal in a cracking clay soil that has been allowed to dry out can lose much water through dead storage and absorption and will certainly take up a lot of water during filling. This can lead to management problems and wastage of water. The conveyance efficiency of such a canal flowing intermittently may be very poor, even though seepage loss as such may not be very great.

Health

If the canal can dry out at intervals, aquatic weeds don't get a chance to establish themselves. Neither do snails, a vital link in the schistosomiasis chain, nor mosquito larvae. So drying out can be a cheap and effective means of improving local health conditions and facilitating maintenance. But if the canal is designed to dry out, it should not have pockets of dead storage that could stay wet. It is often a good strategy to line reaches of canal that pass through a village in order to reduce the health hazards of stagnant water.

3.8   Institutional Management

It is important to consider the institutional side of irrigation management at an early stage in planning. If an existing project is not working properly, there may well be invisible reasons for it which stem from a flaw in the management structure. Putting it right may not cost anything at all. Ignore it at the risk of wasting a lot of money in concrete and steel.

Irrigation projects bring together an enormous range of technical and social disciplines which need to interact properly in order to stand any chance of success. But bureaucracy usually gets in the way. Smallholder Irrigation Schemes suffer more than most development projects from a surfeit of conflicting interests and many different government departments all supposedly co-operating but failing miserably in the attempt. And who loses? The farmer of course, the one who's supposed to be benefiting most.

One management book worth reading is 'Further up the Organisation' by Robert Townsend (Ref. 14). Everyone who directs one or more other people in their work should read it.

Line management

Figure 3.10a is a typical government management structure, familiar to everyone who's ever been involved with any government organisation. On an irrigation project there are numerous different technical disciplines involved in its operation and management. Unfortunately in most governments (and in many large private companies too) these disciplines are separated from each other by the line management system. The intermediate and lower levels of management are left out of figure 3.10a for simplicity, but each of the separate departments will have three or four intermediate levels, from senior managers down to field inspectors who are the ones in contact with the farmers. So the hapless farmer will be dealing with a dozen or so field workers from all walks of life, all reporting to separate bosses through separate lines of the management system which don't interact. It is easy to imagine what happens when, as is usually the case, a farmer comes up with a problem that has multiple causes. It gets tossed around like a ping-pong ball, from one department to the next, and in the end nothing is resolved.

Here are some real-life examples of what can happen with a line management system.

- A field channel remains dry on the day when it should be flowing. The ditch-rider gets beaten up by angry farmers. The farmers breach the distributary bank and disrupt supplies downstream. The downstream farmers beat up the ditch rider again. He is absent for 2 days complaining to his boss who writes a memo to the deputy head of the water management division who writes a letter to the project manager. Meanwhile water management in the field breaks down completely.

- A dispute arises between two farmers, over the precise location of their common boundary. They complain to the extension officer who says it is not his concern. They complain to the ditch rider who refers it up the line and it eventually gets to the project manager who refers them to the survey department. The farmers go to the survey ministry who say they are no longer working in the area and are too busy to see them. The family feud lingers on for years and someone eventually gets injured.

- The management decrees that dry-foot[9] crops shall be grown on upland soils this year. The farmers cannot get seed. The extension officer says he doesn't have the means to organise supplies. The farmers grow rice, as usual.

Unit management

Volvo, a Swedish car company, were quick to recognise the debilitating effects of traditional line management systems. They reorganised their workforce into self-supporting cells in which groups of workers became involved in the entire process of car making from start to finish. Morale and productivity improved beyond all hopes.

In a similar way we can reorganise the line management system into something else that works much better. Figure 3.10b shows a unit management system that is made up of exactly the same bits and pieces as the line management of figure 3.10a. But there are some important differences. First of all the farmers and the project manager are linked directly through one person, called the Unit Manager. Secondly, everything else is kept to the side so it doesn't block up the system. And the farmers are not being bombarded by arrows. There is a single line of communication between the farmer and the project manager. So when the farmer has a problem he doesn't have to chase round any of a dozen different field officers who he might only see once or twice a year. He goes to one person only, whatever the problem. And when the Project Manager wants to communicate with the farmers he goes through the same person, the Unit Manager.

It is a modular system, based on a conveniently sized chunk of the project. When we first tried it out on the Mahaweli project in Sri Lanka the unit covered 250 ha over two distributary channels. As the management system was refined and the teething problems associated with a newly-settled project were resolved, the unit was extended to cover 1000 ha.

With unit management, 90 per cent of all problems can be solved on the spot without delay. The remainder are referred to the relevant quarter by the Unit Manager, and the farmers only have to deal with him.

But suppose the Unit Manager turns out to be not an ideal person and shows favouritism or unfairness toward a particular farmer? There's a built-in bypass for this, via the local village administration. In reality there has to be a safety valve to ensure that the Unit Manager can never become a tyrant.

Here's how the same problems would be handled by the unit system:

- The Unit Manager and his field assistants are responsible for distributing water within the unit. At this level, below level 2, water is considered as another agricultural input. If the current problem is from the outside, the Unit Manager contacts the water delivery agency and then informs his farmers and re-schedules the supply accordingly. The farmers then don't feel the need to damage the canals because they are kept informed of the actual supply situation.

---

9 The term 'aerated root zone' is sometimes preferred to 'dry-foot'.

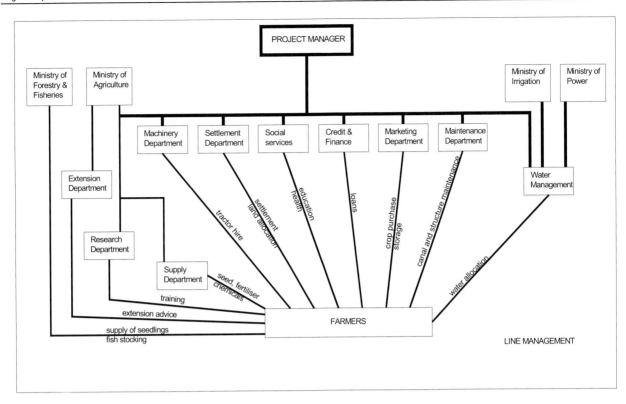

LINE MANAGEMENT

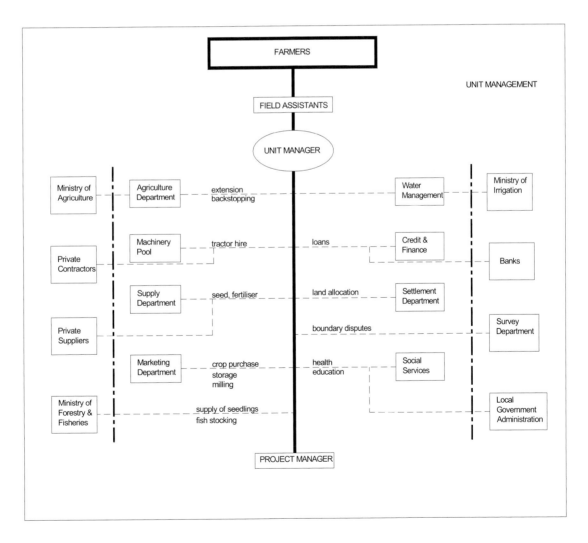

UNIT MANAGEMENT

figure 3.10 *A traditional line management structure, and improved Unit Management*

- Concerning the boundary dispute, which is a very common occurrence on a newly-settled scheme, the Unit Manager either resolves it on the spot or calls in the survey department who re-establish the boundary.

- In obtaining seed and inputs on time, the Unit Manager plays an assisting role. He contacts the supply department or local suppliers to ensure that enough seed is available in good time, and if not to re-schedule the planting programme in a controlled way. He has no hand in procuring the seed other than acting as a go-between.

Authority and assistance - conflicting roles of water managers

Imagine you are a farmer at the tail-end of a supply-scheduled irrigation scheme. How do you see the local gate operator?[10] Is he a friend who gives you water whenever you want it, or is he an enemy who tries to prevent you taking water whenever you want it? The chances are he's young, poorly paid and inexperienced, and you can probably persuade him one way or another to follow your way of thinking anyway. But then, how do you see your local agricultural extension worker? He's definitely a friend, helping out with your problems in the field and passing on training and information from the government technical departments.

So there's a difference, highlighted by the assistance-oriented role of the extension worker and the authoritarian or allocation role of the water manager. It's important to realise the potentially conflicting roles that a manager might have to support. In the unit management system, the Unit Manager is the interface between these two roles. In a line management set-up, there is often no clear delineation of authority, and conflict arises as a result.

Farmer participation in management

One of the recent catch phrases[11] in irrigation circles is *farmer participation*. Get the farmers involved in their own project, instil a sense of responsibility for their own scheme, and many of the management problems that plague the big schemes can be by-passed.

> Here's an example from a developing country. A group of peasants decide they need a pump. They ask the government but the government won't give it to them (why should it? They managed all right up till now without one). So a well-educated outsider gets them together to form an association. They elect a leader. They all agree to paying 50 dollars towards the cost of a pump. (That is to say, nobody openly disagrees in public when the leader proposes it.) The pump is bought, although it is rather smaller than they wanted because not everybody paid up. Then, who is to be responsible for its safekeeping and maintenance? Not me. Not me. Who borrowed the delivery hose and didn't bring it back? Not me. Who'll buy the diesel fuel for it? Not me. And so on.
>
> Which developing country was this, India? Zimbabwe? Sudan? No, it was Scotland. I know, because I was one of those peasants. And non-co-operation in the face of non-absolute necessity seems to be just human nature.

It is usually encouraged with the formation of water users groups or associations, which are made responsible for routine maintenance and water distribution. If this can be done as an adjunct to pre-existing administrative organisations, like the village councils in east Java and the Mandal Panchayats in India, then irrigation management can be absorbed into the normal running of the community like education, health, and religious affairs. This has been achieved recently in Turkey (Ref. 12), where responsibility for irrigation management has been devolved from a central government ministry to local government. Thus although the farmers are still not directly involved in the routine operation of

---

10 *Or ditch-rider, call him what you will, but the man who is employed by the management to control your water.*
11 *Along with 'water management', 'sustainability', 'beneficiaries', 'target groups', 'players' etc. These cliches seem to be regenerated every year and resurface with irritating frequency in the titles of technical papers presented at ICID conferences. I don't like these catchy phrases because it's too easy to use them without thinking what they mean.*

the irrigation system, they at least have the power to elect the local officials who are. The states of the former Soviet Union are similar in that irrigation system management and maintenance is handled by the *Rayon* (regional) and *Oblast* (municipal) local governments.

But beware the unwanted imposition of Euro-social ideas on unsuspecting third world farmers. There is a steady stream of social engineering disasters emanating from uncomprehending ideals of farmers contributing cash or kind toward construction and maintenance of their own canal systems. There has to be a real and urgent need before a farmer will part with his money.

Privatisation

Privatisation started to become politically fashionable in the late 1980s, thanks to a certain British Prime Minister perhaps. In this context privatisation means off-loading government ownership or responsibility for operation into the private sector, either to the farmers themselves or to an intermediate private subcontractor. But is it just a means of off-loading responsibility from a government line management system that can't cope, or is it really to benefit the farmers?

On those East Java tube-wells that had been built 10 or 15 years previously, the ones that worked were those that had been broken down for a while and had eventually been taken over by the farmers at their own expense. They had got fed up with government lethargy and hired a mechanic to repair the pump, bought the diesel fuel themselves and organised the water distribution themselves. They had unofficially been privatised, and they worked. The same thing happened on other schemes, when farmers got fed up waiting for official repair work and rebuilt canals themselves and at their own expense. These schemes were bound to succeed because the farmers regarded them as their own property and made sure that they were properly maintained.

On a big scheme it might be appropriate to consider partial privatisation of some key functions. Years ago in Sri Lanka I once proposed the privatisation of main canal operation and management. There were at the time several highly competent private management companies, with expertise developed on the tea and rubber estates, who could have done the job. They would then have been at the interface between supply and demand, absolving the government of direct responsibility for the chronic breaks in water supply which had such a disastrous effect on downstream water distribution. And they would have none of the complexes induced by the conflicting interests of authority and assistance. The arguments are still valid today.

There is little doubt that private tube-wells and pumps work better than government ones, but only when they have been instigated by the farmers themselves. Privatisation is not some magic wand that is guaranteed to turn a badly-run scheme into a good one overnight. Any attempt to force privatisation on unwilling farmers will achieve nothing. And if more than about 50 farmers are involved the administration is likely to be difficult.

3.9 Water Charges

Water charges are a bone of contention on every small-farmer irrigation scheme in the world.

At one time or another, water tax or water charges have been promoted by governments and funding agencies to recoup funds for maintenance, to instil some feeling of responsibility for the project amongst the farmers and thereby to encourage farmers to maintain their own system, to prevent wastage of water, or simply to pay for the project. Most donor agencies now insist on some form of water charging as a means of capital cost recovery. But the field is fraught with difficulties and even where water charges are successfully collected the alternative aims of more efficient management may be sacrificed.

There are several ways of charging for water, and in the developing world not many of them work:
- by volume of water delivered;
- by crop;

- by area farmed;
- by area irrigated;
- by season;
- by time or duration of flow;
- by number of irrigations (warimetric);
- by hidden taxation;
- by deduction from controlled crop price;
- in kind by forfeiting a percentage of crop (Burma).

By volume

If water is to be sold by the bucketful then the buckets need to be counted. Water in a pipeline can be easily metered, but it is not so easy in a network of open canals, and quite impractical when considering thousands of small farms of a hectare each. Water is too easy to steal and too difficult to police when it gets down to this level. It is common practice on the big irrigation schemes in the USA and Australia where the farms are large and water can be metered at the tertiary canal head by a Dethridge wheel or propeller meter. But for smallholder canal schemes in developing countries, it is impractical.

However, the gradual acceptance of pipeline distribution systems, coupled with developments in cheap electronic technology, may change this. A recent breakthrough in water management for smallholder irrigation has come with the development of the 'Acquacard' in Italy, which is a system of metering, monitoring, controlling and payment for small volumes of water delivered through a pipeline network. It is based on an electronic card (the dimensions of a pack of ten cigarettes) which slots into a weatherproof box containing a closure valve and water meter, located at the field outlets of a pipe network. It is powered by a lithium battery which remains charged for 10 years, so requires no power connections. Each user has an individual card on which are recorded details of the volumetric amount of water he has paid for or is entitled to. All water transactions are recorded on the card and also in the meter box, from which details can be downloaded by managers equipped with master cards, and relayed to a master computer in the accounts office. Trading of water between farmers can only be done via the main office. Farmers are allocated an amount of water commensurate with their size of farm. The amount of water available varies according to the water stored over the winter. An assured amount equivalent to 200 mm of irrigation is allocated for a set fee (currently 0.09 Euros per cubic metre). Water up to 400 mm can be purchased (all in advance) for an increased fee, depending on availability, and above this at double the basic fee.

The project in which this system was pioneered is the Ofanto Scheme, centred on the town of Foggia in the south-east. The numerous rivers that flow into the Adriatic were dammed in the 1960s and provided with off-line storage reservoirs that store most of the winter runoff for use in the summer growing season. Over 120,000 ha in the region are served by the pipeline network. Typical farm size is 1 - 2 ha, and crops grown under drip and microjet include citrus, olives, grapes, peaches and artichokes, with a wide variety of other fruit and vegetables.

As with other smallholder pipe distribution systems, it is neither practical nor cost-effective to provide a point-of-use outlet to every field. A typical farm outlet serves 5-6 hectares, which may cover four or five individual farms. The farmers sharing an outlet have to arrange amongst themselves when to irrigate, since only one can irrigate at a time. This is reportedly acceptable to the farmers and there have been no social problems arising through water-sharing arguments in the first 5 years of operation.

figure 3.11   ***an 'Acquacard' unit controlling water to a hectare of artichokes, Foggia, Italy***

This is a limited-rate demand system, with ultimate control amongst the water users being limited to small and manageable groups. The cost of the regulating device is only about 900 Euros (USD1200), system efficiency down to farm outlet level has been measured at 94 per cent, and crop yields and overall production have greatly expanded since its introduction.

By area

A water charge based on a fixed rate per hectare of farm might be considered fair and equitable. However, the problem with this method lies in the way that the farmer perceives it. Having paid so much per hectare, he feels not only obliged to use the water but also entitled to use as much as he can. Even if he doesn't really need it, the water is his because he paid for it. So he'd rather pour it down the drain than see it going to someone else down at the tail-end. Water charges based on area don't encourage efficient water use. On supply-scheduled schemes which are subject to periodic water shortages, the imposition of a water charge on this basis can actually precipitate a complete breakdown of water management.

In India there is a detailed system of recording and tax collection established by the colonial administration, which is still going strong. The only trouble with it is that the farmers have to rely on the integrity of the individual Lands Officer whose job it is to visit each and every farm and determine the amount of land under each crop. There is a small lift scheme near Karwar in Karnataka, which wasn't cropped at all last year because everyone went off to work on a big construction project nearby. Yet the Lands Officer had recorded a fair proportion of the scheme as being under crops. He was too busy to go out on site so he just filled in the record book whilst sitting in the office. The farmers were not amused when they received their water tax bills. In his excellent book on the Warabandi system (Ref. 5), Malhotra sees this reliance on the Irrigation Booking Clerk as a major drawback.

By crop

On the rain-starved island of Madura, off eastern Java, the tobacco crop is an important source of income for many small farmers. They irrigate by hand using traditional palm-leaf buckets. They are charged for water on the basis of number of plants, and everyone pays the high water price for tobacco because it is economic to do so. It's easy enough to count the relatively small number of plants that each farmer grows, and there is little scope for argument.

In southern India the government water charge for sugarcane is 800 rupees/ha per crop, four times the price of water used for paddy rice. Nobody pays it because in order to justify the cost the farmer would have to ensure that the right inputs went in, like fertiliser and attested seed cane. But these inputs cost money and in yet another of irrigation's vicious circles, the cane crop stays ragged and withering.

So, if the water charge is to be based on per crop per unit area, the price has to be right.

Argentina is using satellite imagery to check the crop area. Modern imagery has a ground detail resolution as fine as 1m, and can also indicate which crops are being grown and what stage of growth they are at. As a tool for central government planning and monitoring it looks set to increase in use.

By time

The tube-wells of eastern Java supplement the hit-and-miss surface schemes with assured but expensive groundwater. Farmers pay for the pumped groundwater on the basis of the length of time that the pump operates. This is OK so long as the water gets to where it is meant to be going. If it gets lost on the way the tail-end farmer pays for more than he really gets. Then he starts fighting with his neighbours who get more for the same price. This was a major justification for expensive canal lining on conjunctive use projects there.

By number of irrigations

The 'warimetric' method, sometimes applied on warabandi systems, is a means of charging for water on the easily-countable basis of number of times that irrigation water is issued in a season. It assumes that all farmers get the same or an equitable amount per irrigation, but it breaks down when they don't.

By season

Most countries have clearly defined cropping seasons, say summer and winter, or wet and dry, in which it is practicable to levy a single water charge per season. The charge may vary from season to season, and from crop to crop. In southern India for example the charge for Kharif (rainy season) paddy is less than that for Rabi (early dry season) paddy, because more water will be used in Rabi.

By manipulation of controlled prices

This is perhaps the most subtle way of getting money from the farmer. If the government is buying the crop at a guaranteed price, this might be set slightly lower than the prevailing local, national or world market price in order that the difference can be offset against the cost of providing water. But it means that the market has to be largely under government control and not easily by-passed by the farmer selling on the side. It's possible with crops that have to be processed at a central facility, like sugarcane or cotton, but difficult with crops like paddy that are easy to store, transport and market.

A method like this relates the water charge directly to crop yield, rather than to area or amount of water used. This means that the efficient farmers pay more than the less efficient ones. To look at it one way, the good farmers are penalised. On the other hand, the bad farmers get a lower income. What is fair?

By forfeit of crop

Payment in kind, in the form of a proportion of the crop, is common for farm inputs such as seed and fertiliser, and sometimes in lieu of an irrigation water charge. In Burma, each farmer has to contribute a certain amount of paddy to the central government. This means that in drought-prone areas, if the irrigation fails all the crop might be commandeered by the government.

In the mid-1980s, the ruling military junta in Burma demonstrated mind-numbing incompetence by demanding that every village contribute a set quota of rice each year to government-controlled stores. This applied even to the dry zone of upper Burma, which has mainly sandy soil and little rainfall, and no tradition of growing wetland rice. In order to avoid heavy penalties, the dry zone villages bought in paddy from southern rice-growing areas, in order to donate their required quota to the government. Burma had at one time been a major exporter of rice, but policies such as this quickly put an end to it.

Nevertheless, payment in kind is often a long-established informal procedure between individuals, and one which may be more acceptable than the use of money.

Free water

In most countries having long traditions of irrigation, water has always in the past been free of charge. This is not to say that there are no water rights, because most irrigation-based societies have developed their own rules for water distribution, either implicitly as in the passing from field to field of the Indonesian rice terraces, or developed out of practical necessity like the first-in-line-takes-all of the west Asian spate irrigators. But the idea of water having a financial price is an alien one to most farmers.

It was very noticeable in east Java that when water was free of charge as on the run-of-river diversion schemes, supply-scheduled rotations were accepted by all farmers. But as soon as the conjunctive use schemes were developed with payments levied for pumped groundwater, supply scheduling failed completely.

The same thing happened in Sri Lanka, where water has traditionally been free and available to all. But once a farmer starts paying for it, the water is his, not everybody else's. Water assumes a monetary value, whereas it was previously considered to be as free as the air we breathe.

There is often a good case for keeping water free. Save on the expenses of collecting water tax. Recoup the cost invisibly in a boosted local economy. Keep the farmers happy. But in order to make it work, invest in some education.

Education

As a tool for increasing water use efficiency or reducing water use, water charges may look good on paper. In practice it may be far easier and cheaper to provide some education, not only to farmers but to field advisors and water managers too. The guiding principle is that it is in nobody's interest to waste water. The assertion too often heard from officials and field advisors is that farmers are ignorant and it is impossible to explain to them the principles of water management. In my experience this is seldom true. Farmers are never ignorant, and usually willing to accept ideas that will affect them in a beneficial way. Some instruction in the use of a simple soil probe for example (Ref. 13) can bring a degree of certainty to the guessing game of when to irrigate. Pointing out the mechanics of water travelling through the root zone is easy - it may just be the case that farmers never thought of it in that way before.

It is a simple matter to present the basic technical justifications for not wasting water. More disturbing perhaps is the lack of readiness of many officials to pass on the relevant knowledge. Knowledge is power. Divest knowledge and you lose your power over people. In my experience it works the other way. Share your knowledge and you earn respect from those people. You also learn from them and increase your own knowledge.

References and further reading for chapter 3

(1) Laycock, A., 'Management aspects of Smallholder Irrigation Schemes' Proc. ICID 12th Congress, Fort Collins, 1984.

(2) Merriam, J. L., 'Flexible Supply Schedule Missing Link for Effective Surface Irrigation and Automation', Proc. ASCE Irrigation and Drainage Division Conf., Hawaii, July 1991.

(3) Chambers, R., 'Managing Canal Irrigation', Cambridge U, 1988.

(4) American Society of Civil Engineers, 'Planning, Operation, Rehabilitation and Automation of Irrigation Water Delivery Systems', 1987.

(5) Malhotra, S. P., 'The Warabandi System and its Infrastructure', CBIP, New Delhi, 1982.

(6) Townsend, R., 'Further up the Organisation', Knopf, 1984.

(7) Abernethy, C.L., 'Performance Measurement in Canal Water Management: a Discussion', ODI/IIMI Network paper 86/2d, 1986.

(8) Sampath, R.K., 'Some Comments on the measures of Inequity in Irrigation Distribution', ODI/IIMI Network paper 88/2f, 1988.

(9) Muller, M., 'Experts cling to Water Myths', New Scientist, 24 June 1976.

(10) Taj ed Din, Owen, Hennesey, 'Water Control Aspects of the Gezira Irrigation Scheme, Sudan', ICID 12th Congress, Fort Collins, 1984.

(11) Smout, I.,'Farmer Participation in Planning, Implementation and Operation of Small-Scale Irrigation Projects', ODI/IIMI Network paper 90/2b, 1990.

(12) Svendsen, M., Nott, G., International Irrigation Management Institute, Short Report on Locally Managed Irrigation No.17, 'Irrigation Management Transfer in Turkey: Early Experience with a National Programme Under Rapid Implementation', 1997.

(13) Merriam, J.L., 'Simple Irrigation Scheduling Technique using a soil probe', ASAE conf on evaporation & irrig scheduling, San Antonio, Texas, Nov 1996.

(14) Cornish, G., Perry, C. J., van Steenbergen, F., 'Charging for Irrigation Services – Guidelines for Practitioners'. DFID report OD 153 and 'Water Charging in Irrigated Agriculture' FAO Water Report 28, 2005.

(15) Smout, I., Gorantiwar, S., 'Improving allocation of Irrigation Water in southwest India', Proc. Institution of Civil Engineers V159/WM2, June 2006.

CHAPTER 4 CANAL OPERATION & AUTOMATION

*Most irrigation schemes are designed by engineers who don't think enough about the way in which water can be moved around a scheme and more importantly, **WHY**. The mechanics of water flow is a useful thing to understand even for non-engineers. Then the need for and functions of certain structures will be apparent. Some structures are better than others at providing hydraulic stability. Incorporating storage at some point in the system is a prerequisite for most schemes that require regulation within the scheme, but there are several ways of doing this.*

Automation is a word that is often misunderstood. It does not necessarily mean complex electronics, computer networks and remote-controlled gates. It can often mean simple structures and ensuring that the canal can work properly with the minimum of intervention by people. Automation and control concepts and techniques are developed and used outside the world of irrigation, and unsurprisingly there are several areas of misunderstanding which contribute to the general confusion. The concept of sensitivity is a better way to describe the performance of a canal and an aid to the design of canals for easy management.

4.1 How Water Flows

The obvious statement that water flows downhill obscures some less than obvious facts about what it does on the way and what it can be made to do. A natural river runs fast and slow, deep and shallow, through lakes and down rapids and waterfalls. A canal will do all these, but in a controlled way and for specific reasons.

Water moves in a canal in one of several states. The most familiar is sub-critical flow, in which velocity is low and the water moves gently without much disturbance on the surface. Any interference in the flow will cause a wave to move both downstream and upstream.

As the velocity increases, due to a steeper slope or a smoother channel or a sudden change in geometry like a fall or step, the nature of the flow can change drastically to super-critical. In supercritical, or shooting flow, the water flows as a fast shallow jet. An interference in the flow will not cause a wave to travel upstream.

In passing from one state of flow to the other the flow goes critical. In hydraulic theory, critical flow is the state of minimum energy, and for all other energy levels there are two possible states of flow, sub-critical and super-critical. Water flowing with a certain energy can be either sub-critical or super-critical, and can change from one state to the other, each having a unique combination of velocity and depth. A change like this is most commonly seen as a hydraulic jump, the standing wave at the foot of a weir where fast-moving supercritical flow changes abruptly to deeper, slow-moving sub-critical flow.

All hydraulic control structures are aimed at influencing the state of flow or keeping it constant. The type of structure possible is influenced by the state of flow upstream or downstream of it.

These are steady states of flow. Some other, non-steady states, are relevant to irrigation channels. The advance wave when a channel is filling is a gradual increasing of flow rate from zero to its full supply rate. The nature and time-to-stabilise of the advance wave can have a profound effect on the ease of irrigation management, likewise as can the recession wave, when a canal reach is shut down or its flow reduced. If either the advance or recession wave is too long, it can be difficult to control or measure the delivery of water.

All these states relate to flow which in hydraulic terms is 'turbulent' as opposed to 'laminar', which doesn't apply in irrigation canals. However, it is more useful to use the term more loosely in referring to macro-turbulence which may need to be controlled by structures, and micro-turbulence which acts at the boundary layer between water and canal bed or sides. The latter has an effect on sediment transport, erosion and siltation.

4.2   Canal Sensitivity and Response Time

When the headgates of a canal are opened, either to increase the flow by an increment or to start up a canal that was previously closed, water advances as a tapered wedge. This advance wave is shallow at its leading edge, gradually building up to normal depth some distance behind. At any particular point along the canal, the advance wave will arrive some time after the headgates are opened, and normal depth will be attained some time after that. The ***response time*** of a canal or system of canals is a measure of the time taken for the effects of control gate operation to travel along the system to where it is required. This is normally at a division structure or an offtake point at the field end.

The response time will be affected by the flow velocity and hence the slope and roughness. It will also be affected by the amount of dead storage in the canal which must be filled before full discharge is attained. Cross-regulation structures will affect the response time in different ways. Open undershot or downstream-control gates will permit a faster response time than fixed weirs or upstream-control gates, in which each reach must be filled in turn before full discharge can be developed in the next. Intermediate offtakes if left open will slow the overall response time. It follows that the quickest way to fill a canal is from the tail-end upwards.

The ***sensitivity*** of the system is a measure of the capability of the system to respond to changes in water demand at the offtakes, and how quickly the required discharge can be delivered to an offtake once it is opened.

High sensitivity is necessary if there are frequent changes in demand at offtake points. A system operated on demand, with irrigation in daytime only, and a variety of crops under the same command, will normally need a sensitive system. A supply-scheduled system operating on proportional flow or monoculture of say sugarcane would not need to be so sensitive.

The sensitivity is easy to assess once a canal is simulated on a computer spreadsheet. There are also clear pointers in an existing canal in the field. The sensitivity is influenced by:

- The velocity, in which channel roughness and slope are prime factors.
- The volume of dead storage, which needs to be filled before the advance wave can proceed.
- The cross-sectional profile. A parabolic is generally more efficient than most other shapes.
- Leakage or seepage, and open outlets which reduce the flow in the parent canal.
- Cross-regulator structures. Overshot or upstream-control structures will pond water and lengthen the response time. Open undershot gates or downstream-control structures will not affect it.

The response curves in figure 4.1 indicate the advance of a release of water along a small trapezoidal irrigation canal, predicted on a computer simulation and then measured in the field. The Bhaire Canal is concrete lined in situ but with a variable cross section, and a variable longitudinal slope. The response curve is timed at the point of advance of full water depth, not the very beginning of the advance wave.

The y axis is time in minutes after letting the water in at the head of the canal. The x axis is the distance along the canal. The longitudinal section of bed level is also shown. The steep reaches of the canal result in higher velocity and a flatter response curve. The response curve is also affected by the discharge, and by the roughness, variations of which are shown on the graphs.

BHAIRE LEFT MAIN CANAL
varying discharge

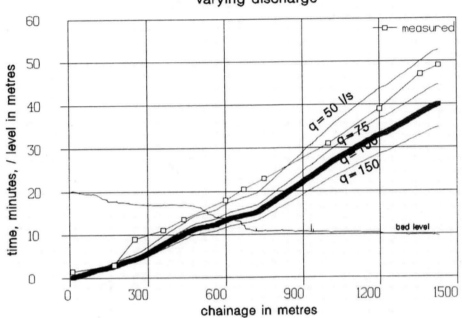

BHAIRE LEFT MAIN CANAL
varying Mannings n

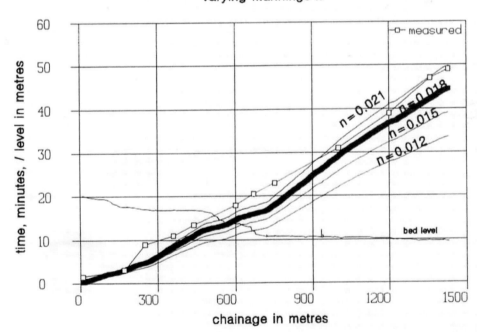

figure 4.1   *Response curves, showing the time for an advance wave to attain normal flow depth at points along the canal*

The response time of small canals is measured in minutes, but on larger systems it can often be measured in days. The Upper Swat Canal before rehabilitation had a response time of 5 days from head to tail. The Merowe Left Bank Conveyor in Sudan would have had even longer, were it not for the designers incorporating downstream control in the lower half of the system.

4.3 Modes of Control

The concepts of *upstream* and *downstream* control are applied to a sequence of canal reaches separated by cross-regulators, or in the case of a pipeline, by pressure-controlling valves. Beware of the control engineers' definitions which refer to the location of the control point within the reach. Irrigation engineers' definitions usually refer to the control point in relation to the controlling structure, but in practice what we really mean is something different again. We mean, in a more holistic sense, that if the *discharge* at the outlet or from the end of a reach is controlled from downstream of the reach, it is *downstream control*, and if the *discharge* at the end of a reach is controlled from upstream then it is *upstream control*. Note that *water level* control does not enter into the definition, although controlling water level or head is usually a crucial intermediate stage in controlling discharge. At the end of the day the irrigation engineer is only concerned with who controls the *discharge*, and where he has to be in order to do it. Unsurprisingly there are several ambiguous definitions in common use.

In a canal we can control the water level and the discharge. In a pressurised pipeline we can control the head or pressure, which is analogous to the canal water level, and the discharge. We can easily measure the instantaneous water level or head, but in the field it is much more difficult to directly measure the instantaneous discharge. Hence the importance of water level in canal control theory. Various combinations of control are illustrated in figure 4.5. Modes of operation of the canal or pipeline system can be categorised as one of these:
- Upstream control;
- Downstream control;
- Mixed control;
- Volumetric control;
- Centralised control.

The principles of upstream and downstream control are not synonymous with *supply and demand scheduling*, although in many cases a supply-scheduled scheme will also be upstream-controlled, and a demand-scheduled scheme will often be downstream-controlled. Many schemes have a combination of control, either in series, with different control logic applied to different levels of the system, or in parallel, as in the case of mixed control. Mixed control involves regulators which react in different ways according to prevailing water levels. Volumetric control aims at maintaining a constant volume of water in each reach. Centralised control co-ordinates the operation of several regulating structures.

All these modes are relevant to both *manual* operation and *automation*, but although traditionally-minded engineers may dream up excuses galore there are not many good reasons to opt for manual control, wherever you are in the world.

Upstream control

Upstream control infers that the discharge in each reach is controlled from its upstream end. Changes in flow emanate in sequence from the headworks downstream through the system. Water levels are controlled at the downstream end of each reach by cross-regulators or fall structures, which if adjustable are operated in response to water levels on their immediate upstream side. Offtakes are preferably located close to and upstream of the cross-regulators, where water levels are most readily guaranteed. There must be a communication link between the control gate at the head of a reach and the outlets, in order that the sluices may be adjusted to give the required combination of discharge and water levels. Long-crested weirs or Amil gates can minimise the problems in maintaining water levels, but can pass only discharge that is not used in the reach. They also imply long filling times for a canal, which has to fill from the top down, feeding any open outlets all the time during filling.

In order to increase the discharge in response to an indent or to reduce it in the case of harvesting or rainfall, the management needs to be informed from the downstream end, whereupon the gate operators will open or close the headworks sluice gates and progressively adjust all gated cross-regulators and offtakes down the main system in sequence in order to maintain water levels at the required elevations. By the time the increased or decreased flow arrives at the point where it is required, a considerable time period may have elapsed, by which time the water may no longer be required and it is wasted. Before rehabilitation of the Upper Swat Canal in Pakistan, this response time of the main and branch canals could be as long as 5 days. And this is not a particularly big scheme; its main system canals are only 140 km long with an average slope of 0.0002 and the headworks inflow is 80 cumecs. The similar-sized Canal Fala de Modolo in Mali had a response time of 7 days on account of its very flat slope of 0.00004. Before its automation with BIVAL, management losses were 50 per cent. (Ref. 3.)

Upstream control does not therefore permit much *flexibility* in water use. Large canal systems have to rely on detailed and inflexible operating rules, which may include shutting down some branches in rotation. Wastage in the form of escape spillage is inevitable except where active automation is effectively employed.

figure 4.2   *Upstream control by a float-operated self-regulating gate, Thailand*
(courtesy Wolfgang Stenzel)

Downstream control

Downstream control means that the discharge in each reach is controlled from the downstream end, either from individually adjusted offtakes or automatic cross-regulators. Changes in flow emanate from the tail-end and work sequentially upstream though the system. Water levels are controlled at the head of each reach by cross-regulators, which open or close in response to water levels on their downstream side. Offtakes are located preferably close to and *downstream* of cross-regulators for minimum fluctuation in parent canal water levels. Low-pressure pipelines are excellent conduits for downstream control and are often used in combination with higher level canals which provide intermediate storage. In a low-pressure pipeline the pressure-reducing valves are analogous to the cross-regulators in a canal.

The great advantage of downstream control is its *short response time*, which can be effectively zero. It therefore offers maximum flexibility and is ideal for demand scheduling. In turn this offers the prospect of minimal wastage through management loss. Downstream control can be achieved by targeting constant water levels at various points along the reach using passive or active automation. Alternative arrangements are shown in figure 4.5.

Passive automation with local control in the form of self-regulating constant downstream level gates requires in-built intermediate storage. This is provided as ***wedge storage*** in level-top canals and hence implies a higher construction cost than other methods. Importantly however, no communication links are required, and this method is commonly used in remote locations or in developing countries having unreliable power supplies.

The location of the sensor can be moved downstream all the way to the next gate, thus removing the need for wedge storage altogether. This depends on reliable communications and reliable electronic control equipment. This particular case gives rise to some of the misunderstandings between irrigation engineers and control logic scientists. The control point is at the downstream end of the reach and immediately upstream of the downstream gate, but it is downstream of the upstream gate which is being controlled, and it is therefore downstream control.

figure 4.3   ***Downstream control gates on the Pehur High Level Canal, Pakistan***

Mixed control

Mixed, or composite control, uses the principles of both upstream and downstream control in the same regulation structure, depending on the prevailing water levels. The aim is to escape a potential problem in downstream control, that of the canal running dry through too much abstraction downstream. Mixed control can be realised by appropriate active control technology, but a special type of self-regulating composite gate (*vanne mixte*) is also manufactured by Neyrtec (now GEC-Alsthom.) This is described in section 4.8 below.

Constant volume control

Active automation with distant remote control reduces the need for wedge storage. The ***constant volume*** method relies on a target water-level sensor in the middle of the reach. A well-known version of this method is known by the acronym ***BIVAL***. It works on the premise that at zero discharge the water surface will be horizontal and at maximum discharge the water surface will be parallel to the bed of the canal but will pivot about the mid-reach water level, which remains constant at all states of flow. It has been used successfully in remodelling large canals with very flat slope and slow response time. Wedge storage is still required but is less than in the previous case since it is only accommodated within the downstream half of the reach. It is sometimes referred to as **volumetric control**, but there is a major difference between it and the **controlled volume** principle which is operated under centralised control as described in the following section.

Centralised control

The above modes of operation are all *serial* - they rely on sequential operation of regulators each in response to the action of the next structure upstream or downstream. Centralised control allows the *simultaneous* operation of all cross-regulators, and as such it has to be actively automated.

Centralised control is applicable to long canals with remote-controlled cross-regulators co-ordinated by computer at a central office. It is of relevance in a supply-scheduled situation of bulk delivery, such as the California Aqueduct. With centralised control all the cross-regulators can be operated simultaneously so there is no sequential time lag along the canal as there is in a conventional upstream-controlled canal. Moreover there is no need for wedge storage in order to meet downstream demands, as occurs with conventional downstream control. Water along the entire canal moves together. The **related level** method of providing canal storage can best be controlled in this way. The **controlled volume** principle, in which volumes rather than water levels in each reach are targeted, can only be done with centralised control, and is the principle of operation of the California Aqueduct. With a length of 600 km and discharge of around 450 cumecs this is almost certainly the largest body of water ever to be moved around so precisely.

Not surprisingly, in order to achieve this degree of control it is necessary to monitor water level, gate openings and discharges continuously and simultaneously at many points along the canal. The canal itself has to be of regular cross section so that the geometry is fully known and volumes can be accurately calculated. Also the water demands have to be known in advance in order that the operation of all gates can be programmed. This is an example of **active** automation, which is described more in the discussions of control logic later in this chapter.

figure 4.4   *A pump station on the California Aqueduct – an example of centralised control*

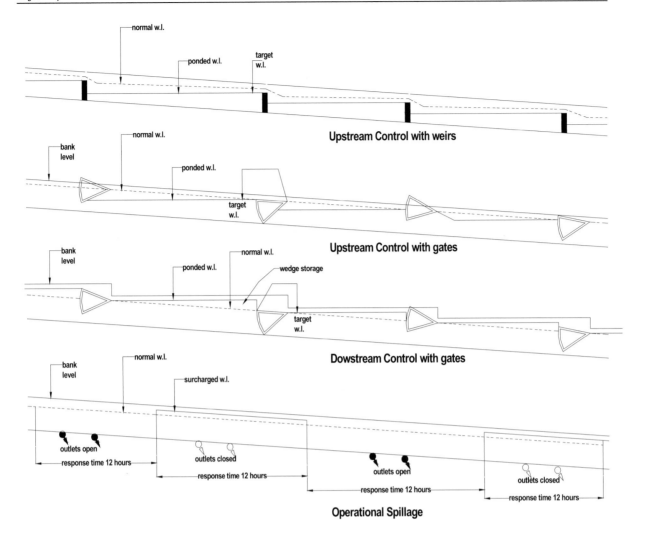

figure 4.5   Modes of canal regulation.

4.4   Intermediate Storage

Intermediate storage is a means to:

- Reduce response time;
- Increase flexibility in operation;
- Increase the hydraulic stability of the system.

> Think of the domestic water supply in your own home. Whenever you want water, you just turn on the tap and there it is. But where does the tap water come from? From the tank in the roof. And before that from the city water tower. All are forms of intermediate storage, which has the effect of smoothing out the fluctuations in demand, and also in supply.

Storage ponds

A canal scheme is made much easier to manage if it has intermediate water storage built into the system. The 500 ha newly-rehabilitated Banavasi scheme in Karnataka State, India, utilises several pre-existing tanks for intermediate storage. These are filled up periodically by the Minor Irrigation Department on a supply schedule, by pumping from the Varada River. Farmers then take water from the tanks via the parabolic canal distribution network on a demand schedule. By this means the scheme is effectively broken down into several smaller schemes which are easier to operate than one big scheme. And if the power supply fails or the pumps break down, the intermediate storage reservoirs act as a safety net and field irrigation is not affected as it would be with a bare canal system.

Indeed the biggest problem before rehabilitation was the unreliability of the power supply. Without intermediate storage, irrigation was only possible when the pumps were running. Fastest response time of the main canal tail-end was 8 hours, and low voltage often interrupted the supply for 10 hours a day. Since rehabilitation, the power supply has not improved but irrigation has, thanks to the simple expedient of linking in to intermediate storage.

And in the case of an existing project with apparently no room to put any new reservoirs, consider the possibility of renting the land. Even though land owners may be reluctant to sell or give up land permanently, they may well be prepared to rent enough land to construct a storage reservoir. This tactic has worked in India and Egypt, as well as the USA (Ref. 2.)

If the water carries sediment, storage reservoirs may suffer from siltation such as that in figure A4.18. This is an on-line reservoir, in which a run-of-river diversion channel runs uncontrolled into a pond, from which irrigation water is drawn off over 12 hours during the day. A more effective scheme would be an off-line reservoir into which clean water could be skimmed from a suitable intake structure. Siltation is considered in more detail in Chapter 7.

Night storage canals

Intermediate storage can also take the form of oversized main and branch canals, which may fill up at night and be depleted during day-time irrigation, but this is normally only a feature of large schemes. The Gezira Project in Sudan has very flat minor canals with a high freeboard to accommodate 24 hours' continuous inflow. But as is clear from figure 4.6, these canals are the first to get silted up and although they are de-silted regularly there's nowhere to put the silt except on the banks. The Gezira is also one of the few irrigation schemes in the world in which the land surface is rising, due to the silt deposition.

Sri Lanka has its own tradition of single-bank canals in which cross-drainage flows are collected in the main canals and extra storage is automatically built in. (So does the Crinan Canal in Scotland, although its purpose is for navigation, not irrigation.)

figure 4.6   ***Silt accumulating in a Gezira minor canal, Sudan. As the silt is dug out, the banks
grow higher each year***

Night storage vs. night irrigation

There is a long-running controversy amongst irrigation academics, about whether a scheme should be run all night or not. On a big project with canal response times measured in days it is impossible to turn the canals on every morning and off every evening. In practice the main system canals at least have to be kept flowing continuously. This implies either continuous flow with night-time irrigation or some form of intermediate storage.

Arguments in favour of night-time irrigation usually cite the fact that intermediate storage costs money, either in the form of oversized canals or of storage reservoirs. This story from Sri Lanka might help to swing the argument:

One year on system H of the Mahaweli Project, 18 farmers died. It was 1979, in the early days of this project, when people were being resettled onto what had over the past thousand years reverted to virgin jungle. Eighteen farmers out of several thousand might be construed as no more than the average death rate, except that they all died of the same cause; snakebite, whilst trying to irrigate at night.

Sri Lanka has many species of snake, but two are deadlier than the rest. The King Cobra moves quickly and will even give chase when it's cornered. The Russell's Viper is, like all vipers, slow and sluggish. It lies in wait for its prey, motionless until something disturbs it. It doesn't take kindly to being trodden on by a farmer in flip-flop sandals. And both snakes only come out at night.

The local ayurvedic remedy for snakebite is a little unsavoury and completely ineffective. First, you go to the doctor. He will invite you to urinate into a cup, which you then have to drink in order to maintain the strength in your body. After that, you have to dance around the room for 10 minutes in order to drive away the evil spirits. Then the doctor wraps some leaves and cow dung around the wound, and then you sleep it off. In fact nine times out of ten you curl up in a corner and die.

One day, one of the farmers I was working with didn't show up for work. Where was he? Round at the doctor's house. At seven the previous evening he had been bitten by a viper and was now receiving the best attentions of the ayurvedic. We found our farmer in the doctor's house, curled up in a corner with a bunch of leaves tied round his ankle and his face the colour of wet concrete. Our conversation proceeded along the following lines:

Me: 'This man must go to hospital'.
The doctor: 'No'. (Because that would suggest to the villagers that the doctor has failed and is no good and nobody will come to him ever again.)

Me: 'This man must go to hospital'.
The neighbours: 'No'. (Because a hospital is the place where people go to die.)

After several hours of confrontation and discussions with all and sundry the upshot was that I could take him to hospital but the doctor had to come along too in order to maintain credibility and make it appear that the doctor himself had suggested it. And worse, I had to promise to the entire village that if this man died I would be the one responsible. We managed to find the snake, which had been killed and thrown in a ditch.

The hospital in Anuradhapura was 35 km away, and we got there late in the afternoon. It was a heaving mass of people, sprawling all over the veranda and choking the corridors. I grabbed the dead snake and held it aloft, and a passage opened up in front of us as everyone shrunk back in fear.

The doctor said yes, they had the serum for vipers. And it works well provided you get the injection within an hour of being bitten. But our man by this time had been bitten 22 hours ago. No chance, said the doctor, but we'll try anyway.
He was 10 days in hospital, but he lived. I was not lynched. And nobody irrigates at night any more.

Level-top canals

Level-top canals can serve the same purpose as night storage canals, but we are here making a distinction between the specifically night-time filling, supply-scheduled idea of night storage canals, and a more automated system of any-time filling and any-time, on-demand irrigation. Level-top canals are associated with more flexible scheduling patterns. They serve the same purpose of intermediate storage, and as the name implies they have horizontal bank tops and longitudinally sloping beds. They are split into reaches at intervals of several kilometres, depending partly on the size of scheme and the terrain but more specifically on the behaviour of transients, waves caused by gate movements which travel back and forth along the reach and can cause spillage or instability in gate operation. Each reach is controlled by an automatic gate which works on the principle of maintaining a constant water level on its upstream or downstream side, and thus each reach is an automatically-regulated intermediate reservoir. They are ideal conveyors for use in an integrated pipeline distribution network. The stored water in level-top canals is often referred to as wedge storage, on account of the prism of stored water that can be mobilised immediately. This is demonstrated in figure 4.6.

The Ghab Project in Syria is 46,000 ha of irrigation now being rehabilitated with level-top canals and a plastic pipeline secondary distribution network. Its primary distribution routes are simple, with a main canal along either side of the flat floodplain of the River Orontes. But its water sources are complex, with gravity diversion from the Orontes and separately from a tributary dam, a dozen inflowing springs along the canal route, some of which are pumped into the canal, and reuse of drainage water from two new pumped-storage dams up in the mountains nearby. There can be no other way to control all this water with any realistic expectation of success, except with level-top canals and pipelines.

The Pehur High Level Canal in the North West Frontier Province of Pakistan is another example of a canal system with multiple sources of water, which can only be regulated effectively with automation. The Swat River supplies water into the Upper Swat Canal system, the tail-end of which is supplemented by the new High Level Canal from Tarbela Reservoir. The 25 cross-regulators on the combined system cannot be co-ordinated through manual control without immensely complex planning and management. But with the downstream control provided by float-operated self-regulating gates no management effort is required at all.

Related level control

Related level control is a means of providing intermediate storage within a canal by co-ordinated control of a series of cross-regulators. It is applicable to night storage when there is a constant flow into the headreach but irrigation is only carried out in daytime, or buffer storage in the case where water is first used for peak hydroelectric generation which does not coincide with irrigation demands. In this case the controller operates the gate so as to maintain a constant difference in (or a similar function of) water level on either side of the gate. Storage is accommodated within the canal freeboard. Level-top canals are required if the downstream sensor is situated at the tail of the reach.

Operational spillage

In a long canal system, in which the response time from head to tail is greater than a day, and where the water is from an unregulated run-of-river supply or can be re-used further down the river system, operational spillage offers a cheap and effective means of partial automation. It simply means that the main canal system is kept flowing at a large enough flow rate to satisfy flexible demands, e.g. daytime-only use. When demand in a minor canal drops off at night, excess water is retained in the parent canal and passed downstream. On a large main canal, this 'spillage' merely contributes to the following day's irrigation demand further down the system, and so it is not really wasted at all. Only at the tail-end of the system is water really spilled. Even then, in many situations the spilled water can be picked up again in other schemes lower down the river system, and the net overall wastage is not great.

The Friant-Kern Canal in California was operated for 50 years on a tightly-controlled arranged supply schedule. But each user was obliged to take water for a full 24 hours at a time. This meant a lot of wastage in the field at night when irrigation went unattended, and a consequent severe drainage problem arising from over-irrigation and low irrigation efficiencies. Recently the canal has been operating on a more flexible demand schedule in which users can shut off the water whenever they like. Operational spillage retained in the main canal system has absorbed all daily fluctuations in demand with only a 6-inch encroachment in freeboard, this being the additional intermediate storage got at zero cost. Irrigation efficiencies have increased, labour costs have reduced and irrigation has become more economic for every farmer who no longer has to irrigate at night.

Conjunctive use of groundwater

Many of the big run-of-river schemes in Java are located in areas of volcanic soils having an extensive aquifer at shallow to medium depths. Tubewells as deep as 100 metres, each serving up to 50 hectares, are linked in to the surface water canal network and provide water on demand at peak irrigation times. Although pumped groundwater is more expensive than run-of-river, its guaranteed availability ensures that crops can be planted on time even if the rains are late. This is important in this overcrowded island where 300 and even 400 per cent cropping intensities on tenth-of-a-hectare farms are commonplace.

In Bangladesh, small tube-wells are used everywhere and serve the double function of controlling the groundwater level and supplementing the canal irrigation supply. Many of these are privately owned and serve single farms only, with open wells and cheap pumps lifting only a few metres. In parts of Pakistan and India open wells are commonly used for conjunctive use, with various water-lifting devices, from Persian wheels to shadufs to petrol-driven or electric pumps.

Low-pressure pipelines

Replacing the tertiary canals with a pipeline will automatically create built-in storage. Not much volume, but enough to balance out the fluctuations in flow caused by farmers taking water at irregular intervals, and enough to provide a zero response time. Only a pipeline system can give zero response time at the field end. Pipelines are discussed in more detail in chapter 8.

4.5   Gate Operation

Canal regulation, i.e. control of the water flow, is achieved through the operation of various types of gate. They may be *operated* manually or under power. They may be *controlled* manually or automatically, and *controlled* locally or remotely. There is a big difference between manual operation and control, and we should beware the confusing semantics of English when translating instructions into another language.

Regulation of an irrigation system involves the following gate operations:

- Discharge control is achieved by opening and closing gates at the head of regulated canal reaches, and at the head or tail of regulated pipeline reaches.
- Water level control is achieved by opening and closing cross-regulator gates to attain target water levels either upstream or downstream of the structure.
- Outlet control is achieved by opening and closing farm, field or watercourse outlets, usually in accordance with a predetermined programme.

Manual gate operation

The simplest form of gate is a steel leaf sliding in grooves at either side. In small canals of about 100 l/sec discharge or less these are light enough to be lifted open and closed by hand. The operation may be facilitated by a ratcheted lever arrangement as in figure 4.7, or a screw mechanism.

Larger gates require mechanical gearing, which can be through a screw-threaded rising or non-rising spindle mechanism, a rack and pinion, a worm drive or bevelled gear arrangement, a cable drum winch, or a counterbalanced cable or chain giving a straight vertical lift. Even when the gates are powered, one of these systems might be provided as a manual override in the event of power failure. Several types are shown in figures A7.04 – A7.06.

figure 4.7   A lever-operated check gate, Imperial Valley, California

Powered or motorised gate operation

Larger gates can be operated through an electric motor coupled to one of the mechanisms mentioned above, or hydraulically. Hydraulic operation entails pumping oil into a cylinder which activates a ram coupled directly to the gate. The pump would normally be electrically operated, and a manual override facility is also possible, using a lever-operated jacking device. Most large gates and all high-pressure gates are operated electro-hydraulically.

Gate self-operation

Gates intended to react in response to water level can be designed with floats and counterweights such that they will open or close according to water levels on their upstream or downstream side. These are usually radial or flap gates having no moving parts except the entire gate itself. They are a means of passively automating a system and are described in detail below.

4.6   Gate Control

Manual control

Manual control in this context means non-automated. Whilst it is often the case especially on older schemes that gates have to be physically operated (i.e. manually operated) by a man or woman winding a handle, most large gates are equipped with electro-hydraulic motors. However, even if the only effort entailed is pressing a button on a control panel 80 km away, this is still manual control, but also remote-controlled.

The implications of manual control are:
- It entails some form of communication, be it telephone, radio, ditch riders, or farmers' committees, between control points and organisational centres.

- Gate operators and support staff have to be employed, often round the clock, and paid for by the canal operating authority.

It will be apparent that even a modestly sized canal system will require substantial inputs from its management in terms of organisation, providing trained operators and running expenses. The level of management input required is nearly always underestimated. As an example, the Upper Swat Canal in Pakistan was rehabilitated recently including the construction of 12 radial gated cross-regulator structures on the main canal. Each of these required a full-time gate operator and assistant, and they had to be available round the clock. So working 8-hour shifts requires 72 people, say 85 allowing for stand-ins for holidays and sick leave. Then the communication system had to be upgraded. The Morse code telegraph network of First World War vintage was replaced by a high frequency radio link. With radio operators and administration staff it takes about 100 people to operate just those dozen structures. The discounted cost of all these people over the life of the project amounted to more than the capital cost of the structures; in effect their cost was more than doubled by the hidden costs of operating them. And another problem soon appeared. The Irrigation Department did not have the resources to provide, train and support these extra operators. All this contributed to the ready acceptance of automation in the next project, the Pehur High Level Canal.

Refusal gates

Refusal gates are a means of allowing the end users in a supply-scheduled scheme to close off the water when they don't need it. A frequent situation arising in rehabilitation of old proportional distribution schemes, which were built with no flexibility at all, is where the farmers do not want water because they are harvesting, or applying a top dressing of fertiliser, but the water comes anyway and they can't stop it. So they turn it into a drain or onto the road, whichever is easiest. Refusal gates at the head of the tertiary or watercourse can be open or shut, there is no intermediate regulation. If they are shut by the farmers the water remains in the distributary canals and is not wasted. The only problem then is to ensure that if too many farmers close their gates at any one time, there is somewhere for the water to be stored or else to spill safely. Usually the excess is stored within the canal freeboard, but additional side escapes may be necessary.

figure 4.8   ***Refusal gates installed on a precast outlet, Pakistan. (left – open, right – closed)***

Remote control and configuration

Remote control is not necessary for all types of automation, although some forms such as active automation of regulator gates depend on it. If it is necessary to press a button to activate a motor which then raises a gate this is ***motorisation***. If the button is located in a control panel away from the gate, this is then ***remote control*** also. The remote control process can be ***automated***, by having a sensor linked to a electronic controller which activates the gate when, for instance, the water surface reaches a certain level. The term *remote control* is something of a loose cliché, and we should describe it more accurately in terms of the *configuration*. **Configuration** describes the relative locations of control devices, sensors and controllers, and can take several forms:

- **Local control** refers to the sensor being located in the same canal reach as the device being controlled. This is also referred to as **distributed control**, in that local controllers are spread through the system.
- **Local close control** applies when the sensor is adjacent to the device, such as the float in a self-regulating gate.
- **Local distant control** refers to a sensor in the middle of the reach downstream (BIVAL) or at the downstream end of the reach (EL-FLO, PID.)
- **Semi-local distant control** is sometimes used by control engineers to describe de-coupled controllers that rely for some of their input information on other, usually adjacent, controllers in the canal system.
- **Remote distant control** is when the controller is not in the vicinity of the device, and is normally the same as
- **Remote centralised control**, in which the action of all devices is determined by a single controller, which could be a computer or a person sitting in a central office.
- **Hierarchical centralised control** applies where two or more levels of control are combined. For instance local or semi-local controllers might handle individual operation of regulator gates, but these would in turn be linked to a remote central controller which updates target water levels and assembles input data. This is also referred to as **master supervisory control**.

SCADA

SCADA is the acronym for 'Supervisory Control And Data Acquisition'. A SCADA system includes collecting information, relaying it back to a central site, analysing the data and displaying it on a console, and if required, carrying out any consequent control operations. Control may be automatic or initiated by operator commands.

A SCADA system is composed of:
- Field instrumentation, such as water level transducers, gate opening indicators, pump flow meters.
- Remote stations, situated at pump stations, cross-regulator structures, headworks.
- Communications network, which may be hard-wired, satellite, microwave, telephone landlines or mobile phone network.
- Central monitoring station, with computer processing to present data in a useable form and to relay automatic instructions back to the remote stations if required.

When designing or specifying a SCADA system, it is important to bear in mind the speed at which information technology is advancing. On a large irrigation project the lead time from design through tendering and construction to installation of the control system is likely to be 5 years or more. It is not often possible for the project designers (mainly civil engineers) to predict what computers will be available or what telecommunication systems will be in place when the civil works are complete and the water is flowing. So a performance specification must be drawn up and the detailed design left to the specialist IT contractor who comes in at the end of the project.

Examples of recent projects which involved SCADA are the Pehur High Level Canal, and Merowe Irrigation Project in northern Sudan. The Pehur High Level Canal is downstream-controlled by self-regulating gates all the way to the tail, and so needs no communication links along the canal system. It does however, depend on a SCADA system to operate the headworks (figure 4.9), which comprise a bank of five Howell-Bunger valves at the termination of the Gandaf pressure tunnel. These operate under a head of up to 120 m releasing water from Tarbela Reservoir into the canal at such a rate as to maintain a target water level at a certain point in the canal headreach. At this point is the remote station with a device for measuring water level. This is linked by buried cable back to the headworks. The central monitoring station is computer programmed to open or close the valves in a sequence which depends on the rate of rise or fall of water level at the remote station. (More details are given in Ref. 14.)

The Merowe project in Sudan, due to commence construction in 2007, has a main canal system 500 km long. The lower part of this is automatically downstream-controlled by self-regulating gates, but the upper part is upstream-controlled by a series of ten radial-gated cross-regulator structures. There are also 7 pump stations supplementing canal flow by pumping from the Nile at peak demand periods only. The aim of management is to balance the outflow from the Merowe Dam (a 1200 MW structure on the 4th cataract of the Nile) with the irrigation demand at the start of the downstream-controlled section 300 km away, whilst providing for 30 intermediate irrigation offtakes on the way. As with the Pehur Project, any water spilled unnecessarily means power wasted at the damsite, since water not run through the canal system would otherwise go through the turbines. The SCADA design is aimed at monitoring the gate openings and water levels at each cross-regulator, the water levels and operating status of every pump, and measuring the flow discharge entering the canal system from the dam. All the remote stations will be manned by pump or gate operators, to whom operating instructions will be relayed by mobile phone from the central monitoring station.

4.7   Why Automation

Automation means a system being able to activate, operate or regulate itself without any intervention by people. Automation can be done for one or both of two basic reasons: either to do a difficult job more easily and with less chance of error, or to save on labour costs and effort. An entire canal system can be automated, or only parts of it, say an outlet structure or a cross-regulator.

Automation can be passive or active. *Passive* automation can be achieved without resorting to computerisation and electronic controls. It can be achieved merely by installing a structure such as a bifurcator with no moving parts. *Active* automation usually implies an electro/electronic control system with circuitry based on concepts of *feedback*, *feed-forward* or *fuzzy logic*.

In many developing countries the merest suggestion of *automation* strikes terror into the minds of irrigation engineers, who are already struggling to maintain the simplest equipment, let alone complex electronic circuitry, in an operational state. But with passive automation at least it will be possible to improve the canal system at very little cost without inviting extra problems of maintenance.

Do not confuse the terms *automation* and *motorisation*. Very often water control gates are equipped with electric or hydraulic motors to make them easier to operate. In turn these motors might be operated by a switch on the wall of the structure, or by an operator sitting in front of a fancy control panel a hundred kilometres away, pushing buttons to open and close gates by *remote control*. But none of this is *automation*. So long as it requires a person to activate the process, it is not automation.

Automation should not be considered until the required mode of control of the irrigation system is determined. Much wasted effort can go into the active automation of supply-scheduled, upstream controlled systems when a better alternative would be downstream control with demand scheduling.

Automation to save labour

Figure A4.7 is an example of some of the most labour-intensive irrigation in the world, at the Jember sugar cane estate in East Java. The soils are fertile volcanic loams with an even and gentle slope, ideal for long-line furrow irrigation. Water is fed into long furrows, but the cane is grown on cross-ridges 5 m long which are irrigated by women standing in the main furrows with buckets. It's a perpetuation of the old Dutch *renosso* system of cane growing, and it certainly generates employment. But in most countries labour doesn't come that cheap any more.

Automation is entwined with availability and cost of labour. But it is not only advanced countries like the USA who have to automate. Even Pakistan and India, with burgeoning populations where labour is supposedly plentiful and cheap, can have labour shortages on many irrigation schemes. In Chapter 3 we looked at the Gezira scheme and saw how the shortage of labour has made farmers reject the modes of water management that the scheme was originally designed for. Semi-automatic field irrigation is possible there because of the black cotton soil. Even partial automation such as the installation of intermediate storage reservoirs can drastically reduce the labour input and just as importantly utilise the convenience of labour. In California, Merriam cut his farm labour costs by 80 per cent just by building a storage reservoir and getting the flexibility to operate his irrigation distribution system properly.

Automation for easier operation

So automation does not necessarily mean high technology that developing countries can't afford or manage. Even in a labour-rich area, an automated scheme that works is better than a labour-intensive one that does not. And automation should not be looked upon as merely a means of saving labour. The strategies for automation or partial automation, say, by using float-operated cross-regulator gates, can often be justified by considering the difficulties in manual operation, and the consequences of not overcoming these difficulties.

More and more large schemes are moving away from their original concepts of supply-scheduled proportional distribution. They are entering their first rehabilitation phase with their end users clamouring for more water, delivered to a more flexible schedule. The long-established management practices are often not equipped or attuned to the implied operational changes, and automation offers the only realistic chance of management adapting to the changing demands upon it. The Pehur High Level Canal together with the Upper Swat System is an example of a complex system that needs automation to adapt to evolving development of the system. This is an old canal system that has recently been extended and supplemented with water from a second source. The supply-scheduled, continuous flow principles of its original design are becoming less relevant as the physical canal network becomes more and more complex. In fact it would be impossible to operate the system at all without some form of automation. This view was not initially shared by all those involved with the design and management of the new parts of the project. Old traditions die hard. However, since 2003 the scheme has been operating well.

Automation and control

The way in which its delivery system is operated has a huge bearing on the success or failure of an irrigation project. In its simplest format, the canals might run uncontrolled over the whole project. If the source is regulated from a dam or reservoir then the ensuing continuous flow may yet be easily managed by individual farmers as necessary. The extreme uncontrolled case of wild flooding is a viable option in some countries where flash floods are rare but welcome. Here there is no control on either the canals or the source. It is therefore also automatic, in that the system works by itself.

However, most schemes need some control over their water. Although on paper a diagram of steady flow in a canal delivering water to many outlets seems simple, in practice there is no such thing as a steady-state canal. There are always continuous tiny variations in flow which have a knock-on effect along the system and which need to be controlled. They are caused by farmers opening or closing offtakes, by seepage or leakage outflows, by inflows from rainfall, by fluctuations in supply from a river, or by pump stoppages. So there is a need for gates to control the discharge down each branch, weirs to control the water levels, intermediate storage to smooth out fluctuations in the system. But once gates are introduced then someone has to operate them. And people are not the most reliable of operators, because they are subject to all kinds of non-technical influences such as not understanding the correct procedures, bribery and corruption, and just plain making mistakes. Then the more gates there are, the more room there is for error. And the more errors there are, the more the chance of the system breaking down. See Chapter 11 for some horror stories.

So by making the system, or at least a part of the system, operate by itself, automation can significantly reduce the potential for human error. And straight away more complex modes of operation become possible. Traditional viewpoints shun any kind of automation in developing countries, on the grounds that any breakdowns will not be repaired. But there are some subtle ways to automate, without getting involved in anything too mechanical.

figure 4.9   *SCADA-controlled outlet valves supply the Pehur High Level Canal*

The simplest methods of automation involve fixed structures such as weirs and splitters, which are often not even recognised for the automatic structures they are.

Float-operated self-regulating gates have been used successfully for the past 50 years in developing countries, to control canal flow according to water levels upstream or downstream of the gate. So canal discharges can be automatically and continually adjusted 24 hours a day, depending on the water use. To do so manually would require an army of gate operators in constant touch with every farmer and each other.

Pipelines offer a straightforward means of partial automation. Replacing a canal with a closed or semi-closed pipe means that the flow into it becomes self-regulating, because a pipeline cannot overflow.

In pumped schemes, the simple expedient of installing cut-out switches activated by rising or falling water levels can remove at least one degree of freedom-to-make-mistakes from pump operation. Using submersible pumps that don't require priming is another.

So automation can take many forms. In the more sophisticated developments taking place in the USA today, some surface furrow irrigation is being refined with pulse and surge methods. These start with the low operating costs of surface irrigation and are aimed at economising on water use and reducing wastage of water in the field, by varying the flow rate to match the intake rate of the soil with the depth of water to be applied. Spile pipes with servo-operated valves that are radio-controlled are another state-of-the-art idea that may be of little use to small farmers in developing countries but could over the next half-century find more application in countries where in the year 2006 such concepts are unthinkable.

Partial Automation

It is often the case that only part of a scheme will be manually controlled, and the rest automated in some way. The open flume outlets (see figure 7.8) of Pakistan and northern India for instance are designed to distribute a fixed proportion of the parent canal flow into the watercourse. The intention is that if the discharge varies in the parent canal, then the discharge through the outlet will vary by the same amount. In practice they only work effectively over a very limited range of parent canal water levels, but they are nevertheless an example of a passive automatic structure. Upstream and downstream of this though, the system is manually controlled. The distributary headworks upstream have manually-operated gates and the farm outlets along the watercourse are opened and closed by farmers on a warabandi schedule.

It is also possible to have partial automation in an individual control structure. Gated cross-regulators should always be designed with an overspill weir, which can also be the crest of the gate leaf. Minor changes in discharge will pass over the weir without significantly changing the water level. The cross-regulator gates need only to be adjusted for large changes in flow.

4.8 Passive Automation

Devices for passive automation are acted upon *by* the water; they do not act *upon* the water. They include proportional dividers, float-operated gates, counterweighted gates for upstream control, baffle distributors, long-crested weirs and proportional outlets. They require no computerised control and no human intervention beyond their initial adjustment. They are designed for a variety of jobs:

- Controlling upstream water levels;
- Controlling downstream water levels;
- Controlling division of discharge;
- Controlling outlet flow.

Long-crested weirs

This is the simplest form of automatic structure, even though it may not be recognised as such. It does the job of controlling water level upstream close to a pre-set level even though the discharge varies. The discharge over a weir is proportional to the crest length times the head raised to the power 1.5. By lengthening the crest the head can be kept small. Hence a large change in discharge can be handled by a small change of head, so that water levels are made insensitive to the discharge.

figure 4.10   *Labyrinth weir used as an emergency overflow structure*

Several configurations are in common use. Figure A4.8 shows a side weir in a small canal. Figure A4.20 shows an oblique weir in a larger canal. In the first case an offtake immediately upstream has been closed and the weir's function is to safely by-pass without damage the check gate which is also closed. In the second case the function of the weir is to maintain a certain minimum upstream water level at all stages of discharge in order that the offtake immediately upstream may operate. The duckbill configuration of figure 7.10 is another common arrangement, which is effective at dissipating energy downstream of the crest. The Labyrinth weir of figure 4.10 is a less common but equally effective structure, being a combination of several duckbills in parallel.

Using a long weir in combination with an undershot gate is an effective means of partial automation if the objective is upstream control. Any small increases in flow will pass over the weir without any need to adjust the gate, and the water level upstream will remain almost constant. The gates then only need to be operated to accommodate large variations in discharge. This means that gates do not have to be attended constantly and there is a better chance of the canal operating satisfactorily. Undershot radial gates can also be designed to overtop safely in the event that upstream target water level is exceeded.

Self-regulating float-operated gates for constant water level

These are hydro-mechanical gates of which the best-known are those manufactured by the French company Neyrtec (now part of GEC-Alsthom.) They are usually referred to by their trade names Avio, Avis and Amil, and we will persist with the tradition here since many irrigation engineers will be familiar with these. Avio and Avis gates are for downstream control, respectively for orifice and free-surface flow. Amil gates are for upstream control. They are also referred to as constant downstream level, or constant upstream level gates. A special composite gate is available for situations requiring a combination of upstream and downstream, or relative level control. The basic research and development was done 40 or 50 years ago and since then there have been numerous attempts to copy them with varied success. Figure 4.11 shows an interesting Soviet-designed variant of the Amil gate, for parabolic canals.

The advantages of self-regulating gates are:
- They require no operators.
- They require no computation of gate opening and closing times.
- They respond instantaneously to fluctuations in water demand and flow downstream.
- They require no communication system between gates or along the canal.
- They prevent canal wastage and spillage by closing automatically.

The principles of operation of these gates are shown in figure 4.12. A target water level for zero discharge and gate closed is set at the design stage, and the float causes the gate to move up or down until the target level is attained. In order to avoid the destabilising influence of turbulence, the float of the constant downstream level gate is usually housed in an open chamber through which water flows through a small orifice. The water level at zero flow is higher than that at maximum discharge by a small amount known as the *decrement*. For the largest gates this is about 150 mm.

During gate installation the counterweights, which comprise a cylinder filled with steel reinforcing bars, are adjusted to give the exact degree of sensitivity required for the gate to operate at all water levels. The gates are trapezoidal in end elevation to avoid binding against the cheek plates during operation. The constant downstream level gate is set with its trunnion axis at the target water level.

These gates can handle discharges up to 40 cumecs. Avios can cope with a head difference of up to 10 metres. For large canals it is common practice to install at least two gates in parallel, so that one can be taken out of service during low demand periods to facilitate maintenance.

figure 4.11 ***Parabolic self-regulating upstream control gates awaiting installation, Maskane Project, Syria***

The composite gate (*vanne mixte*) acts as a constant downstream level gate but only within a pre-set range of upstream water levels. When the upstream level drops below the lowest limit, the gate closes and acts as an Amil. Similarly if the upstream water level exceeds the upper limit, the gate opens to permit excess water into the downstream reach. This operation is achieved by locating two operating floats in individual side chambers having piped connections into the canal. The water levels in the chambers are controlled by weirs and orifices which are adjusted during installation to give the required operating range. The gate also acts in its mid-range on the principle of ***related level control***, since forces from the two floats will balance each other over rising or falling water levels and the upstream and downstream water levels will tend to have a constant difference. However, this is not the primary purpose of this type of gate.

Radial gates using the same float-operated principles but with a more complicated system of cables and pulleys are sometimes used in river control work in the UK. The tilting gate is used for fine control of upstream water level, usually in combination with undershot radials which are opened to pass floods.

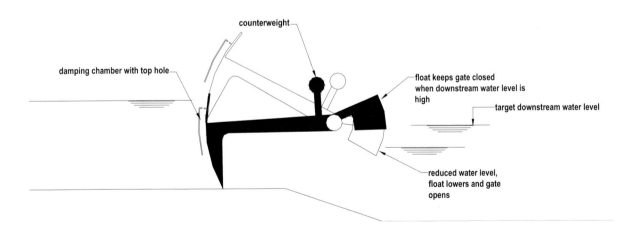

figure 4.12 ***Principle of operation of downstream control self-regulating gates***

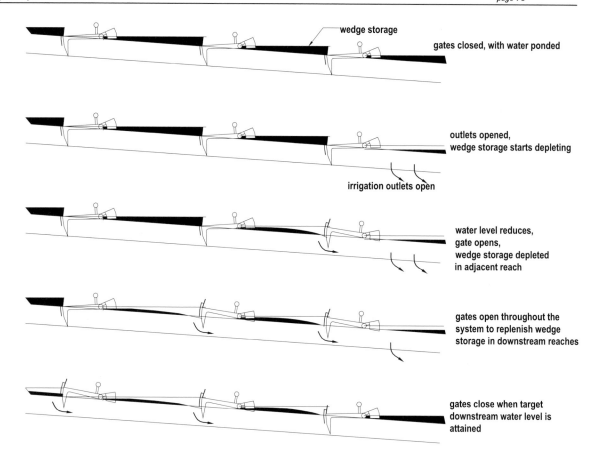

figure 4.13   *Downstream control in adjacent canal reaches*

Hunting and transients

One problem associated with this type of gate is that of *hunting*, in which the gate movement overcompensates in response to fluctuations in water level, which in turn affects the next gate along the system. The gates operate out of phase with each other, creating transient waves up and down the reach, which may cause damage and overtop the banks. This tends to happen if the reach is too short. Some manufacturers give guidelines for the minimum amount of wedge storage in the reach. Gate stability can also be checked by means of a transient analysis, a hydraulic computation of non-steady flow. For the 30 cumec Pehur High Level Canal, the minimum reach length was around 5 km. The bed slope too is restricted to a practical range, generally less than about 0.00025, above which the cost of constructing level-top canal banks increases substantially.

During commissioning of the Maira Branch Canal gates (figure A4.11a-f) severe hunting occurred. This was corrected by reducing the size of the inlet to the float chamber. The main float is located inside a steel chamber, which has an opening into the surrounding water. The purpose of the chamber is to isolate the float from the effects of turbulence or rapid changes in water level. More damping is afforded by constricting the size of the opening.

Counterweighted gates for upstream control

There are several well-known versions of a simple automatic flap gate that starts to open when the pressure of water on its upstream face exceeds the opposing force of a counterweight tending to keep the gate closed. An example is shown in figure A4.12. They can be made very cheaply but as is the case for most self-regulating gates they are not watertight so cannot be used as isolating gates.

Proportional dividers

Proportional dividers are useful in schemes which rely on proportional division at some or all levels of the canal system, for instance the protective schemes of India and Pakistan, or the run-of-river schemes in Java. Here the function of the system is to spread the available water around as equitably as possible, meaning that water shortages are also spread evenly. Hence when the discharge in a parent canal fluctuates, the discharge in all offtaking canals varies in the same proportion. The basic structure is a weir with a splitter wall or vane which divides the flow in a pre-set ratio.

figure 4.14   A proportional divider abstracts a fixed proportion of the flow in the parent canal

The French company Neyrtec (now GEC Alsthom) manufactures a division structure with a movable steel vane which can be initially adjusted to any desired split ratio. In Pakistan we developed a family of bifurcators and trifurcators based on a Crump weir and fixed splitter wall. The example in figure 4.14 shows a watercourse offtaking from a distributary canal on the Upper Swat system. As the flow passes over the weir crest, the splitter wall divides it in the required proportion. The nose of the splitter wall is located at a point which splits the crest length in the same proportion as the required discharge. This offtake takes about one-tenth of the parent canal flow, so the width of the offtaking flume is one-tenth that of the whole weir. Limitations arise when the offtaking flow is proportionately smaller than this. If it were say one-hundredth of the parent canal flow, the weir would be ten times as wide if we were to keep the offtake the same width. There is a practical limit to the width of the offtake; if it is too small then construction methods cannot meet the close tolerances required, and wall friction at the sides of the flume will interfere and choke the flow.

These limitations can be countered by a double bifurcator (figure 7.6), in which the flow is split again by a second Crump weir. It uses a little more head but in many cases gives a more compact structure. Difficulties in construction were experienced as the local contractors were unused to working to close tolerances, but these problems were largely overcome by precasting. The design is described more fully in chapter 7.

These structures are very simple yet do a job automatically. That is, to divide the flow in a fixed proportion no matter what the incoming discharge. They replaced the old open flume outlets that were designed to do the same job, but in practice were never very accurate, only worked properly over a narrow range of water levels in the parent canal, and were susceptible to error caused by siltation or weed growth in the parent canal. It was easy to cheat with the old flumes; a favourite trick among some farmers was to bathe their buffalo in the distributary canal immediately downstream of their offtake so that the water level rose and an unfair proportion flowed into their watercourse! It was somewhat surprising then that some of the first bifurcators were smashed by farmers who perceived the Crump weir as an obstruction to flow rather than as a means of guaranteeing equitable distribution.

Flumed outlets for proportional discharge

The early years of the 20th century were a period of great innovation in India, where irrigation development was taking place on a grand scale. Notions of water management other than supply scheduling and proportional flow were never given any thought, for at that time protection against famine for as many people as possible was the prime concern. One of the many new types of structure to emerge from the free hand that was given to many of the design engineers of the day was the proportional outlet.

The open flume proportional outlet and its developments the Adjustable Proportional Module (APM) and Adjustable Orifice Semi-Module (AOSM) were developed as a means of abstracting a fixed proportion of the parent canal discharge into a watercourse. When the parent canal discharge varied, so did the outflow into the watercourse by the same percentage, so that at least the outlets downstream received a fair share. They were therefore passive automatic structures. Their advantage over split-weir types is that they can be used to feed a small amount of water from a large canal, in situations where a split weir would be too cumbersome and expensive. Their disadvantage is that they only perform well over a limited range of water level. More details are given in section 7.4.

Baffle distributors for constant discharge

Baffle structures operate on the principle of flow constriction. They are designed to deliver a constant outlet discharge even though the upstream water level varies. The best-known baffle distributors are manufactured by Neyrtec (figure 4.15.) They operate as a throttled orifice. The throttling effect is generated by an inclined plate which directs the flow backward and downward above the entrance to the orifice. As the upstream head increases, so does the throttling effect, and the discharge remains more or less constant (i.e. to within +/- 5 per cent of design) over a range of upstream heads. In an unthrottled orifice the discharge would increase in proportion to the square root of the head.

figure 4.15   *Modular distributor*

The basic element is a steel plate inclined into the flow. The rectangular orifice can be closed with a steel shutter or left open to give a discharge related to the width of the orifice. Although frequently used as a single offtake giving a fixed discharge into a minor canal, it is also commonly used in a combined set of different sizes as in figure 4.15. This gives controlled regulation over a range of discharges using any desired permutation of orifice openings. A modular range of widths is manufactured, and used in combination there is no limit to the discharge which can be handled by any one structure.

The operating head is in the range 0.1 - 0.2 m for the smallest sizes and 1.0 - 1.4 m for the largest. A variation using a double baffle plate increases the maximum operating head by about 50 per cent.

In supply-scheduled situations which require tight control over discharge the durability and effectiveness of this type of structure is well proven. However, it is not cheap. It pays to be completely clear from the outset of design about the mode of water management that will be applied.

4.9 Active Automation

There are clear advantages in using passive automation where possible. With passive automation there are no cables, computers or controllers involved. No decisions are expected from operating staff. Little maintenance is required. No communication between regulator structures is called for. However, there is a cost associated with this, for instance in raising the canal banks for downstream control. We can also get downstream control with active automation but the trade-off is between a saving on civil works and the extra cost and maintenance of complex computerised control networks.

Some canal systems are unsuitable for passive automation, particularly if the longitudinal gradients are steep. The Froude number at design flow has been suggested as an indicator of suitability for automation. If the Froude number is less than 0.3, passive automation is feasible, and a Froude number greater than 0.5 implies suitability only for feed-forward control. In most irrigation projects, canal Froude numbers are typically less than 0.2.

Active devices act *upon* the water, and are not acted upon *by* the water. They are usually gates or pumps which come into operation in response to a water level rising or falling to a certain target level. They may also be operated manually, as in the case of a simple sluice gate. When the operating procedure is automated, the thought process of the manual operator has to be emulated by the control system:

- *'When the water level gets too high, I must close the gate.'*
- *'When people downstream want more water, I must release more from the reservoir.'*
- *'It is raining, so the people downstream will want to shut down their water supply, so I had better start closing down the cross-regulators now before the canal overflows.'*

Now it is necessary to transcend some language barriers, between the language of the farmers, that of the irrigation engineer, and that of the control theorists!

Control theory

Control theory is a science in its own right, with practical applications in every branch of applied science. Concepts of logic are an active, albeit jargon-ridden, area of research, and recent advances, especially in fuzzy logic, offer some exciting prospects for the automation of irrigation systems. But first, an explanation of some of the jargon.

The ***system under control*** might be water level in a single reach in a canal or it might be volumes in every reach of the entire irrigation scheme, or discharge in just some sections of it. It is what we are trying to control in an automatic process, and we therefore need to be specific in terms of water level, discharge or volume.

The ***Controller*** contains the software or applies the mathematical theory that does the job. It may work independently for a specific device or it may be linked in with other controllers.

Input is the supply of information and data required to be processed and used on the system under control in order to attain a required ***output***. Input data is also referred to as the ***control action variable*** and might comprise a single gate opening value (Single Input), or an array of discharge measurements and gate openings from sensors located all over the system (Multiple Input.) Raw input data may be converted into more meaningful commands through an input function, e.g. water levels integrated in space and time to give a discharge, or discharges integrated in time to give a volume.

Output is the ultimate action on the thing we are trying to control, the ***controlled variable***. It may be a water level, a discharge, or a volume in a reach. The controlled variable will have a target value which may be constant or changing in a known way. Output function might be an electric signal to a gate motor to raise or lower the gate.

Multivariable control infers more than one input or outputs, and is usually defined in the name assigned to the process. Hence MIMO means multiple input, multiple output; MISO means multiple input, single output; SISO stands for single input, single output, etc.

Feedback is the principle used in all self-regulating control systems, in which information on the result of an action is used to repeat the action in a modified way or to a modified extent. Feedback systems are subdivided into closed-loop and open-loop systems. A ***closed-loop*** feedback system is fully computerised or automatic and allows no intervention by the operator. An ***open-loop*** feedback system can respond to intervening control signals from an operator.

Feed-forward is the ultimate ***open-loop*** principle in which relevant information from an external source is applied into the process in order to achieve a certain objective. It is used in ***predictive controllers*** and ***gate stroking***, where the gate operations required to control water levels are anticipated ***in advance*** of the water levels actually falling or rising. It relies on accurate estimates of what actions will be required in the short-term future. The estimates might be based on past experience, or information on weather conditions or crop harvesting, on changing discharges from outlets, or from a combination of various sources. The feed-forward processing might be done through a computer simulation model of the canal system, or by ***fuzzy logic***. The term ***feed-forward*** is descriptive but confusing. In practice it still needs ***feedback*** in the form of monitoring target discharges and levels in order to assess whether it is working correctly or not. So it is not an alternative to feedback, but rather an augmentation of it.

When two logics are used together, they may be either coupled (interdependent) or decoupled (acting separately as in the following description of dynamic regulation.) Decoupling often speeds up the whole process of canal control, leading toward ***real-time*** control.

Fuzzy logic

What enables a person to ride a bicycle? What enables water master Gary Sloan to operate his upstream-controlled canal system, 160 km long with many offtakes and 36 cross-regulator gates in series, so that the water levels never vary by more than 2 centimetres? Both these are examples of unstable systems that would fall over or fail if left to themselves, but which can succeed with expert human intervention.

The human brain is a master of fuzzy logic. In the case of the bicycle, the statically unstable machine is kept upright by a great number of tiny shifts in weight of the person riding it, which compensate for its tendency to fall over. In the case of the Santa Rosa Canal in Arizona, the gates are 'tweaked' every few minutes in such a way that any tendency of water levels to rise or fall in any of the 36 reaches is anticipated and compensated for in advance. In both cases the person in control relies on past learning experiences and the skills both developed from it and inherent in the human brain. These are not easy to replicate in a computer program.

Until the science of fuzzy logic was developed, it was very difficult to program a robot to ride a bicycle or reverse an articulated truck around a corner. The reason for the problem lies in the difficulty of creating an exact mathematical simulation which accurately describes the anticipated motion and also deals successfully with unforeseen perturbations such as might be caused by a grain of gravel in the road, or a gust of wind on the side of the truck. You can do it with standard feedback logic, but you need dynamic regulation using both open-loop and closed-loop control which gives a rapid series of correcting adjustments to the steering wheel.

Fuzzy logic allows an imprecise, human-like way of thinking to be applied in the programming of computers. Conventional computer logic relies on unambiguous statements like yes/no, true/false, and black/white. Human logic can happily process vague statements such as *rather hot* or *quite cold*. In canal operating terminology: '*The water level is probably going to increase so we had better close the gate a bit*' can now be formulated mathematically and processed by computers.

Fuzzy control is applicable to complex processes which have no simple mathematical model, and highly non-linear processes. It is being successfully used in control of hydroelectric dam gates, camera aiming for the telecast of sporting events, planning of bus time-tables, earthquake prediction, cancer diagnosis, improved fuel-consumption for vehicles, and now in irrigation canal system operation.

For a crash course in fuzzy logic, log in to a good web site (Ref. 8), because I'm taking you no further here except to describe a recent project in the USA (Ref. 11) which was converted to fuzzy control.

The Dry Gulch Canal is upstream-controlled from a reservoir and required to be operated to give a constant water level at the tail of the canal, from which there are numerous turnouts. The total offtake demand fluctuated from 0 to 7 cumecs, and control gate operations were complicated by the accumulation of debris which partially blocked the trash screens and influenced water levels in an unpredictable way. The canal is old, unlined and irregular in cross section, and no survey data were available for it. Using sensors on the regulating gates and existing Parshall flume measurement structure it was possible to maintain tail-water levels within a tight range.

There is a huge practical advantage in fuzzy logic: you don't need to know the exact geometry and dimensions of the canal. It is not necessary to measure all the offtake flows. Furthermore, a fuzzy controller can accommodate changes in the system without affecting the result; in control logic terms, it is *robust*. And it doesn't require a lot of calibration to tie in the theoretical results with what actually happens in the field. So you can apply the principles of controlled volume to an old unlined canal with irregular banks and prone to periodic change through erosion, weed growth and siltation, which would take a lot of time to survey accurately.

Active automation of a canal system

It is becoming rare for an irrigation design engineer to be working on an entirely new canal system. Most design scenarios involving automation concern an existing canal system which was originally designed and built long before automation was possible and which is now due for rehabilitation. So having decided on some form of automation the prime concern is how to convert from manual to automatic, not how to build it automatic in the first place.

Gate stroking is a method of centralised feed-forward control, in which a large number of small gate movements are anticipated in advance in order to reduce the response time of a canal to changes in flow. The controlled variable is usually water level at the end of each reach. Starting at the downstream end, the water level and discharge in each reach are calculated using one of several mathematical algorithms.

If water levels are off target, gate adjustments are calculated as a function of time. The actual gate movements are complex and require the use of variable speed motors. Although these are expensive, their starting current is low so there is a power saving over conventional fixed-speed motors.

Centralised controlled-volume operation targets the volume of water in each reach which is varied according to a pre-determined schedule. Typically the discharge requirements at all outlets are updated twice a day, and in the case of the California Aqueduct, pumping is scheduled to take advantage of off-peak power, so that the upper reaches serve as buffer storage during peak pumping times. Being a typical case of upstream controlled supply scheduling, it does not offer much flexibility and water levels are not constant. However, the response time of the system is effectively reduced to the response time of a single reach, since all gate operations are simultaneous and the water is shifted along the canal *en bloc*. It is more suited to large conveyors rather than in a distribution system feeding small outlets.

Centralised dynamic regulation of a canal system utilises a combination of both closed-loop and open-loop control, for optimising the management of all water passing through the system. A target discharge is estimated from forecasts of demand. This is a feed-forward or open-loop process. A correction discharge is then applied from a closed-loop determination of actual volume in each reach. If there is a shortfall in volume then extra discharge is let in, and if there is too much volume in a reach a cutback in discharge is made. Examples of canals controlled in this way include the Canal de Provence in France, and the Canal Rocard in Morocco.

Centralised real-time control aims to provide water on demand in an upstream-controlled canal system. It thus emulates a passive system of level-top canals and downstream constant-level gates, but without involving expensive civil works. It requires an estimate of the demand at every outlet, and the processor calculates necessary discharge and flow conditions in each reach, taking into account the response time of each reach. The actual, *real-time*, conditions over the whole system are fed into the processor and repeatedly updated at intervals as frequent as 10 minutes. A closed-loop calculation of the required gate settings is made (emulating a self-regulating gate operation) and transmitted to individual regulator structures in the field. Examples of large canal systems using this version of control are referred to as CAGC and SCP, both French canal management companies.

PID is an acronym for Proportional, Integral, Derivative, which describes the mathematical simulation algorithm of a type of controller used in early active systems for *local distant downstream control*. The water level at the end of each reach is targeted as a constant by operating the gates at the head of the same reach. A similar method is EL-FLO, developed by USBR, and PI.

CARDD is an acronym for Canal Automation for Rapid Demand Deliveries, which speaks for itself. It is also applied for *local distant downstream control*, but in an attempt to reduce overall response times it takes into account water-level data from several points along the reach. It was also developed by USBR.

PIR uses feed-forward logic in a further refinement of *local distant downstream control*. It is a predictive controller which estimates the future target water level based on the effect of gate adjustments already made.

SCADA (Supervisory Control And Data Acquisition) system refers to the combination of telemetry and data collection. It consists of collecting information, transferring it back to a central site, carrying out necessary analysis and control, and then displaying these data on a number of operator screens. The SCADA system is used to monitor and control plant or equipment. Control may be automatic or can be initiated by operator commands. A SCADA system is composed of Field Instrumentation, Remote Stations, Communications Network, and Central Monitoring Station.

Most of these methods rely on a mathematical algorithm based on an unsteady flow simulation model of the canal system or at least accurate knowledge of its geometry. But in the real world canals are not of regular prismatic section. Bring on the fuzzy logic.
There are some well-documented controllers which enable a standard gate to do the job of a self-regulating gate controlled by a target water level. The Littleman and Zimbelman controllers were developed in the USA for local downstream control.

Instrumentation, communication and motorisation

Instrumentation covers the sensors which measure the controlled variables such as water level, gate openings or discharge, the type of controller and the method of communication between the two.

The measured variables in canal control are one or more of water level, discharge and gate position. Available sensor types for water level include piezoelectric pressure transducers, floats and float switches, and ultrasonic sensors. Floats are in commonest use in developing countries as they don't require a power supply. Discharge can be measured directly by ultrasonic flow meters, or indirectly by water levels at measuring weirs, or indirectly by water levels and openings of gates. The latter method is however, inaccurate, especially when the gate is near the extremes of its travel or when the outflow is submerged. Gate position indicators can be mechanical levers, gears or cables linked directly to the sensing device, or an electronic shaft encoder which converts the position of the gate operating shaft into an electronic signal.

Communication links are necessary for all configurations other than float-operated gates in which the floats are a part of the device. The communication may be provided by means of a ditch rider on a bicycle, or by telemetry which includes radio signals, telephone line, electronic impulses via fibre-optic or electric cabling. Telemetry requires power at the sensing point and although it is usually only a few watts it has always been a discouragement to remote control in developing countries where theft and maintenance are problematical. Solar power is increasingly being employed in remote locations in order to recharge small batteries running instruments.

Radials are the commonest type of large gate selected for motorisation, since they require less effort to raise and are less prone to jamming than vertical slide gates. Roller gates and wheel gates are still in common use in Pakistan and India for large structures. These are large vertical lift gates with roller bearings or wheels which support the horizontal thrust, but they require a heavy supporting frame and usually also counterweight ballast. Tilting gates can also be used, such as the Obermeyer gate which is activated by pumping air into an inflatable bladder. On account of their low cost small slide gates are used everywhere for passing discharges up to a few cumecs. It is a common situation in rehabilitation work to retro-fit existing gates with motors and remote control.

A pair of hydraulically-operated radial gates in Morocco are controlled by radio from a central office 50 km away. This is part of the dynamically-regulated Canal Rocard system, and is an example of active automation.

figure 4.16   Motorisation

The actuator or lifting device can be an electric motor driving either a winch with cables or a screwed gear, or a hydraulic cylinder. Small to medium-sized radial gates of the size used in canals are usually cable-operated with the winch on a horizontal axis above the gate. Small vertical lift gates on a rising spindle can be raised or lowered by turning a horizontal wheel supported on a thrust bearing. This arrangement is not suitable for heavy gates which require reduction gearing and are better suited to hydraulic operation.

Hydraulic operation, in which a piston is activated by a small electric pump moving oil between the cylinder and a reservoir tank, is always used for high-pressure gates such as bottom outlets of dams, since the large breakout force required can be applied directly to the gate without relying on mechanical linkages for which maintenance is difficult. It is increasingly used for smaller canal gates where it presents fewer problems of accidental damage, theft and maintenance.

4.10 Evolution from Manual Protective to Automated Productive - a case study

Continuing the story of the Upper Swat system from section 2.2, this presents an example of a large scheme, which is in gradual metamorphosis from manual to automated. The whole process will probably take the better part of the 21st century, but this inexorable change has already begun with the rehabilitation of the old system and the construction of the Pehur High Level Canal. Looking into both the past and future the programme of development will probably appear something like this:

- 1901 - the Upper Swat Canal and Benton Tunnel is conceived as a means of bringing protective irrigation water to an area of 100,000 ha, in order to guard against famine and widespread crop failure.

- 1914 - the canal system commences operation as a manually operated, supply-scheduled, proportional flow scheme. No cross-regulation is provided on the main system. Open flume **proportional outlets** and APMs are used.

- 1925 - additional distributaries are constructed to extend the scheme into areas of rough terrain.

- 1935 - a 20 MW power station, and the Burkitt Tunnel are built to utilise available head from the inter-basin transfer.

- 1950 - a second power station and the Dargai power canal are constructed. Variable power demand begins to take priority over irrigation and canal discharges start to vary. Buffer storage is constructed using related-level cross-regulators, but proves difficult to control. The main system continues without cross regulation.

- 1960 - tail-end users suffer as head-end farmers diversify and grow more crops which require more flexible and more frequent water supply. Several weir-type cross-regulators are built on the main canals to increase supply into some distributaries, but this also exacerbates the tail-end problem.

- 1978 - Tarbela Dam is completed on the River Indus. The Pehur High Level Canal is conceived as a means of taking Indus water to supplement supplies in the Upper Swat command.

- 1985 - rehabilitation of the adjacent Lower Swat Canal commences. The main canals are designed for intermediate storage and minor canals are gated with the aim of demand-scheduled operation. Neither the farmers nor the canal operators understand the changing principles water management and the scheme continues to operate on a proportional supply schedule.

- 1989 - a small pilot scheme is established within the Lower Swat command to demonstrate automated operation of a **semi-closed pipeline**. It continues to work well but few in the Irrigation Department understand its principles of operation.

- 1993 - rehabilitation of the Upper Swat Canal system commences with the intention of doubling the canal capacity and water duty. It is designed for supply-scheduled proportional flow but manually-operated **cross-regulators** are installed for upstream control on the main system. **Proportional dividers** are introduced to cope with variable discharges at lower levels in the system. However, additional capacity is built in to the freeboard of all canals in order that higher discharges pertaining to a demand-scheduled operation may be accommodated at a later date.

- 1995 - design of the Pehur High Level Canal commences. The scheme is justified on its capability of guaranteeing winter flows to the tail of the Upper Swat system and permitting extension of the Upper Swat canals into 30,000 ha of previously unirrigated lands. The High Level Canal and the tail of the Upper Swat main system, the Maira Branch Canal, are designed for automated **downstream control with self-regulating gates**. No practical non-automated alternative is deemed possible to cope with the complex requirements of minimising wastage and power foregone from Tarbela whilst topping-up a variable run-of-river supply from the Swat. Distributary canals are still to be operated on proportional division with upstream control at their head.

- 1996 - design is completed for the previously-unirrigated Topi Area for 4,000 hectares of low-pressure **semi-closed pipeline** served by the level-top Pehur High Level Canal. The design is met with scepticism by the Irrigation Department and a compromise is reached under which the main delivery pipelines are retained with orifice-type outlets into open watercourses. The scheme can thus be readily converted at a later date into full pipeline distribution at minimal cost.

- 2000 – construction of Topi pipeline network is complete. The Harris valves are not installed because there is no water in the main system due to delays in tunnelling. The Harris valves are held in storage by the contractor, but when water becomes available 2 years later the government refuses to pay a few thousand dollars for storage and the valves are not yet installed. The pipes come into operation as closed gravity canals.

- 2003 - construction of Pehur High Level Canal is completed. The self-regulating gates work well. This is their first application in Pakistan.

- 2009 – the Topi Harris valves are re-discovered and a bright engineer reading this book manages to get them installed and operating properly.

- 2012 - a third power station, Malakand 3, is constructed, placing more constraints on water availability.

- 2015 - **active automation with computerised control** installed on the existing Upper Swat Canal cross-regulators. The entire main canal system is now **automated**.

- 2020 - Tarbela Reservoir silts up and flows into PHLC become less reliable.

- 2030 - small reservoirs added at head of distributary canals to enable **flexible demand scheduling**.

- 2040 - Topi converted to **full pipeline distribution**.

- 2050 - **pipelines** adopted over much of the Upper Swat command area. New extension areas irrigated by pumping into level-top canals and semi-closed pipelines.

It will have taken a century and a half and six generations to make the sea-change transition from manual to automatic; from confrontation to harmony; from subsistence to prosperity; from water wastage to efficient water use. And the greatest barrier, perhaps the only barrier, to this change is the prevailing attitudes of engineers and politicians who sometimes only look backward and see what we did before, rather than look forward and see what our sons and daughters might achieve.

References and further reading for chapter 4

(1) Ven Te Chow, Open Channel Hydraulics, McGraw-Hill, 1959.

(2) Merriam, J.L. and Styles, S., 'Use of Reservoirs and Large Capacity Distribution Systems to simplify Flexible Operation' - XXVII IAHR /ASCE Conf, San Francisco, Aug. 1997.

(3) American Society of Civil Engineers, Planning, Operation, Rehabilitation and Automation of Irrigation and Water Delivery Systems, 1987.

(4) Cemagref, International Workshop on Regulation of Irrigation Canals: State of the Art of Research and Applications, Marrakech, April 1997.

(5) Goussard, J., Automation of Canal Irrigation Systems, ICID, 1993.

(6) Malaterre P.O. 'Regulation of irrigation canals: characterisation and classification'. International Journal of Irrigation and Drainage Systems, Vol. 9, No.4, November 1995

(7) Malaterre, P.O., http://www.montpellier.cemagref.fr/~pom/perso.htm

(8) Peter Bauer, Stephan Nouak, Roman Winkler, http://www.flll.uni-linz.ac.at/pdhome.html

(9) Thomas-M. Stein, http://fserv.wiz.uni-kassel.de/kww/irrisoft/irrisoft_i.html

(10) Plusquellec, H., Burt, C., Wolter, H.W., 'Modern Water Control in Irrigation', World Bank Technical Paper 246, 1994.

(11) Stringham, B. L. and Merkley, G. P., - 'Field Application of a fuzzy controller for an irrigation canal in Roosevelt Utah', International Workshop on Regulation of Irrigation Canals: State of the Art of Research and Applications, Marrakech, April 1997.

(12) Parrish, J. B., Automation of sloping canals, 'International Workshop on Regulation of Irrigation Canals: State of the Art of Research and Applications, Marrakech, April 1997.

(13) Parrish, J. B., http://www.iihr.uiowa.edu

(14) Laycock, A., Swayne C. G., Marques, J., - 'The Pehur High Level Canal, Pakistan', Proceedings of the Institution of Civil Engineers, v158/WM3, September 2005.

CHAPTER 5   IRRIGATION WATER DEMANDS

Some basic knowledge of crops and their physiology is essential in order to design an irrigation canal. The designer needs to know the amount of water flow that is required at all points along the canal. That requires an understanding of the crops that are being irrigated and the way in which they use water, and also the soils and the way in which water is taken into the soil. It is also necessary to know how the system will be managed, and when and how water will be directed along the canals. Recent developments in irrigation planning philosophies lead to the concepts of flexibility and congestion, which require a radical change in accepted design procedures.

Water always gets lost along the way. It leaks, it seeps, it overflows, it is spirited away by animals, human and others. It is wasted, stolen, or lost to the ravages of nature. A great deal of water loss can be controlled or prevented, once we know what to look for. And if it cannot be prevented, we need to estimate the losses in order to design a system that will work.

5.1   *Estimating Irrigation Requirements*

Why plants need water

Plants absorb water through their root system. The *root zone* of the soil needs to be replenished with water either by capillary rise from the groundwater zone below, or by downward infiltration from rainfall or irrigation.

A plant makes use of the water in several ways. Firstly, the water carries dissolved nutrients such as nitrogen, potassium, phosphorous and many trace elements. These nutrients are utilised by the plant in growing, that is, in building new cells through photosynthesis. Water moving through the plant distributes these nutrients to the places where they are needed.

Secondly, water fills the individual plant cells and keeps the plant rigid. If the plant's water content is reduced, the internal water pressure in the cells is reduced, the cell walls start to collapse and the plant starts to wilt. Wilting is one indication of the need to irrigate, being a sign of water stress in the plant.[1]

When the water has deposited its nutrients in the plant, any excess water that is not required to fill new plant cells is evaporated into the atmosphere through stomata, which are small pores in the underside of its leaves. The stomata are controlled by a pair of guard cells, which inflate with water to open, and collapse when deprived of water to close.

Evapotranspiration

The physical process in which water travels through the plant from its roots upward into its leaves and out into the air is called evapotranspiration.

The plant has some control over its rate of evapotranspiration. If water is in short supply, wilting causes the stomata to close and the rate of evaporation is reduced. If there is a continuous supply of water such as in paddy rice, the plant may pump water through at a faster rate than it actually requires in order to grow. This is known as luxury consumption.

Different crops use different amounts of water. High-yielding rice uses a lot. Groundnuts and pulses need less. Sorghum, a crop famous for its resistance to drought, needs little. Many crops can have a crop factor in excess of 1 at some stages of growth, because the plant acts as a pump, ejecting water vapour into the atmosphere through a leaf area which is considerably larger than the ground area on

---

1 *Plants will recover from wilting, until they reach permanent wilting point, the point of no return, which varies between species.*

which it grows. Waterweeds such as *Salvinia* and Water Hyacinth can have crop factors as high as 2 or 3, and are notorious for wasting water.

A crop needs varying amounts of water during different stages of growth. Hence the crop factor varies over the life of the crop. Some stages of growth are more critical than others, in terms of the plant's ability to withstand moisture stress without the crop yield being affected.

The crop factor is a measure of each crop's water demand from month to month. Multiply it by the reference evapotranspiration ET_0 to get the water demand. Representative figures are given in table 3.1 but no self-respecting agronomists will be entirely in agreement over these and different crop varieties will have different figures. It can also be calculated over 10-day periods (decades) or even 5 day periods (pentades), but for practical design purposes monthly figures are usually good enough. However, modern spreadsheets mean that any extra work involved in processing more figures is minimal, so the rule is to utilise whatever data are available.

Crop	Duration	Crop factor Kc at months after planting										
	days	1	2	3	4	5	6	7	8	9	10	11
groundnuts	120	0.5	0.9	1.1	1.0							
groundnuts	150	0.5	0.7	1.0	1.0	0.6						
pulses	100	0.5	0.8	1.0	0.4							
pulses	60	0.7	1.0									
cotton	170	0.5	0.6	0.8	1.1	1.1	0.6					
sorghum	150	0.5	0.9	0.9	0.8	0.6						
HYV rice	120	1.3	1.1	1.2	1.0							
local rice	170	1.2	1.0	1.0	1.0	0.7						
wheat	125	0.6	1.1	1.0	0.6							
maize	105	0.5	1.0	1.1	0.4							
maize	120	0.5	0.9	1.1	0.9							
sunflower	120	0.4	1.0	1.2	0.8							
sesame	90	0.4	0.8	0.9	0.7							
vegetables	continuous	1.0										
alfalfa	continuous	1.0										
sugarcane, mother crop	300	0.5	0.7	1.0	1.0	1.0	1.0	1.0	0.8	0.8	0.5	
sugarcane, ratoon	350	1.0	1.0	1.0	1.0	1.0	1.0	1.0	1.0	0.8	0.8	0.5

table 5.1   *Crop factors*

Climatic conditions also affect the amount of crop water use. Hot temperature, strong winds and low humidity all increase the evapotranspiration. The intensity of solar radiation and the albedo or reflectivity of the crop also affect it.

The Penman Formula, or the Penman-Monteith derivative[2] of it, is often[3] used to calculate the Potential Evapotranspiration of a reference crop, based on the prevailing climatic conditions. The reference crop is short grass, and its crop factor is 1. This potential evapotranspiration or ET_0, is the yardstick for calculating the water use of all other crops. It is also referred to as the open water evaporation, on the premise that a large body of water such as a lake will have a similar evaporation rate as the reference crop. The crop water requirement is simply $ET_0 \times$ the crop factor, in millimetres per day or per month. The pain of calculating Penman figures is now removed by an excellent piece of software named CROPWAT which is available on application from FAO's Land and Water Division in Rome.

---

2 *Also known as the Modified Penman formula.*
3 *And often too much reliance is placed upon it; it was developed in a temperate climate and modified for tropical conditions, and can result in an over-estimation of water requirement..*

Other methods are also used, the most common being derived from measured open pan evaporation figures, with ET_0 commonly taken as $0.8 \times$ the pan evaporation. However, this method too can be unreliable, because the true pan factor is uncertain and often closer to 0.5, depending on the location of the pan and the climate. Other methods such as Thornthwaite's and the Blaney-Criddle formula, base evapotranspiration estimates on temperature records. These formulae can be found in Ref. 6.

Penman formula, modified by Doorenbos and Pruit (Ref. 9)

$$ET_0 \, mm/day = \{W[R_s(1-\alpha) - \sigma T^4 (0.34 - 0.044 \, e^{0.5})(0.1 + 0.9n/N)] + [1-W][0.27(1+0.01u)(e_s-e)]\}C$$

where :

Δ = slope of saturation vapour pressure curve

γ = psychrometric constant

$W = \Delta/(\Delta+\gamma)$

R_s = incident solar radiation

a = albedo or reflectivity (= 0.25 for crops)

σ = Stefan Boltzman constant

T = mean temperature in degrees absolute

n = number of sunshine hours

N = maximum possible number of sunshine hours

e_s = saturation vapour pressure in millibars

e = actual vapour pressure in millibars

u = wind run, km per day

C = a correction factor dependent on ratio of day to night wind run, relative humidity and solar radiation

---

Crop water demand

The crop water demand described above may be satisfied by rainfall or soil moisture. Whenever the amount of available water drops below the crop water demand, irrigation is needed. The irrigation requirement is thus the difference between the crop water requirement and the effective rainfall:

Irrigation requirement = $ET_0 \times$ crop factor - effective rainfall

Some crops, notably rice, may require extra irrigation for the purpose of land preparation. Some crops may be pre-irrigated, so that the soil is already moist at planting time. And others may be pre-irrigated a month before planting in order to germinate dormant weed seeds that will then be left to die off through drought. All these extra irrigations must be taken into account in calculating overall irrigation demands.

An example spreadsheet calculation of irrigation demand using the cropping pattern of figure 5.1 is given in table 5.2. The diversion requirements are worked out on a 1000 hectare basis, which was a convenient unit for this particular 50,000 ha scheme in Nigeria. It is a straightforward matter to extend the calculation to include say pumping costs or reservoir volumes, by adding extra rows at the bottom.

Estimating rainfall

When planning an irrigation scheme, the fundamental data come from historical rainfall records for the area in question. Major uncertainties arise from the fact that rainfall varies greatly from year to year. For designing the capacity of a scheme it is normal to take a 20 per cent low rainfall year, that is, a year in which there is a 1 in 5 (20%) chance of the annual rainfall being as low as this. For calculating average use or long-term pumping costs, use mean rainfall figures.

Monitoring an on-going scheme can take the actual rainfall daily or month by month, calculate the theoretical water use for the actual cropped area, and then compare this with the actual water use to estimate the scheme efficiency.

row		unit	calculation	may	jun	jul	aug	sep	oct	nov	dec	jan	feb	mar	apr
1	Evapotranspiration ET0	mm		223	189	174	158	168	183	147	140	136	185	195	228
2	rainfall, 20% low year	mm		43	80	201	243	115	24						
3	effective per centage	%		70	70	70	70	60	50						
4	Effective rainfall	mm		30	56	141	170	69	12						
5	**COTTON, 30%**	ha			300	300	300	300	300	300					
6	crop factor				0.5	0.6	0.9	1.1	0.8	0.5					
7	crop water requirement	mm	row 6 * row 1		95	104	142	185	146	74					
8	net deficit	mm	row 7 - row 4		39			116	134	74					
9	field application efficiency				0.7	0.7	0.7	0.7	0.7	0.7					
10	diversion requirement	mm	row 8 / row 9		55			165	192	105					
11	conveyance efficiency				0.85	0.85	0.85	0.85	0.85	0.85					
12	system head requiremnt	mm	row 10 / row 11		65			195	226	124					
13	ditto, volume per month	Mm3 1000 ha	/ row 12 * row 5/100,000		0.19			0.58	0.68	0.37					
14	**MAIZE, 10%**	ha			100	100	100	100							
15	crop factor				0.5	1	1.1	0.4							
16	crop water requirement	mm	row 15 * row 1		95	174	174	67							
17	net deficit	mm	row 16 - row 4		39	33	4								
18	field application efficiency				0.7	0.7	0.7	0.7							
19	diversion requirement	mm	row 17 / row 18		55	47	5								
20	conveyance efficiency				0.85	0.85	0.85	0.85							
21	system head requiremnt	mm	row 19 / row 20		65	55	6								
22	ditto, volume per month	Mm3 1000 ha	/ row 21 * row 14/100,000		0.06	0.06	0.01								
23	**GROUNDNUTS 15%**	ha			150	150	150	150							
24	crop factor				0.5	0.9	1.1	1							
25	crop water requirement	mm	row 24 * row 1		95	157	174	168							
26	net deficit	mm	row 25 - row 4		39	16	4	99							
27	field application efficiency				0.7	0.7	0.7	0.7							
28	diversion requirement	mm	row 26 / row 27		55	22	5	141							
29	conveyance efficiency				0.85	0.85	0.85	0.85							
30	system head requirement	mm	row 28 / row 29		65	26	6	166							
31	ditto, volume per month	Mm3 1000 ha	/ row 30 * row 23/100,000		0.10	0.04	0.01	0.25							
32	**GROUNDNUTS 20%**	ha			200	200	200	200	200						
33	crop factor				0.5	0.7	1.1	1	0.6						
34	crop water requirement	mm	row 33 * row 1		95	122	174	168	110						
35	net deficit	mm	row 34 - row 4		39		4	99	98						
36	field application efficiency				0.7	0.7	0.7	0.7	0.7						
37	diversion requirement	mm	row 35 / row 36		55		5	141	140						
38	conveyance efficiency				0.85	0.85	0.85	0.85	0.85						
39	system head requiremnt	mm	row 37 / row 38		65		6	166	164						
40	ditto, volume per month	Mm3 1000 ha	/ row 39 * row 32/100,000		0.13		0.01	0.33	0.33						
41	**WHEAT 45%**	ha								450	450	450	450		
42	crop factor									0.6	1.1	1	0.6		
43	crop water requirement	mm	row 42 * row 1							88	154	136	111		
44	net deficit	mm	row 43 - row 4							88	154	136	111		
45	field application efficiency									0.7	0.7	0.7	0.7		
46	diversion requirement	mm	row 44 / row 45							126	220	194	159		
47	conveyance efficiency									0.85	0.85	0.85	0.85		
48	system head requiremnt	mm	row 46 / row 47							148	259	229	187		
49	ditto, volume per month	Mm3 1000 ha	/ row 48 * row 41/100,000							0.67	1.16	1.03	0.84		
50	**VEGETABLES 5%**	ha		50	50	50	50	50	50	50	50	50	50	50	50
51	crop factor			1	1	1	1	1	1	1	1	1	1	1	1
52	crop water requirement	mm	row 51 * row 1	223	189	174	158	168	183	147	140	136	185	195	228
53	net deficit	mm	row 52 - row 4	193	133	33		99	171	147	140	136	185	195	228
54	field application efficiency			0.7	0.7	0.7	0.7	0.7	0.7	0.7	0.7	0.7	0.7	0.7	0.7
55	diversion requirement	mm	row 53 / row 54	276	190	47		141	244	210	200	194	264	279	326
56	conveyance efficiency			0.85	0.85	0.85	0.85	0.85	0.85	0.85	0.85	0.85	0.85	0.85	0.85
57	system head requiremnt	mm	row 55 / row 56	324	224	55		166	287	247	235	229	311	328	383
58	ditto, volume per month	Mm3 1000 ha	/ row 57 * row 50/100,000	0.16	0.11	0.03		0.08	0.14	0.12	0.12	0.11	0.16	0.16	0.19
59	system head total reqt.	Mm3 / 1000 ha		0.16	0.60	0.12	0.03	1.25	1.15	1.16	1.28	1.14	0.99	0.16	0.19
60	ditto, 24 hr flow	l/s/1000 ha		63	230	47	11	482	444	448	495	441	384	63	74

table 5.2   ***Calculation of irrigation demands***

The **effective rainfall** is only the proportion of rainfall that can be utilised by the crop. During a heavy rainstorm or if the soil is already saturated a high proportion of rain will be lost as runoff. Very light rainfall after hot weather may evaporate without entering the soil. For the conditions of coastal Karnataka in India for instance, effective rainfall averages about 60 per cent of the total. This is an estimate based on the tenuous premise of discounting monthly falls in excess of 800 mm or less than 20 mm and assuming a 20 per cent loss on the rest. All this is more than a little subjective, but the vagaries of actual rainfall make it impossible to predict accurately in advance. Beware of over-precision, for much of irrigation is more art than science. There is little point in calculating theoretical evaporation to within a millimetre if the effective rainfall is at best only an educated guess.

Empirical relationships like this have been developed in many countries, and FAO have chronicled many of them (Ref. 10.)

Field irrigation requirements

The amount and frequency of water to be applied on the field will depend on several things:

- The irrigation requirement estimated from the evapotranspiration, rainfall and crop factors.
- The rooting depth of the crop.
- The type of soil and its moisture-holding capacity.
- The method of irrigation and its field application efficiency.

The root zone

The rooting depth of the crop will vary according to its stage of growth and soil conditions. A high water table or hardpan caused by puddling for paddy rice can restrict the rooting depth, whilst a free-draining loamy soil will permit roots to attain their maximum depth. Typical rooting depths for mature crops are shown in table 5.3. These figures refer to active depths from which most of the water is taken up. Maximum depths may attain double these figures. Crops such as sesame, safflower and sorghum known for their drought resistance can abstract water from a depth of several metres, although under normal irrigated conditions this would not be the case and for optimum yields a lesser root depth would be assumed in irrigation scheduling calculations.

Crop	active rooting depth
	cm
groundnuts	80
pulses	70
cotton	120
sorghum	200
HYV rice	100
local rice	120
wheat	120
maize	100
sunflower	130
alfalfa	200
sugarcane	200
tomato	70
sugarbeet	80
soya	130
grapes	150
citrus	200
banana	75
tobacco	75

table 5.3   ***Rooting depths***

The actual rooting depth can be drastically affected by the method of irrigation. Drip irrigation can often create a tight ball of roots around the emitters and inhibit the full utilisation of nutrients in the surrounding soil. Over-irrigation of dry-foot crops can encourage shallow root development which leaves the plant vulnerable to sudden drought or water shortage.

The rooting system is normally denser near the soil surface and the roots here tend to be more efficient at abstracting water from the soil. So the soil should not be allowed to dry out completely before irrigation; a common rule of thumb allows for two-thirds of the available moisture in the root zone to be replenished at each irrigation.

Available water

Not all the moisture in the soil is available to the plant. The available moisture is the amount that can be extracted by the roots. Some moisture is held in tiny pores in the soil where it can only be extracted at very high suction pressures. (In clay, some water is held very tightly to the microscopic clay particles by strong molecular forces.) When a soil is saturated, some water will drain out under gravity, either to the drainage system or to the subsoil as deep percolation.

After a soil has released all its drainable water under gravity, it is said to be at field capacity. And the function of irrigation is to restore the soil moisture content to its field capacity.

The available moisture content of a soil at its field capacity varies, for example:

Sand	8-12 per cent of the soil volume
Loam	10-16 per cent
Clay	15-20 per cent

A soil at field capacity contains air in the larger pore spaces and water, held by surface tension, in the smaller ones. Most plants need some air around the roots and do not survive totally saturated conditions for long. Aerobic bacteria, some of which fix nitrogen from air to soil, may have a symbiotic relationship with the plant roots, and obviously require air. Paddy rice is an exception, along with reeds such as *Phragmites* which have a useful application in reedbeds for sewage treatment.

The moisture-holding characteristics of a soil can be measured in the laboratory and expressed in the form of a pF curve, the moisture content plotted against the logarithmic value of the suction force required to extract the water from the soil matrix. This should be a routine test for all soils being irrigated for the first time.

Cropping patterns and intensity

As a precursor to designing the whole scheme, water consumption data of the crops to be grown can be conveniently aggregated through a cropping pattern. It is often represented graphically as in figure 5.1, although normally it would also be tabulated in spreadsheet format. The graphical pattern is useful to show harvesting and planting periods, represented by the tapering sections in the graph, and to highlight potential problems such as labour or machinery shortages which might occur when for instance several harvesting periods coincide. The aggregated pattern can simplify the calculation of scheme irrigation requirements, but with computers it is just as easy to calculate the water requirements of individual crops, then add them to get the scheme requirements. A crop calendar diagram is similar, but without the scaling to indicate percentage area covered by each crop.

The *cropping intensity* is derived from the overall cropping pattern and is used in feasibility studies to describe the likely impact of a proposed irrigation scheme over the pre-existing situation. A single crop grown over the entire area of a scheme in one season only would imply an intensity of 100 per cent. The pattern of figure 5.1 has an intensity of 130 per cent over the year, which includes two cropping seasons. This may appear low for an irrigation scheme but in this case the pre-existing intensity was only about 30 per cent. Many places in the world have localised intensities in excess of 200 per cent. In many of the conjunctive-use schemes in eastern Java 300 per cent (i.e. on average three crops per year, including a pulse crop of 2 months duration) is common. However, as a measure of success or failure the cropping intensity is irrelevant, because it takes no account of yields or agricultural production levels.

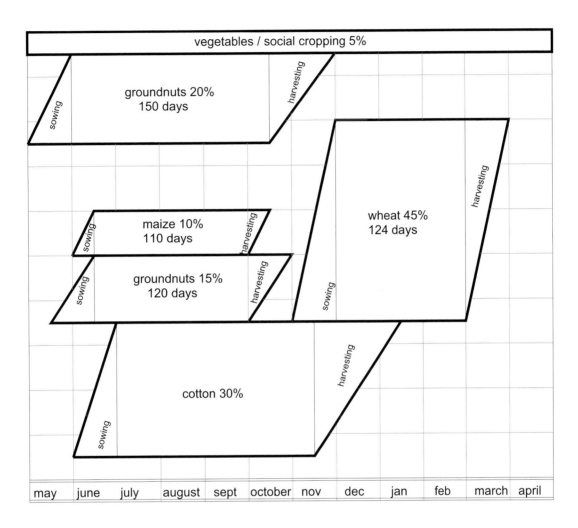

figure 5.1   *A typical irrigated cropping pattern for north-eastern Nigeria*

Calculating field application rate

Ideally, the plant's water supply should be replenished at exactly the same rate that it is used up by the plant, resulting in zero stress. This state of affairs can be met fairly closely by centre-pivot or linear move sprinklers, intensive drip or micro-irrigation. However, most of the large schemes around the world are based on surface irrigation or other methods necessitating an intermittent application of water.

Consider the example of irrigating groundnuts in a sandy loam soil having an available moisture content of 12 per cent. The rooting depth is 80 cm and therefore the total available moisture to the crop with the soil at field capacity is 12 per cent of 80 cm = 9.6 cm or 96 mm.

We will be aiming to irrigate when two-thirds of this is depleted. That is, 2/3 × 96 = 64 mm which is therefore the design irrigation application. After allowing for a field application efficiency of 70 per cent, the amount needed on the field is:

64 mm / 70% = 91 say 90 mm

How often must we irrigate? Go back to the irrigation requirement calculations in table 5.2, taking as an example the long-duration groundnuts planted in June. The estimated ET_0 is 189 mm/month, crop factor 0.5, which gives a crop water requirement of $189 \times 0.5 = 95$ mm per month. Suppose the expected rainfall doesn't arrive on time but we wish to plant anyway[4], in which case this is also the irrigation requirement. And we can apply 64 mm at a time, so the maximum irrigation interval is $64/95 \times 1$ month which is about 20 days, say 3 weeks. (Engineers should note the agricultural degree of precision!)

Local field conditions and the stage of growth of the crop can modify the design frequency. Very hot weather might lead to excessive wilting which would indicate that a more frequent application is required. Within the first month after planting, the root system has not developed fully and the plant cannot utilise water in the mature root zone. Nevertheless at the end of a dry season such as Northern Nigeria's it is normal with the initial irrigation to establish a reserve of moisture in the full root zone even though it may not be fully utilised by the crop at the outset. Legumes such as groundnuts rapidly develop a tap root which follows the moisture downwards through the soil as it dries out.

The method of irrigation can also modify these ideas. Whenever moisture is depleted from the soil, the plant will be under some degree of stress which will reduce yield. Trickle irrigation, by delivering water at very frequent intervals such as every day, can eliminate water stress entirely. Some sprinkler systems such as centre pivots can have the same effect. But if we are constrained to some form of rotational distribution system, then some moisture stress is inevitable.

Stream size

Most irrigators if given a choice of flow rates will try to take the maximum that they can handle, because that will give the shortest time for which they will have to irrigate. The shorter the time spent irrigating, the cheaper the cost of labour. The shorter the time spent irrigating, the more time is left for other things in life. The maximum amount of water that can conveniently be handled by one farmer or irrigator is a subject of some speculation, but for most situations of small-farmer, manual control it is in the range 25 to 50 l/second.

5.2   Water losses

Not all the water that enters the canal at its head gets to where it is intended, i.e. the crop. Some is lost on the way. It is all too easy to let slack management interfere with the flow of water, resulting in wastage that can attain alarming proportions. The ways in which water can be lost, and the efficiency factors used to describe it, are essential knowledge for all people involved in the canal from its initial design through construction to operation, management and maintenance.

> In the English city of Manchester, the water reticulation network is 150 years old. It leaks so badly that the people there have to pay for twice as much water as they actually use. It's not so different in the developing world.

Irrigation water can be lost along the canal and irrigation system in several ways:

- Seepage
- Leakage
- Management loss
- Dead storage
- Absorption
- Evaporation
- Deep percolation in the field
- Runoff from over-irrigation

---

4 This is a very common situation for which irrigation is essential in many countries even well-endowed with rainfall. Planting on time makes optimum use of climatic conditions and reduces problems of seasonal labour shortages and machinery utilisation.

Seepage

Seepage from the canal can be heavy if the soil is permeable and the canal is unlined, poorly compacted, or badly lined. In designing a canal the aim is usually to limit the seepage loss to no more than 5 per cent of the canal flow, either by lining or using a suitably impermeable soil with good compaction. In practice it is possible on small canals with high quality lining to reduce seepage loss to less than 1 per cent. However, too often the lining is badly constructed and seepage loss is high, especially where the canal runs in embankments. There have been plenty of cases when lining has actually increased the seepage loss, a situation which is entirely predictable if inappropriate methods are used.

soil type	seepage rate, steady state[5]	infiltration, dry soil	compacted
gravellysand	5 - 10		
medium sand	1 - 5	>1	0.6
fine sand	1 - 3		
sandy loam	1 - 3	0.4	0.3
loam	0.5 - 2	0.5	0.2
cracked clay	0.5 - 2	>0.5	<0.05
clay loam	0.02 -0.2	0.2	0.15
dense clay	<0.02	0.15	0.06
raw mineral soil (fill)[6]	0.2		
raw mineral soil (in cut)	0.1		

table 5.4   Indicative water movement rates in various soils, m/day

Estimating potential seepage loss from an unlined canal is an uncertain science, but some indicative figures are given in table 5.4. These have to be modified by a feel for the soils on site and some field seepage tests. Most soils will compact denser than their natural state, and their permeability can be significantly reduced by even a small amount of compaction. A dry soil will have an initially rapid intake of water, especially if it has open surface cracks. In estimating seepage loss or absorption loss from a canal that is initially dry we should not only consider the steady-state permeability rate but also the dry-soil infiltration characteristics.

The practice of expressing seepage loss in terms of per centage of flow is not very useful, especially in a long canal. The Yeleru Left Bank Canal in Andra Pradesh (see figure 5.3) is 150 km long and leaks like a sieve. A recent project to improve its conveyance capacity was based on the supposition that the present losses were 70 per cent, and after rehabilitation the losses should be reduced to 32 per cent. In practice this turned out to be meaningless. A detailed site inspection soon showed that heavy seepage was occurring in some reaches in which the embankments had not been properly compacted, and that there was also leakage through structures and theft of water by farmers along the way. Furthermore the canal was partially blocked with soil slipped from the steep banks and rock that had not been fully excavated at the start. These caused the banks to overtop in places, at discharges well below the design level. Instead of glibly talking of reducing the loss by 40 per cent, we did a more comprehensive analysis based on measured and targeted permeabilities of the banks after rehabilitation of earthworks. When plotted reach by reach as conveyance curves (fig. 5.2) a more exact picture emerged. The measured flow in Rabi, when there was very little irrigation, indicated an existing loss of 76 per cent of the flow entering at the head. Most of the loss was in the upper reaches where the banks were in filling and had not been properly compacted. In some places the seepage rate was measured at over 2 m/day. The target values after rehabilitation were based on a maximum seepage loss of 0.2 m/day, leading to the middle and upper curves in figure 3.2 for the two planned phases of the project. On this basis the overall projected loss in Rabi would be reduced to 28 per cent and 19 per cent for the ultimate phase of development. Applying the same criteria to the irrigation season in Kharif, the overall losses would fall from 72 per cent at present to 18 per cent and 8 per cent respectively. So the

---

5 From Ref. 1.

6 Measurements from Upper Swat Canal, Pakistan, sealed with silt carried in suspension for 80 years.

overall loss is wholly dependent on the incoming flow at the system head, and it can be entirely misleading to talk about per centage seepage losses without looking at all states of flow in the canal.

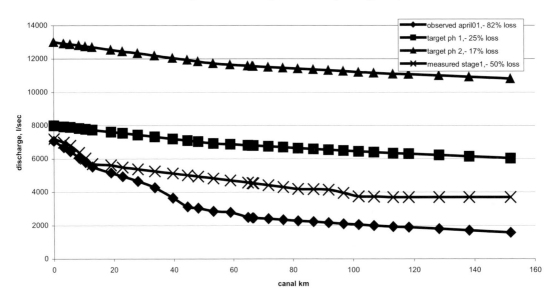

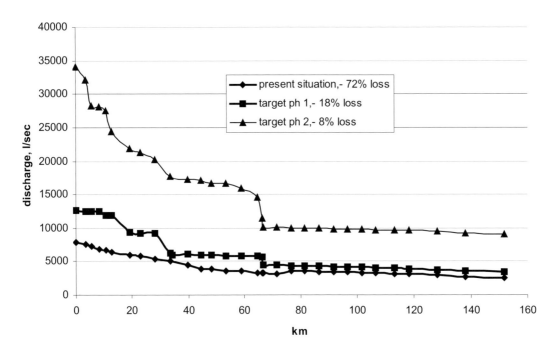

figure 5.2   ***Conveyance curves for the Yeleru Left Bank Canal***

figure 5.3   Yeleru Left Bank Canal

Leakage

Leakage through gates, structures and animal holes can be significant if the canal is badly designed and constructed or poorly maintained. Regular maintenance is the key to reducing leakage losses through animal activity. However, good design and construction is the key to avoiding them in structures. Use vibrated concrete in preference to stone masonry, and ensure that gate frames and seals are properly embedded.

It is useful in practice to differentiate between seepage and leakage: seepage is a natural steady process, which depends on the natural condition of the soil and occurs over a wide area. Leakage occurs at discrete points and is more easily preventable.

Management loss

Management losses, deriving from incorrect operation of the canal system, are potentially the biggest source of water wastage, and also the cheapest and easiest to correct, *provided they are recognised.* Gates and regulators left open when they should be closed can lead to massive water losses. The wastage shown in figure 5.4 was from a tank in a drought-prone area of India, when farmers opened the tank sluice too much and wasted an entire season's supply of irrigation water through ignorance of the need to conserve water. Figure 5.5 shows an improvised turnout causing flooding of the road.

The response time of a canal system needs to be taken into account in management. Even on a small scheme it may take several hours for water to run along the length of the canal system. Any water still running after it is needed will be wasted, unless it can be stored. So gates and pumps in the upper reaches of the canal need to be closed in advance to allow the water in the canal to travel down to where it is required. Unless the canals are designed for rapid response, such as the parabolics in figure 9.11, then water management can be extremely imprecise, especially if the delivery schedule is on demand.

The management of a scheme can often be hampered by deliberate damage to canals and structures caused by farmers, either in attempts to steal more water or to prevent their neighbours from getting it. This type of problem can have its root cause in social conflicts or in design faults, but can usually be nullified by sensitive management. Water lost through theft or deliberate damage is therefore treated as a management loss.

figure 5.4   ***Wasted water from Arshingeri Tank, Karnataka***

figure 5.5   ***Management losses from Zaida Minor flood the road in NWFP, Pakistan***

Dead storage

By dead storage is meant a volume of water that cannot be utilised. The term is more commonly applied to reservoirs in referring to the pond that remains below the lowest draw-down level. Dead storage losses occur when water is let into a canal and it takes time for the dead storage of the canal to be filled before water flows along it. Unlined canals that are badly maintained can have large pockets of dead storage in which water is ponded. Due to their roughness and in order to keep velocities low to limit erosion, unlined canals have a larger cross-sectional area than lined canals, and careless maintenance can enlarge the section or over-deepen parts of the canal. The response time of an unlined canal can be seriously lengthened by its excessive dead storage. And when the canal is shut down, water will collect in pockets of dead storage and be wasted to evaporation or deep percolation. On some warabandi-operated schemes in India, the dead storage losses are recognised as a reason for adjusting the water charges: *Bharai* is the time allowed to take up dead storage in a watercourse reach, and *Jharai* is an allowance for the recession of flow rate after a watercourse is shut down.

Absorption loss

An unlined canal in a clay soil will take up a lot of water into cracks in the soil every time water is let into it after a period of drying out. This is effectively dead storage, which has to be replenished before water can flow any further along the canal. Any dry soil has a rapid infiltration rate to start with, before attaining a steady seepage rate. And the soil doesn't even need to be dry, the same thing happens in a canal that has only been closed for a day. All this water that is absorbed by the canal before it really starts to flow can have a drastic effect on water management.

Absorption has a proportionately more drastic effect on small or less-than-design discharges than on the full supply discharge. Repeated filling and emptying of a canal, say on a short rotation of 12 hours, can result in big losses to absorption and dead storage. Filling a canal quickly is then of prime importance and is best achieved from the tail-end upwards. That means keeping outlets closed until all reaches are filled and is an argument for undershot or notch-type regulators rather than overfall weirs.

Evaporation

Evaporation loss from a small canal is never very significant. Take a typical small canal flowing at 100 l/s with a surface width of 1 metre. Assuming an open water evaporation rate of 5 mm/day, the evaporation loss would be only 0.06 per cent per km. It takes a very long and slow-moving canal to lose much by evaporation.

It can be important in the field however, if small amounts of water are applied to a young crop which does not cover the ground fully. And if sprinklers are used, they can be affected by wind and evaporation. A common problem in rehabilitation of a canal system occurs when the farmers are familiar only with basin irrigation for rice and, having dry-season water for the first time, they attempt to grow a dry-foot crop. If flat basins are used for young upland row crops the water gets spread unevenly and much can be lost to evaporation from the soil surface.

Deep percolation

Deep percolation losses occur in the field when irrigation water goes below the root zone and is lost to the plant. When applying water by surface irrigation methods, it can be difficult to get an even distribution of water across the field. In order to get enough water to the end of the field it may be necessary to put too much on at the start of the field, and excessive percolation loss is the result. Similarly, an irrigation application discharge that is too small can lead to excessive percolation loss due to the long travel time across the field. For this reason, the design duty of field channels needs a lower limit of about 20 l/s in order to get a practical size of field application. Furrow and border strip applications are often given a heavy initial application which is then cut back by reducing the inflow when the advance wave has reached about two-thirds of the length of the field.

Runoff

Runoff from over-irrigation in the field can be a serious source of wastage. Surface irrigation of dry-foot crops is often imprecise and difficult, unless the land is very carefully prepared beforehand. It's quite common to find the drains running as full as the canals, so it's a good idea to look at the drains first in any inspection. If they are full then look for problems of over-irrigation or careless canal operation.

5.3   Irrigation Efficiencies

These sources of water wastage can be quantified in terms of **Irrigation Efficiency**. The overall efficiency can be subdivided into various components, but definitions vary and it is easy to get entangled in semantics. I find the following the most practical because they are categorised according to their means of rectification. If the problems can be isolated, then they can be put right.

***figure 5.6**   Conditions leading to heavy water use. A hot, arid climate, persistent winds, sandy soils, poorly constructed canals and an inappropriate method of field irrigation. The Multaga project in Sudan*

Conveyance efficiency

The conveyance or transmission efficiency is a measure of the water lost through seepage, leakage, absorption and dead storage from the canal system. A figure of over 90 per cent is normally considered acceptable for a lined canal, but a well-constructed canal on a small scheme should achieve better than 95 per cent.

The conveyance efficiency may well be sensitive to discharge. An unlined canal flowing at half capacity will lose proportionately more water to combined conveyance losses than when flowing full, because more of its profile is in contact with the wetted perimeter of the canal. And in leaky canals most of the water is lost around the base of the canal sides, so small discharges get lost quickly.

Considering only the steady state flow, that is neglecting any spillage or slug[7] losses that can be attributed to management, the conveyance efficiency can be formalised as:

$$\frac{\text{sum of steady state discharge reaching fields}}{\text{steady state discharge released at system head}} \times 100 \text{ per cent}$$

Or the conveyance efficiency can be considered on just a single canal or a sequence of canals. If the conveyance efficiency is low, check out the physical state of the canal for the effects of bad maintenance, poor construction and leaking structures.

---

7 *A 'slug' being a short surge of water being sent down the wrong canal or anywhere else it wasn't supposed to go, usually because of gates being incorrectly operated.*

Management efficiency

The management efficiency can often be both the poorest efficiency and the one that is easiest to rectify, once the problems have been identified. It is a measure of the management losses incurred through spillage, wastage, incorrect allowance for response time and travel time, and incorrect gate operation in the canal system.

Considering the non-steady state flow of surges and incorrect spillages which are excluded from steady leakage and seepage, management efficiency over a period of time (e.g. 1 month) can be defined as:

$$\frac{\textit{volume flow reaching field}}{\textit{(volume released at system head - conveyance losses)}} \times \textit{100 per cent}$$

with units of measurement in cumec-days.[8]

If the management efficiency is low, say less than 80 per cent, then take a critical look at people and training, and at the design and location of control structures in the canal system, and indeed at the entire strategy of main-system water management.

However, it may be necessary to look beyond the single scheme and consider the institutional set-up and procedures of the government organisations responsible for the scheme. Often there may be several different government departments involved (agriculture, irrigation, the deputy commissioners office) whose traditional line management structure does not lend itself to efficient problem solving in the field. It is often difficult for separate departments to communicate at the field level. There is then a clear need for a change in management structure such as the Unit Management system which was successfully developed in Sri Lanka (see Chapter 3.)

Application efficiency (in-field)

The field application efficiency is a measure of the water wasted in the field, being the amount of the irrigation demand divided by the amount of water needed to satisfy the demand across the whole field. It takes into account the water lost to runoff, evaporation and deep percolation, and in the case of sprinklers, wind drift also. On a well-run scheme it could be as high as 80 per cent.

Field application efficiency can be defined as:

$$\frac{\textit{volume of crop irrigation demand}}{\textit{volume used to satisfy demand}} \times \textit{100 per cent}$$

Low application efficiency can be improved by land levelling, or some adjustments to flow rates in furrows or alterations to the field distribution system design.

Distribution efficiency

The distribution efficiency is sometimes considered separately, being a measure of the evenness of water distribution across the field. But for most practical purposes it is enough to include it as part of the application efficiency. To get a distribution efficiency of 100 per cent it would be necessary to spread the water to an exactly even depth over the entire field, or to be more precise, give an identical amount of water to every plant. It is easy to measure with sprinkler irrigation, just by collecting the water in a grid of containers. But with surface irrigation it is more difficult. There is a need to probe the soil all over the field to determine the depth of the wetting front.

Note that a very low application efficiency could apply even with 100 per cent distribution efficiency, if for example twice as much water were put on as the crop required. But on the other hand, assuming the irrigation demand to be evenly distributed across the field, 100 per cent application efficiency would be impossible without 100 per cent distribution efficiency.

---

8 *This is a convenient unit for a volume of irrigation water. 1 cumec-day = 86,400 cubic metres =1.18 cusec-months, etc.*

Distribution efficiency can be defined as:

$$\frac{minimum\ water\ depth\ applied}{average\ water\ depth\ applied} \times 100$$

figure 5.7   Poor distribution in furrow irrigation

figure 5.8   Plastic vanes retard the rate of advance in furrows as a means to improve distribution efficiency

Overall efficiency

The overall efficiency of a scheme is a measure of the total losses. It is the amount of water needed by the crop divided by the actual amount of water put into the scheme, i.e. the irrigation demand divided by the actual amount of water put into the scheme at the system head at the pumps or released from the tank. A small scheme might realistically be designed for an overall efficiency of 70 per cent, but in practice a badly-run scheme might operate at less than 30 per cent. At 30 per cent efficiency a scheme is using more than 3 times the amount of water that it should be. If it is a lift scheme, the power cost will therefore be 3 times as high as it should be. And if it is a tank scheme, with the limited amount of water stored in the tank only one third of the potential area can be irrigated.

The overall efficiency is an amalgamation of the rest:

$$overall = conveyance \times management \times application$$

expressed as a percentage.

And the overall water loss is:

$$system\ head\ diversion \times (100\% - overall\ efficiency)$$

To take a real-life example, a typical scheme might have a conveyance efficiency of 90 per cent, a management efficiency of 70 per cent, and an application efficiency of 70 per cent. That gives an overall efficiency of only $0.9 \times 0.7 \times 0.7 \times 100 = 44$ per cent.

Always the cheapest and quickest way to improve on this is to shake up the management efficiency first. Raise that to 95 per cent and the overall efficiency goes up to $0.9 \times 0.7 \times 0.95 \times 100 = 60\%$ which is about as good as can be expected for a big surface-irrigated canal scheme.

Other efficiencies

Mention should be made of two other ways in which the success of an irrigation scheme can be quantified; the *production efficiency* and the *economic efficiency*. Whilst the overall efficiency described above really considers only the engineering side, these provide a more global view of the project although perhaps one that is not so easy to assess. The target crop yield and returns must be known.

The production efficiency (sometimes called the *biological efficiency*) is the crop yield related to the volume of irrigation water used. How much water does it take to produce 1 tonne of rice? It's no good optimising the engineering efficiency if the crop is still poor. An example comes from a 500 hectare tank scheme in India rehabilitated a few years ago. Within three cropping seasons the entire area was covered in green, well-irrigated rice. All the top officials came to see and there were congratulations all round. Unfortunately nobody got out of the jeep and walked around the fields. If they had, they would have seen some of the worst paddy rice in the world. Bad seed, no fertiliser, no weeding, bad land preparation. The irrigation efficiency had jumped up from 20 to 80 per cent. But the production efficiency had plummeted. Probably because the tail-end farmers were so surprised to receive water they didn't really believe it and so they didn't bother cultivating properly.

The production efficiency is:

$$\frac{crop\ actual\ yield}{target\ yield} \times \frac{theoretical\ water\ use}{actual\ water\ use}$$

The economic efficiency is the financial or economic crop return, in relation to the unit cost of water. Is the cost of water justified by the value of the crop produced?

The economic efficiency is:

$$\frac{crop\ return}{target\ return} \times \frac{theoretical\ cost\ of\ water}{actual\ cost\ of\ water}$$

5.4   Canal and system duties

The overall scheme demands need to be worked out on a suitable assumed cropping pattern, but how much water must be sent down the canals and when? There are two ways to work this out. They can be called, for the sake of provocation, the engineers' way and the farmers' way. At first sight there may not appear to be much difference between the two, but they have a profound influence on the design of the distribution system. Having calculated the field irrigation water requirements, they cannot just be extrapolated to the canal capacity by multiplying by the command area of each canal. It is necessary to take into account rotation between canals, intermediate storage, and degree of automation throughout the system.

Canal duty

The duty of a canal or pipeline is a measure of its design capacity. It can be expressed in several ways:

- As discharge related to unit command area, e.g. 9 cusecs per 1000 acres, or 0.7 l/s/ha. In South Asia this is termed the *water allowance,* and relates specifically to the head of the watercourse or tertiary canal.
- As an area that can be irrigated per unit of water, e.g. 250 acres per cusec. This is the commonly accepted format in use in India and Pakistan, and defined in the ICID dictionary (Ref. 16) as the *Full Supply Factor.* It is the inverse of the water allowance, with appropriate modification to allow for system losses. This is a little confusing, since in this case a lower duty actually means a higher canal discharge.
- As a depth of water to be applied on the field, e.g. 8 mm. This is John Merriam's favoured method for pipelines since it is directly understood by the farmer, and if the engineers require it in terms of litres/second they can easily calculate it themselves with an appropriate congestion factor thrown in.

Provided the units and the context are clear it does not matter much, and my own preference is to use duty meaning discharge per unit area, with the context or location within the scheme clearly stated.

On a large scheme the duty will vary along the system, depending on rotational management, system losses, and whether intermediate storage is provided.

Conveyance losses and management losses have to be added on to the field irrigation demand. To get the overall irrigation duty at the system head, take the calculated crop water requirements and divide by the overall efficiency. The figures in table 5.5 are typical of a scheme in which rotational management plays an important role. Normally the main canals will be flowing continuously, 24 hours a day. On the Upper Swat scheme, rotation is introduced at distributary level and the canal duty is therefore greater than in the main canals. At the tertiary or watercourse level the duty is even higher because they operate intermittently and need to cater for a worst-case cropping pattern.

canal	area commanded, ha	design duty, l/s/ha
watercourse	0-100	1.4
minor	100-500	1.4 - 1.0
distributary	500-5,000	1.0 - 0.85
branch or main	5,000-50,000	0.85 - 0.7

table 5.5   Canal duties, Upper Swat System, Pakistan

Water management affects canal duties

Think about the way in which water is going to be managed. There is a minimum practical discharge, below which it is practically impossible to irrigate in the field with a small amount of water. For surface irrigation in poorly levelled fields this is around 10 - 15 l/s, depending on the field and soil conditions. And because not everyone can irrigate at the same time, the canal duty increases down the canal system toward the tail.

In a proportional flow, supply-scheduled system, the water will be divided in proportion to the commanded area, down to tertiary or quaternary level. Beyond this, proportional division will result in flows that are too small for practical use, so some degree of rotation will be necessary. The Warabandi system of northern India and Pakistan works in this way, by rotating between farms along the watercourse.

In a demand-scheduled system, the duty will be a lot higher toward the tail of the system, to make allowances for several adjacent farmers taking water at the same time, and growing different crops. Theoretically, a completely free-for-all demand schedule would need to cater for the extreme case of all farmers growing a thirsty crop such as paddy rice and all farmers wanting to irrigate at once. This may happen in a very small group, but in practice its likelihood rapidly diminishes as the group gets larger. Availability of labour and machinery is always limited; one tractor can only plough so much in a day. The labour market tends to be self-regulating; as the demand for labour goes up, so does its scarcity and its price, until the threshold is reached at which some other crop, which uses peak labour at a different time, becomes more attractive. The sales market also becomes self-regulating (more or less, depending on the extent of government intervention); if everyone produces the same crop at the same time, prices will go down until some other crop becomes more profitable. And then there is the human factor; people have different preferences, different skills, different reasons for growing a particular crop, and different resources with which to go about it.

The demand envelope

An interesting study was made in Pakistan, to see just what happens on a mature, intensive irrigation scheme at various levels along the canal system. The result is the demand curve of figure 5.9. Actual crops being grown on each watercourse were determined from government revenue department records and analysed for theoretical water requirements. The total for each watercourse is shown as a point on the chart. Maximum command area of a watercourse is about 100 ha, above which watercourse demands were aggregated to get demand figures higher up the canal system. For small command areas, there is a very wide spread of watercourse demands, because some grow nothing but paddy rice, whilst others grow oilseed or in some cases have a tail-end problem which reduces their cropping intensity. Localised extremes of water demand cancel each other out and diminish with distance toward the head of the canal system, until the envelope of water demand merges into a straight line. It is the top line of this envelope that a demand-schedule must be designed for. In this case the demand rule was established at a maximum 19 cusec/1000 acres for command areas up to 250 acres, and thereafter by the curve $0.057 \times area^{0.8}$, which reduces to 9 cusec/1000 acres after 10,000 acres cumulative command area.

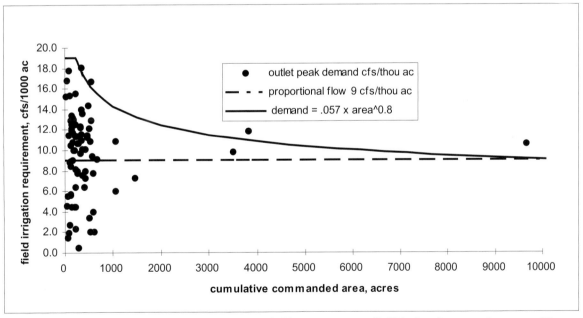

figure 5.9   ***Demand curves of commandable area versus field irrigation requirement , Upper Swat Canal, Pakistan***

It is interesting to note that the irrigation scheme which provided these data had been operating for 80 years as a proportional flow supply-scheduled system, yet despite this the diversity of cropping is huge. In effect it demonstrates a rather inefficient system, which either does not work as it was intended, or else wastes water in horrendous amounts in some places and imparts unduly high crop water stresses in others. (All of these were true to varying degree. It would be interesting to do a similar study using actual crop yields rather than theoretical water requirements; the envelope ought to be much flatter.)

Cropwise irrigation demand - the engineer's approach

Beware the typical engineering mindset: this much evaporation divided by that much area equals so much water. So the canals have to be this big. And this can lead to canals so small at the tail of the system that the water they are supposed to carry soaks into the ground before it gets half way across the field. It is the same mindset that has resulted in almost every major irrigation scheme in the world being designed for supply-scheduled operation, simply because it never occurred to engineers to think any other way. 'We'll build the scheme, everyone will get 0.5 l/sec/ha and be grateful for it...' This may be acceptable for a centrally-managed monoculture scheme such as sugarcane in an African nucleus estate or cotton in the Aral Sea Basin, but for the majority of smallholder schemes more flexibility is increasingly being demanded.

Going back to the Nigerian scheme in table 5.2, the aggregate discharge requirements for the assumed overall cropping pattern are given in the bottom row. The maximum required discharge is 495 l/s per 1000 hectares in December, after allowing for conveyance efficiency and assuming continuous 24 hour/day operation. So if the scheme area is 50,000 ha the discharge capacity at the head of the canal system is $50 \times 0.495 = 25$ cumecs. And likewise down the system, the canals getting progressively smaller as their command area reduces, until rotation is introduced at some point in the system, say at tertiary level.

The unit stream - the farmer's approach

The farmer's approach is less precise but eminently practical. 'We may never be able to predict exactly how much water each field needs, so just make sure I can get enough water when I really need it....' Such vagueness may be anathema to many engineers, but its fuzzy logic is inescapable to anyone who has ever been a farmer.

The alternative is based on an entirely different approach, the farm, not the area to be irrigated. Looking at it from the farmer's viewpoint, we introduce the concepts of the ***unit field***[9] and the ***unit stream***.

The ***unit stream*** is the *probable* discharge required to irrigate the ***unit field*** in a single day. It should be based on prevailing conditions of soil, cropping patterns and potential evapotranspiration, field application efficiency, and an 8-hour day. A day might acceptably be stretched to twelve hours for a short period in which a heavy peak water demand occurs, such as in Pakistan planting wheat in November or maize in June. In many countries a 10- or 12- hour working day might be considered the norm, but the scheme should not be designed too tightly; flexibility depends on having adequate capacity built in to the distribution system. A project built now should still be operating after a hundred years, by which time a 12- hour working day may well be outdated.

The unit field is of such an area that it can be easily irrigated in a working day, given an appropriate stream size. There are no hard and fast rules for ascertaining the size of the unit field. In the USA 8 ha or more might be appropriate (Ref. 3) compared with only 1 or 2 ha in a developing country employing manual labour only. In Pakistan the Topi area was designed for 2 ha units.

The unit field has no relation to the actual size of farm. In large farms, each unit field may in reality be no more than part of a single field. Where farm sizes are very small the unit field will represent an aggregate of several holdings. This is the case for instance in parts of Pakistan where land holdings have become so fragmented through generations of handing-down that there may be 60 holdings to a hectare.

The method of irrigation will also affect the size of unit field. Taking border strips and furrows as average, basins can utilise a higher than average flow rate. Sprinklers or microjets would use a lower rate.

---

9 Sometimes called unit farm, or unit farm area.

Unit stream size is one that can apply an adequate depth to a unit field in a day of 8-12 hours. It should be as large as the available labour can manage. The larger the stream, the less labour time is required for irrigating. The less time required, the more time is available for doing other things. The less labour that has to be hired, the cheaper it all becomes. In practical terms for most developing countries the unit stream is likely to be in the range 50-100 l/s (1.5-3 cfs.) This stream size is quite manageable by a single farmer who will often split it into smaller sub-streams during irrigation.

Consider the previous example slightly differently, from the farmer's viewpoint, growing the same crop of groundnuts on the same sandy loam with furrow irrigation. The crop still needs an application of 90 mm every 20 days. Assume the unit field is 2 ha, and we want to irrigate no more than 8 hours/day. That is

$$90 \; mm \times 2 \; Ha \,/\, 8 \; hours, \; = .09 \times 20000/8/3600 \times 1000 \; l/sec = 60 \; l/sec$$

which would be the unit stream for a frequency of 20 days.

However, we need to look at the worst case, that is the combination of crop, soil and climatic conditions that could occur to give the maximum water demand. So suppose we look at sugarcane, with a maximum crop factor of 1.0 and rooting depth of 2 m. It will need for the period in question 189×1.0 mm/month and the replenishment would be:

$$2/3 \times 2m \times 12\% = 160 \; mm$$

and its maximum irrigation interval would be

$$160/160 \times 1 \; month = 1 \; month, \; or \; 30 \; days$$

In this event the unit stream for an 8 hour day, 30 day frequency and 70 per cent application efficiency would be:

$$160 \; mm/70\% \times 20000/8/3600 = 160 \; l/sec$$

Particular vagaries of the crop will modify the unit stream. Some crops may require heavy irrigations prior to planting (e.g. wheat in Pakistan) or for land preparation (e.g. rice in Indonesia.) Furrow irrigation at its most efficient may use a cut-back stream so the initial application rate will be higher than the mean.

Stream size can only be based on field evaluation and experience, of which not much is yet documented anywhere. Since given sufficient flexibility the farmer can adjust his own stream to any size up to the maximum available at the outlet, the design stream size is not crucial. Hence the term '*probable*' in the opening definition. It matters little if the actual value chosen turns out to be slightly wrong. If it is too small the irrigator may have to work 9 hours instead of 8, and vice versa. Abnormal weather conditions will also modify the length of time the unit stream is applied. **The important thing to note is that such adjustments are possible with a flexible system, whereas in a fixed rotational supply schedule they are not.**

Flexibility and congestion

All irrigators if given the choice will irrigate at a time that suits them. This will normally be in daytime. If they have guaranteed access to water, they will take water only when their crop needs it. A system which gives farmers flexibility in choosing when to irrigate and how much water to apply is the basis for a successful demand-scheduled scheme.

Consider a canal which is operated on a strict rotation 24 hours/day such as a warabandi system. Each farmer has to take water at his allotted time, be it in the middle of the night or the heat of the day, and on his allotted day, be it his birthday or Friday or Sunday or daughter's school speech day, and irrespective of the crops he is growing or whether or not the crop needs water. There is no flexibility to switch the time of irrigation to something more convenient or more conducive to crop growth. There are no free time slots to move to. There is 100 per cent **congestion**.

Remove some congestion. Build in some flexibility. Arrange it so that the farmer can irrigate at a time that suits himself and the crop. Get the congestion down to around 60 per cent and the scheme becomes workable. Do all this by designing more capacity into the supply system.

As with designing the unit stream or unit field size, there are no firm rules yet on what level of congestion to aim for. In years to come, feedback and research in the field will contribute more useful data, but for the moment we have to back up our theory with intuition and such experience as we can bring to bear. Much of the original thinking on the subject comes from John Merriam (Ref. 3, 4, 5) who considers 60 to 70 per cent congestion to be the right order for lower laterals. Higher up the system congestion can be increased to 85 per cent on the mains, and at the farm distributors it should be reduced to around 50 - 60 per cent.

$$congestion \% = \frac{number\ of\ unit\ farms}{number\ of\ unit\ streams \times days\ irrigation\ frequency} \times 100$$

In Pakistan a statistical analysis assuming a binomial distribution of abstraction showed that on laterals with more than 20 unit fields the design capacity could be reduced by some 25 per cent without affecting the congestion at farm distributor level. This increase in allowable congestion at upstream levels of the system can be allowed for in the calculations by a suitable reduction factor.

Table 5.6 is a fragment of a spreadsheet for a pipeline irrigation scheme in Pakistan. The area is divided into distributor groups typically comprising between 10 and 20 unit fields, numbered from the tail-end upwards. The operating rule is to restrict the number of farmers irrigating at any one time within a group to whatever number results in an acceptable level of congestion. Hence the number of unit streams per group, and consequently the duty of the pipeline at various points, are calculated.

row 1	distributor group	1	2	3	4	5
2	area, ha	54	46	34	26	20
3	no. farms	27	23	17	13	10
4	no. streams	4	4	3	2	2
5	nominal frequency, days	10	10	10	10	10
6	congestion	68	58	57	65	50
7	optimum streams	4	4	3	2	2
8	cumulative farms	27	50	67	80	90
9	cumulative streams	4	8	11	13	15
10	per cent reduction	0	0	0	25	25
11	reduced streams	4	8	11	10	12
12	stream size, l/s	60	60	60	60	60
13	cum. area served, ha	54	100	134	160	180
14	pipeline capacity, l/s	240	480	660	600	720
15	duty, l/s/ha	4.44	4.80	4.93	3.75	4.00

table 5.6   ***Calculation of congestion***

The target congestion rule adopted is 50 per cent to 70 per cent for individual groups. The key routine in the calculation is in determining the optimum number of streams in each group to give the target congestion. This is an iterative procedure between rows 4 and 7. In row 4 an initial number of streams per group is estimated. In row 6 the congestion is calculated for the given frequency which is 10 days in this example. In row 7 a test is applied to see if the calculated congestion is within the targeted limits, and if not an adjustment is made and referred back to row 4.

Rows 8 and 9 are cumulative farms and streams, working from the tail-end of the scheme upwards. Row 10 applies a reduction factor to the number of streams, in effect allowing a higher congestion as we go higher up the system. The reduction factor applied in this case is 25 per cent when the cumulative number of unit fields exceeds 70.

If this method seems a little vague, it is important to bear in mind the practical implications of an error in estimating the number of unit streams and hence the pipeline capacity. They are not very great. Suppose we change the reduction factor so that the 25 per cent is knocked off only when the number of unit fields exceeds 100. In this example the theoretical pipeline diameter serving group 5 would change only from 484 to 497 mm. In practice it would remain unchanged at 500 mm. This is demonstrated more fully in Chapter 8.

References and further reading for chapter 5

(1) Smedema, B., and Rycroft, D., 'Land Drainage', Batsford, 1983.

(2) Merriam, J.L., 'Irrigation System Evaluation and Improvement'. California State Polytechnic, 1968.

(3) Merriam, J.L. and Styles S., 'Use of Reservoirs and Large Capacity Distribution Systems to simplify Flexible Operation' XXVII IAHR /ASCE Conf, San Francisco, Aug. 1997.

(4) Merriam, J.L. and Styles, S., 'Design and Congestion Considerations for Flexible Irrigation Supply Systems' XXVII IAHR /ASCE Conf, San Francisco, Aug. 1997.

(5) Styles, S., 'Alleviation of surface and subsurface drainage problems by Flexible Delivery Schedules' XXVII IAHR /ASCE Conf, San Francisco, Aug. 1997.

(6) Withers, B. and Vipond, S., 'Irrigation Design & Practice', Batsford, 1974.

(7) Zimmerman, J., 'Irrigation', Wiley 1966.

(8) FAO Irrigation & Drainage Paper 1 Irrigation Practice and Water Management, 1971.

(9) FAO Irrigation & Drainage Paper 24 Crop water requirements, 1974.

(10) FAO Irrigation & Drainage Paper 25 Effective rainfall, 1974.

(11) FAO Irrigation & Drainage Paper 33 Yield response to water, 1979.

(12) FAO Soils Bulletin 42 Soil Survey Investigations for Irrigation, 1979.

(13) Plusquellec, H., Burt, C., Wolter H.W., 'Modern Water Control in Irrigation', World Bank Technical Paper 246, 1994.

(14) ILRI, Wageningen, Fieldbook for land and water management experts, 1972.

(15) Jurriens, R. and Mollinga, P., Scarcity by Design: Protective Irrigation in India and Pakistan, ICID journal V45 No2, 1996.

(16) International Commission on Irrigation and Drainage (ICID), Multilingual Technical Dictionary on Irrigation and Drainage, ICID, 1967.

PART 2 - DESIGN

It is always a temptation for designers to repeat what went before. It is easier that way, and cheaper too. Their actions don't have to be justified, because that is the way it's always done. There are no time-consuming discussions or arguments about how to do it, or what type of structure to put where. The consultant does the job as before and takes his money from the client. The client gets on with issuing the construction contracts and takes his money from the financing agency. The scheme eventually gets built, a bit late and more expensive than planned, but this is normal. The farmers move in, and there is an immediate tail-end problem because of the management learning curve, and because successive generations of designers never cared about it. The politicians get on the bandwagon and curry favour with voting farmers by pressuring the management to release more water than they should. There is not enough money or equipment to do proper maintenance, which is necessary even though the canals are new, because traditional materials of stone masonry were used in construction....

It's the same old story. After 10 years the system needs rehabilitation because maintenance has failed and water management has broken down. Back come the consultants, along comes another financing agency, and we start all over again. And still we stick with design precepts and construction methods that are two centuries old.

Do you the designer really derive any pride from being involved in this? Are you just in it for the money, or did you become an engineer because you wanted to build good things? Things that people will look at in awe a century from now and say, '...just look at that canal, still working well after all this time'. Just look at that aqueduct in figure 1.1. Just look at the Roman Nouria in figure A1.1. If you would have felt no pride at building these beautiful things, maybe you should have studied accountancy after all.

So having searched your soul and found in there some primeval element of engineering pride, some inexplicable desire to build things that are good and held in reverence by future generations, then open your eyes. Start thinking laterally. Start thinking big. Start thinking of the future. Think alternatives. Think farmer. Think water manager. Think people.

Engineers are too readily belittled by bureaucratic procedures, browbeaten by superiors reluctant to pass on their knowledge or reluctant to admit they don't actually have the knowledge to shift their stance away from that of their forefathers. Engineers constantly complain that architects unfairly get a high public profile and take undue credit for designing things. Is it only architects who are permitted to possess flair and imagination? You don't usually get architects designing irrigation projects. The engineers have to perform what architecture there is. It takes a good deal of courage to step outside the engineering norms, but if ever there was a time when engineers needed courage, it is in irrigation now.

CHAPTER 6   CANAL ARCHITECTURE

This chapter describes the planning of a typical canal system and its principles of operation. It introduces concepts of canal shapes and modes of operation, and describes principles behind the adoption of various types of structure.

We consider 17 reasons for lining, including an in-depth study of many significant reasons not often thought of. Reasons for not lining are also given. The criteria for selecting a particular type of lining may vary from project to project. The strategic arguments for lining are presented in a way that demonstrates logical decisions that should be followed through in a typical project situation.

6.1   Canal Layout and Water Delivery

The diagram in figure 6.1 shows a typical lift irrigation scheme, in which water is pumped from a river up through a short pipeline and discharged into a canal system. The main canal runs along the command line, at the highest level to feed as much good farmland as practicable. Branch canals run along the topographic ridges of high ground. Minor canals run down the line of major slope, more or less in the positions required to serve each farm.

Command

The command area is the land that can physically be irrigated by a canal or scheme. It might be commanded under gravity (when water can only run downhill as in a canal) or under pressure as in a pipeline system, when water can be made to flow uphill. So the gravity command level for a particular canal will follow the water level in the canal, less the head loss required in getting the water out of the canal. Because a canal has to flow downhill, it follows that the command line will also move downhill, slightly off the contour at best. In a scheme with good land at a high level, we may wish to maximise the command level by having a main canal with as flat a slope as possible.

Head

The availability of head is the principle criterion in planning or designing a scheme. Head is described and measured in units of height of water, i.e. feet or metres. In a scheme with plenty of available head, the canal structures can be cheap and simple even though they generate a large head loss in operation. In the case of a scheme like the Gezira in Sudan where there is not much head available then the emphasis in design is on conserving head by adopting structures that don't use much head. A pipeline distribution system which uses a lot of head might be appropriate in rugged terrain with steep ground slopes, where there is a big fall in height across the scheme.

Canal hierarchy

Depending on the water management system and size of the scheme, the canals will become smaller in size as their capacity reduces toward the tail-end of the scheme. There is a confusing array of terminology in use. At suitable places the main canal divides into secondary or branch canals, and on larger schemes these may subdivide again into distributaries, tertiaries or field channels. On large schemes, main and branch canals might be termed primaries. Where one canal branches from another, the one from which it branches is often called the parent canal.[1]

The actual terminology used varies from country to country, but this does not matter so long as the function of the canal is clear. Table 6.1 lists some typical nomenclatures, labelled on a scale of 0 to 4 according to which level they might be used on an irrigation scheme. Level 0 canals are inter-basin or

---

1 *In Sudan the minor canals are given names like 'abu sitta' (father of six), and 'abu ishreen' (father of twenty), according to the number of smaller canals they serve.*

bulk-issue canals which deliver water to the scheme from outside. Within the scheme the canals are designated level 1 down to level 4 in decreasing proportion of the total command area which each one serves. Not all levels necessarily exist on a scheme. A small project might only have a main canal (level 1) feeding directly into quaternaries (level 4.)

level	functions	names in use	typical Q m3/sec	typical slope
0	Carries water from outside the scheme	Inter-basin Trans-basin Conveyor, Feeder	50 - 500	0.0001
1	Conveys water from source reservoir or river or from the head of the scheme to distributor canals On small schemes may feed quaternary channels Include intermediate storage, and may be designed for intermediate storage or operational spillage	Lead canal Header Primary, Principal Main, Major Major	10 - 100	0.0001
2	Intermediate conveyance between levels 1 and 3 Intermediate storage or to feed storage reservoirs	Branch Secondary	2.0 - 10	0.0002
3	Often intermittently flowing, feeding level 4 canals On small schemes may feed direct into the field	Distributary Tertiary, Minor	0.1 - 3.0	0.0005
4	Feed a group of farms Feed individual fields, via farm turnouts, siphons or bank breaching	Watercourse Field/farm Channel Quaternary	0.025 - 0.05	0.005

table 6.1   Canal Nomenclature

The tail-end, and associated problems

The start of the canal is usually called its head, and the system head is the offtaking point of an entire canal system, usually at a dam or a river offtake, the headworks. The end of the canal is called its tail.

The tail-end problem bedevils many large irrigation canal systems. Usually this means that the tail-end of the system gets not enough water, and in many schemes no water at all. It is important because it is an easily recognised manifestation of a variety of deep-rooted problems in the scheme. The first thing to look out for on any irrigation scheme that may not be working well is the extent of its tail-end problem.

The tail-end problem can arise in many ways. For simple engineering reasons, the canal system may be broken or in bad condition so that water cannot travel along its full length. For natural climatic reasons there may simply be a shortage of water at the headworks, and farmers near the head naturally try to take all the water they can, by fair means or foul. For social reasons, the strongest and more powerful farmers may be at the head and insist on commandeering the lion's share of the water. For political reasons, there may be divisions along the way in the form of districts, villages, tribes or ethnic groups who have rivalries with their neighbours downstream. For purely economic reasons, it may even make sense to induce a tail-end problem (see section 3.3.) For managerial reasons, the tail-end problem may be impossible to avoid, especially if the canal system is not designed with potential operating difficulties in mind.

Whatever the reason, the tail-end problem is instantly recognisable in the field. Feeble crops, dry watercourses and irate farmers, when their neighbours upstream are doing well, all indicate that the management needs to act decisively. And it also indicates that the designers should learn from their early mistakes and avoid them next time.

6.2 Planning a canal layout

Canals necessarily have to flow downhill, but usually to fulfil their prime function they need to maintain head either for gaining irrigation command or for storage in the case of level-top canals. So the route for a new main canal is usually fairly obvious. But planning the layout of a minor canal system needs care and understanding. The following is a talk-through of the steps in producing a layout design on paper.

Drainage lines

Irrigation blocks served by gravity are always delineated by drainage lines. Given a block of land that is billiard-table flat, drainage lines can be created on a rectangular grid, but that is a rare situation indeed. (Although there are examples from Cambodia and elsewhere in which central planning has contrived to impose rectangular grids on undulating topography with scant regard for physical laws.) More likely is the case of somewhat irregular topography having a network of natural drains around and between which the irrigation layout has to be fitted.

Starting with a contour map of the area to be irrigated, use a blue crayon to sketch in the natural drainage lines. These can be straightened slightly, depending on how much earthmoving and land levelling can be afforded. The result should be a network of blue lines all leading into the arterial drains which are natural rivers.

Ridge lines

Use a red pencil and mark in all the high spots, ridges and spurs. This should give a pattern complementary to the drainage lines, and the red lines should all lead back to link in with the main canal route. These are the natural routes for secondary and level two canals.

Major and minor slopes

The steepest possible orientation of a canal will cross the contours at right angles and hence follow the *major slope*. This is often the most convenient orientation for level 3 canals, but if the major slope is too steep to avoid erosion then the canal will either have to be lined or designed with drop structures. For surface irrigation the nature of the soils will dictate the preferred *minor slope*. Unless the topography is very flat the minor slope is usually the orientation of level 4 canals or field irrigation works such as furrows, which will be constrained by the field boundaries determined during blocking out. The long-line furrows of figure 7.23 are oriented down the major slope, and the level 4 head ditch from which the siphons are abstracting water runs along the minor slope. Typical longitudinal slopes for fieldworks and larger canals are given in table 6.1, but these are indicative only and can be varied over a wide range according to local conditions and with suitable engineering work.

Blocking out

Blocking out is the process of planning farm and field layouts on a new or modified scheme. Initially the design holding size must be known. This can apply either to farms, individual fields, or the notional *unit fields* described in section 5.4. It requires a modicum of artistic talent, and a soft pencil. After locating the block boundaries and the route of the level 3 canals, the blocks are subdivided into fields or irrigation units. The fields are sketched in initially as quadrilaterals with their upper and lower boundaries parallel to the minor slope (which may not necessarily be parallel at top and bottom of the field.) A sample unit of the target size is drawn as a guide, and with some practice it is straightforward to sketch in the fields to within a tolerance of +/- 10 per cent in area. Figure 6.1 shows a block of 20 ha holdings designed for surface irrigation. For surface irrigation the point of water delivery has to be the highest point of the field, after any land levelling. This applies both to pipeline delivery systems and open canals, the difference being that with pipelines there are fewer constraints on routing the distributors and laterals. However, in the case of a pipeline distribution system the unit fields are grouped in a different way to tertiary units of a canal system, as explained in section 8.2.

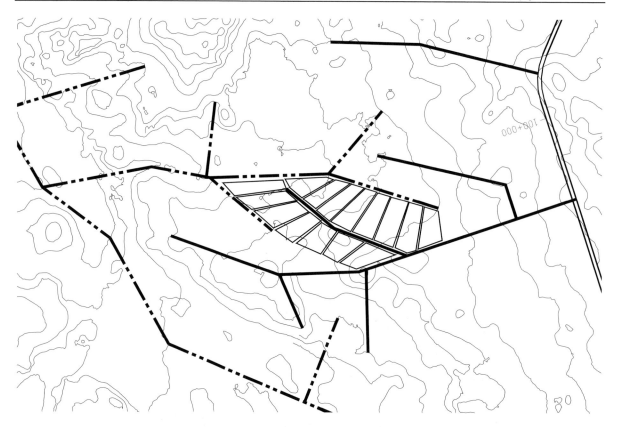

figure 6.1 *A typical scheme canal layout, showing command area, drainage lines and irrigation blocks*

6.3 Canal Architecture

The size, shape and form of a canal is determined by a variety of rules, some more scientific than others, and also by the whims of the designer. Most canals are unimaginatively designed to a standard trapezoidal cross section, because that was the way the last generation of engineers did it. We can do better now for less expenditure of mental effort, thanks in part to computers replacing the slide rule.

Trapezoidal is the commonest for a newly-constructed canal but also the least imaginative and for many purposes not the best. Most engineers are unwilling or unable to

> *A canal cross-section can be any shape, and natural channels often have very complicated shapes which cannot be easily modelled by engineering formulae. But it is sensible to choose a profile that is easy to construct and does the job of carrying water for the least cost and with the best practical hydraulic efficiency. That limits the choice to a few standard sections.*

contemplate any other shape, although with computers today there are no excuses for not designing the best. Parabolic is for many situations the best practical shape for a variety of reasons, explained in Chapter 10. Elliptical closely describes a natural channel or matured earth canal. Rectangular is the most efficient at conveying a heavy bed load. Other shapes including triangular and circular might be dictated by available materials or available construction machinery.

An unlined canal will in time create its own profile depending on the soil type it passes through and the siltyness of the water it carries. This is shown clearly in figure 6.2 which shows the Yeleru Left Bank canal drained down for its annual inspection. This was built to a trapezoidal section in 1991, mainly unlined, but with the intention of lining it later as funds became available. Within a few years it had reverted to its natural shape, which is so close to a parabola as makes no difference. Even the reaches which had been lined, with trapezoidal concrete, had acquired silt deposits in the sharp corners which effectively made it a parabolic section anyway. Parabolic lining is a natural choice.

Rigid linings such as concrete are subject to external stresses from imposed loads and soil, especially clays which shrink and swell with moisture changes. Finite element analyses will indicate stress concentration points in trapezoidal canals especially at the junction of sides and bed. These stresses are minimised with parabolic profiles, a major advantage of parabolics which is seldom recognised even by experienced engineers.

figure 6.2   ***Yeleru Left Bank Canal, a natural parabolic***

Parabolic

The parabola is a steadily and infinitely steepening curve defined by the formula $y = ax^2$, where x is the horizontal co-ordinate, y the vertical, and a is any constant.[2] In practice it is better defined by the only dimensions which can easily be measured in the field, the top width and overall depth. These give a unique shape from which the constant a is calculated and fed back into the formula to calculate the hydraulic parameters for any depth.

It is the best shape to use for a small canal from several points of view. The beauty of the parabola lies in its inherent structural strength, especially if the sides are unsupported. It is easy to design the profile of varying thickness so that the strength is at the root of the cantilevered sides, where it is needed. This is especially significant for precast concrete segments.

Another feature in favour of the parabolic shape is its hydraulic characteristics with a fluctuating discharge, as is often the case in practice. It retains a higher velocity at low discharge than almost any other shape, and hence has slightly less of a tendency to deposit sediment. Similarly its greater depth at low discharge enables it to carry floating and semi-floating debris more easily than a flat-bottomed canal. Conventional wisdom has it that parabolics are superior at carrying sediment, but this is untrue in the case of bed load comprising coarse silts and sand.[3] Since bed material can only move when it is in contact with the boundary layer, the larger the boundary layer the higher the capability of

---

2 In the Former Soviet Union can sometimes be seen canals designed as a second-order parabola $y = ax^3$, or a third-order, $y = ax^4$. Both these give increasingly flatter and U-shaped profiles, and therefore lose much of the advantages of a first-order parabola in maintaining high velocities at low discharge.

3 Some recent research on compound profile channels in India, where round-bottom canals were silting up at a great rate, found them to be poor carriers of heavy bed loads.

transporting sediment. In a wide flat-bed channel the boundary layer in contact with bed material must cover a wider area than in a parabolic. Rectangular channels are more efficient at transporting heavy bed loads.

Perhaps the best feature of the small parabolic is the reluctance of cattle to walk in it, and of cyclists to ride in it. Both these animals can be extremely damaging to small conventional canals.

The theoretical best hydraulic section is with the top width T = 2.06 × depth.[4] This is approximate and varies slightly with longitudinal slope, but in practice it is more convenient to deviate from this rule depending on the method of construction. Small precast segments should have a narrower width than the best section, to increase strength and ease of handling. Cast-in situ concrete should have a wider top width in order to flatten the side slopes and make it easier to construct the canal without formwork. In weak or sandy soils it may be impossible to form the earthwork to a stable slope which is any steeper than 1:2 or about 30°. And for safety reasons, a large canal with steep upper sides is not desirable – people or animals have great difficulty getting out of a smooth lined canal with sides sloping steeper than this. In either case the forfeit in cost of extra concrete is not very significant. Table 9.1 gives a range of profile co-ordinates for manufacturing precast parabolics. All have been made and tested in the field using 35 kN/mm^2 concrete (4500 psi) and the dimensions have been optimised by trial and error.

Parabolics have been used in various countries with variable popularity. The countries of the Former Soviet Union use them as precast canalettes almost everywhere. When introducing parabolics to other countries the main opposition comes usually from the irrigation engineers who have a vested interest in not upsetting the status quo, and who all too often derive a healthy income from inferior quality design and construction. In Pakistan their acceptance by farmers was established through pilot projects using locally-made precast segments, amidst a great deal of opposition from the local irrigation authorities and traditionally-minded foreign consultants. Long since disillusioned with brick channels and watercourses that collapsed after a season and were never repaired, the farmers appreciated their durability and strength and the quick response time. In order to ensure success the initial segments were manufactured properly, with good quality moulds and proper concrete vibration. Too many similar innovations fail because not enough care is taken at the outset to ensure that local standards of construction are improved accordingly.

Larger canals can also be easily built as parabolics, with a saving in concrete volume of around 5 per cent over the equivalent trapezoidal together with gains in structural strength and aesthetic beauty. The Pehur High Level Canal in Pakistan was designed as a parabolic level-top canal to carry 30 cumecs. It promised to be cheaper and easier to build than the equivalent trapezoidal, shown for comparison in figure 6.3, provided it was built in the recommended way using rotating striker tubes for concrete placement. For a canal of this size the greatest advantage is its structural strength. A finite-element analysis showed no stress concentration points and a very low overall stress. A similar sized canal, the Genil-Cabra in Spain, also found greatly reduced stresses in comparison to the equivalent trapezoidal (Ref. 7, and figure A1.12.) This was vital since the canal passed through swelling soils, which would create severe stresses.

With a trapezoidal section, stress concentration can be expected at the junction of base and sides where there is an abrupt change in the profile. This was borne out in previous research and is frequently manifested in practice, when cracking occurs at this point due to settlement or soil movement.

During construction, placing concrete on trapezoidal side slopes requires care, and often results in either a dry mix that cannot be compacted or a wet mix that slumps under the weight of concrete upslope and leads to tension cracking. With the parabolic profile the base of the side is at a gentle slope that is markedly less prone to slumping and cracking. The predominantly flatter slopes facilitate concrete placing overall.

---

4 Not 2.83 × depth as sometimes quoted.

Volume of concrete can be reduced by tapering the thickness. The parabolic section in figure 6.3 has a taper imparted by the fact that an identical parabolic curve is used for the underside and topside of the lining. The two curves are displaced vertically by 100 mm, but this translates to a thinner section with increasing distance up the slope. The trimming equipment shown in figure A9.9 could hence be used with minimal modification for both the subgrade and the finished concrete surface. An even taper results in no stress concentration points. Whilst the sides of a trapezoidal section could also be tapered, this is not normally done and would save nothing over the base. In the case of a level-top canal with continually varying depth, the upper sides of the parabolic steepen as the depth increases and hence the sides shorten in comparison to the trapezoidal sides, which keep a constant side slope.

Setting out for construction can be easier with a parabolic than with an equivalent trapezoidal. Because the shape is defined by a single curve, setting out only requires a single point on the profile, such as the top of one side, or (less usefully) the centreline, to locate it in the field.

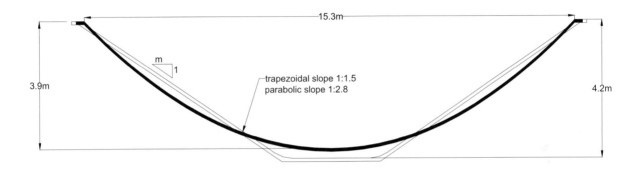

figure 6.3   Equivalent parabolic and trapezoidal profiles for the Pehur High Level Canal designed for 30 cumecs

Trapezoidal

A trapezoidal profile can be defined by its side slopes, depth and bed width. It is the commonest profile and is still universally used because most engineers prefer to work with straight lines rather than curves. The trapezoidal shape is defined by three curves: the base, the left side, and the right side, albeit straight lines.

The theoretical best hydraulic section is a half hexagon with side slope 60 degrees, $m = 0.58$, where m is the horizontal ratio of the slope 1V:mH. However, for all but the smallest channels this is too steep a slope to construct insitu. A practical maximum side slope is 1 on 1 except for small precast segments. On steeper slopes plastered masonry or ferrocement can be laid easily but in situ concrete needs to be shuttered. If the canal is deeper than about 1 metre, laying concrete on the sides can be difficult because the lower levels of the sides tend to slump under the weight of wet concrete above, even on slopes of 1 to 2 or flatter. The standard remedy in the field to combat slumping is to mix the concrete very dry, but then it cannot be compacted and it ends up porous, honeycombed and useless as a water retaining structure. This alone is enough to justify the use of in situ parabolics, in which the slope of the sides progressively flattens out with depth and the chance of slumping and tension cracking is greatly reduced.

Filleted trapezoidal

This is a compromise between the preferred parabolic and the unjustifiably popular trapezoidal, especially for large canals. By radiusing the junction of bed and sides, the stress concentration problems of the trapezoidal shape are largely reduced. For construction of level-top canals in unstable soils, the upper side slopes of a parabolic are variable, and may in some reaches become too steep for the excavated and filled soils to maintain stability long enough to install lining. Suggested radii for various canal sizes are shown in table 6.4.

In the design of the Merowe Project, level-top parabolics were initially proposed throughout the scheme, but then changed to filleted trapezoidal. Since the soils were extremely weak, with sands and dispersive silts, the upper side slopes at the downstream end of level-top reaches would have been too steep for easy construction.

Triangular

Triangular profiles are not common, although machine-graded ditches and grassed waterways are often made with this shape. A triangular canal section is the same as a trapezoidal with zero bed width and is defined by just the side slopes and the depth.

The best hydraulic section is with side slopes of 1 on 1, but in practice this has little practical meaning, since lined canals would normally be given a trapezoidal profile, and the side slopes of graded ditches could be considerably flatter for long-term stability.

Rectangular

Rectangular sections are commonly used in situations requiring limited land take, as conveyance flumes across cliff faces or channels in urban areas for instance. Rock cuts are often easier to construct as a nominally rectangular profile. But beware the temptation to design vertical sides in brick or masonry with a concrete base (as in figure 6.4.) It might be easy to construct the walls, but it is almost impossible to ensure that it is properly bonded at the base of the walls, and incurable leakage is the result, not to mention extreme weakness against horizontal forces from tractor wheels, roots and swelling soils. The best hydraulic section is one of bed width twice the height.

figure 6.4   *Rectangular brick canal, Pakistan, with a 'pucca nucca' outlet*

Circular or half-round

A full semicircle is theoretically the best hydraulic section of all, that is to say, the one that gives the most cross-sectional area for the least wetted perimeter. But this only applies to a theoretical case with no freeboard, which in practice is never designed for.

Circular pipes or half-pipes are available everywhere in one form or another. Figure A.9.72 shows a typical application. They can provide a quick and easy solution to many conveyance problems. But they are often unnecessarily expensive, and are structurally inferior to parabolics when used as segmental lining.

Large in situ circular sections have been built in Spain and Latin America (Ref. 10), using specialised but simple machinery for excavation, trimming and finishing concrete. Zero-slump concrete was used for these large canals, both in order to reduce shrinkage cracking due to its low water/cement ratio, and to permit placement without slumping on the near-vertical sides. The machinery shown in figure 6.6 comprises a rail-mounted gantry straddling the canal having a swinging boom pinioned over the centreline. The tip of the boom holds an earth trimming bucket or a concrete moulding plate with vibrator. A heavy vibrating effort is required for the zero-slump concrete, 1200 kgf per metre at 50 Hz is suggested in Ref. 9. It is pertinent to note the opening comments of Manuel Barragan, the author in this reference, when referring to the scepticism which greeted this innovative idea when it was first tested in the late 1960s: '*... one is always required to overcome not only indifferent physical difficulties but also the atmosphere of scepticism which is always a heavy burden...*'. A degree of scepticism is no doubt useful in rooting out impractical ideas, but it is unfortunately far easier for engineers to be sceptical and do nothing rather than to be adventurous and extend the frontiers of technology.

figure 6.5   Half-round canalettes, Morocco

Other shapes

There are several other shapes occasionally used in small canals:

An ellipse, or semi-ellipse, is sometimes used in corrugated steel culverts to give a greater width and hence greater discharge for a given height of culvert. It also approximates fairly closely to the profile of a mature earth canal that has been flowing for several years and has stabilised to a natural cross-sectional shape.

A hydrostatic catenary has been used for temporary channels of flexible membrane suspended from the top of each side. This is described in Chow (Ref. 10) but is rarely used in practice.

An ovoid, or egg-shape, used to be common in brick sewers, in order to keep the solids flowing at low discharges. In open-channel terms this is close to a parabolic profile.

In The Punjab a compound section of sloping sides and circular arc bed is sometimes used in the mistaken belief that the circular bed will increase sediment transport capability. In practice this is a false assumption, since circular and parabolic shapes are not the best for a heavy bed-load.

figure 6.6   ***Constructing a large circular canal, Spain (courtesy New Civil Engineer)***

Compound channels

Compound channel profiles are composed of more than one standard shape. It could be a trapezoidal with filleted sides, a rectangular U shape or triangular with circular bed, or a trapezoidal with flatter side slopes at the top, or with a berm. There are several instances with larger canals when a compound channel can be useful:

In a main drain which carries a small discharge for most of the time but needs a large capacity to accommodate infrequent flood flows, the small service discharge can be concentrated in a small channel in the bed of the main flood channel. This will keep velocity high enough to prevent siltation and heavy weed growth, or at least contain the perennial weed growth within a small strip where it can be easily maintained.

In an earth canal in partial cut and fill, the side slopes in cut may be more stable than those in fill, so a flatter slope is required in the top half of the canal. The same might be true for a canal cut through layered soil.

For a large drainage channel a flat berm at an intermediate depth can facilitate regular cleaning with machinery such as draglines.

A flood relief channel might have flood banks a long distance away from the main channel, to confine unusually high floods within the floodplain. The flood channel is then a compound one with very wide and shallow side berms. If large floods are infrequent, then the berms can be utilised for grazing or amenities such as playing fields

6.4   Canal lining philosophy

To line, or not to line?

To line or not to line the canals is one of the perennial questions facing engineers and project economists on all canal irrigation schemes. Whether it is a new scheme or an old one due for rehabilitation, the decision to line or not can have serious implications on cost and economic viability of the project. Lining can be very expensive. There are cheap methods also, but they may not last long. Sometimes the answer will be clear from the outset: it may not be necessary to line, or it may be essential. But more often the proper course of action is obscure. In this event there are a multitude of different lining methods to choose from, and a similar number of external factors which can push the decision-making process in one direction or another. As always, individual preferences and prejudices of the designer play a big part, but there are still a lot of lessons to be learnt. The reasons range from a general conservatism amongst engineers and a reluctance to try anything new, to more sinister implications of bribery and corruption. The die-hard engineer sees lining as a panacea for all irrigation maladies. The die-hard socio-economist sees it as an expensive waste of time. They are both wrong.

Reasons for canal lining

Canal lining may be required for a number of reasons, besides the obvious one of preventing seepage losses. The reasons outlined in the left column in table 6.2 are given in order of importance for a particular conjunctive use project in Java. Note that in this case prevention of seepage loss is not a high priority. In the right column are reasons for lining a major project in Sudan, for which the criteria are ranked quite differently.

Java – Madiun Conjunctive Use Scheme	Sudan – Merowe Irrigation Project
1. Fast response time	1. Maintain integrity of cross section
2. Reduced pumping costs	2. Avoid erosion
3. Land tenure	3. Improve bank stability
4. Reduced land take	4. Maximise command level
5. Maintain integrity of cross section	5. Prevent dispersion of sensitive soils
6. Prevent animal damage	6. Prevent seepage in high embankments
7. Control encroachment	7. Reduce siltation
8. Reduce health risks	8. Reduce weed growth
9. Ease maintenance	9. Reduce maintenance
10. Ease management	10. Increase velocity to carry wind-blown sand
11. Prevent seepage out	11. Ease management
12. Prevent seepage in	
13. Limit siltation	
14. Prevent erosion	
15. Limit farmer damage	
16. Structures simplified	
17. Increase discharge	
18. Improve bank stability	

table 6.2   Reasons for canal lining

Rapid response time

On east Java tube-well schemes the distribution of pumped irrigation water is made according to a demand schedule. Farmers pay for water on the basis of hours of pumping time. A typical tube well scheme may cover an area of up to 100 ha, and involve tertiary (level 3) canal lengths of up to 1 km. A farmer situated near the end of such a length of canal may not only have to wait a considerable time from pump start-up to receiving his water, but also, if the canal is unlined, he may lose a large proportion of it on the way. He thus ends up paying for more water than he receives, and may require extra pumping time to compensate for the losses, thereby disrupting the supply to other farmers.

The response time or sensitivity of the canal system is affected by the dead storage volume in the canal (which has to be filled before normal discharge can be attained), by the flow velocity, and by the amount of seepage and management loss. Unlined canals typically have a larger cross-sectional area and a greater volume of dead storage than lined canals. Velocity can always be increased by lining. Seepage can be prevented but only if the lining is appropriately designed.

The first time parabolic segments were used in Pakistan, the response time of the watercourse dropped from 45 minutes to 7 minutes as a direct result. That meant a significant saving in labour as well as a saving in water. Effective canal lining can therefore be an important contributor to good water management, by enabling water to be transmitted around the canal system with maximum speed and with minimum management losses.

Pumping costs reduced

Groundwater pumped from an aquifer 60 m below the surface is not cheap, neither in terms of capital cost of well and pumping equipment, nor of fuel cost in pump operation. Few of the pump installations in Java are yet electrified, and reliance is placed on diesel prime movers which are more expensive to operate and maintain. It follows that any water wasted in distribution is tantamount to throwing money down the drain. The 50 per cent distribution losses which are commonplace mean that the farmer pays twice as much for water as he needs to.

In central India electric pumps are commonplace on small irrigation schemes and pumped water is not so expensive, but there is still good reason to conserve power which is in perpetually short supply.

Land tenure problems reduced

In existing farmland not previously irrigated, such as the limestone areas of north-eastern Java and Madura, farm boundaries are frequently complex and disjointed. It becomes necessary to route canals along existing boundaries wherever possible in order to avoid complicated problems of land re-allocation. The situation is made more difficult by the need for surface irrigation on the steep and uneven topography.

Lined canals require a much narrower reservation width and can be turned through sharper angles than unlined canals, thereby fitting more easily into existing land-holding patterns. But pipelines should always be considered as a possible alternative.

In a recent project to rehabilitate the Kalpani Distributary on the Lower Swat System in Pakistan the designers were faced with trebling the capacity of an existing earth 2-cumec canal on high embankment without encroaching on the intensively farmed land on either side. The only way was to line it, thereby saving several metres in the overall width and in fact accommodating the increased discharge within the old reservation width.

More land available for cultivation

By reducing the canal reservation widths through lining, extra land is made available for cultivation. Some areas of Java have such intense population pressure that farmers are forced to plant crops on canal banks and sometimes in the canal itself in order to utilise as much land as possible. Farms may be as small as one tenth of a hectare, hence the need to plant as many as four crops a year in places.

A typical unlined tertiary canal has a reservation width of about 3 m. This can be halved with lining. The construction of a new unlined canal will often take soil from an overall reservation width of up to 6 m, and many farmers are reluctant to part with such a large piece of land, even if they are paid for it. Appropriate lining methods can largely avoid such disruption of cultivable land.

figure 6.7   ***Population pressure leads to monumental feats of engineering such as irrigating
the slopes of a volcano, Lawu, Eastern Java. The main canal runs at top left***

Integrity of cross-section maintained

Unlined canals, particularly in the volcanic clay soils of East Java or the black cotton soils of India and
Africa, can be notoriously unstable in cross-sectional profile. The swelling and shrinkage properties of
these soils serve to hasten the natural slumping of the sides that tends to form a wider and shallower
section than originally designed. During routine maintenance, weed and silt removal tends to enlarge
the canal through over-excavation and removal of bank soil along with weed roots. The result of
increasing the cross-sectional area is to increase the dead storage and hence the response time, to
decrease the velocity and hence encourage further silting and also slow the response time, and to
encroach upon cultivable areas.

Lining not only restricts the natural processes contributing to destruction of the canal, but also makes
it easier to maintain without altering the cross-section.

Animal damage prevention

Land crabs and rats can be highly destructive in unlined canals. Crabs are able in a matter of minutes
to burrow completely through tertiary canal banks, and in a very short time a canal can become
completely unserviceable through crab damage. The crab problem is particularly severe in heavy soils
such as volcanic clays which tend to be wet for the greater part of the year. Unless crab holes are
continuously blocked, seepage losses can become so excessive that water management procedures can
breakdown completely, in a soil that would normally be considered impermeable. Crabs are not
deterred by weak mortar, and many Javanese canals lined with stone masonry are perforated by crab
holes in mortar joints.

Whilst crabs are active in submerged conditions, rats tend to burrow above the water line but in so
doing weaken the banks and deposit large amounts of soil into the canal. They thus hasten the process
of cross-sectional deterioration and contribute to the rapid breakdown of the canal system.

figure 6.8 *In Eastern Java intensive land use means every available piece of soil is utilised. But cassava grown on canal banks soon destroys the lining*

Lining can keep these animal excavators in check, but unless the lining is properly designed and constructed it can be susceptible to damage which is both difficult to control and expensive to repair.

Much animal activity goes on unseen beneath a canal lining. Small rodents, worms and insects can riddle the bank with holes that act as conduits for any water leaking from a semi-porous lining such as brick or masonry. Because these holes are hidden beneath the lining, they are not closed up or filled with silt and the lining actually leaks more than the unlined canal did.

Cattle and buffalo can destroy small canals that are unlined or badly lined with a weak material. But they will not enter parabolic canals.

figure 6.9 *Cattle can wreck small canals, especially in India where cows are sacred, and canals are not*

Crop encroachment prevention

Severe pressure on land space gives rise occasionally to farmers encroaching on unlined canals in order to plant rice. This is particularly noticeable in some of the larger conjunctive use schemes in east Java, in which tertiary and secondary discharges are prone to wide fluctuations. Some canals lose capacity through farmers planting crops in them, and become by-passed when it is necessary to discharge larger flows. The construction of lining results in a clearly defined course which is less prone to encroachment. However,, unless the lining functions properly the entire canal can become redundant and it is not uncommon to find expensively lined channels full of paddy rice.

Crops planted on canal banks are sometimes inevitable due to population pressure on land, and in this event an appropriate lining method can prevent undue damage to the canal system. The damage caused by cassava on masonry lining in figure 6.8 can be avoided with precast parabolics.

Health

Lining can be a major contributor to improved hygiene. Natural health hazards include snails, mosquitoes, rats and snakes, which are either vectors of disease or a direct hazard to people working in the fields. All these are at home in the damp conditions which may persist with unlined canals. Even though the canals may be operated intermittently, it can take a long time for an unlined canal to dry out, and there may be many pockets that remain continually wet.

Schistosomiasis, or bilharzia, is known throughout most of the tropics. Snails as its potential vectors are to be found everywhere, and it may be only a matter of time before this disease becomes a serious problem in every country having irrigation.

A canal lining that can be completely dried out at frequent intervals will greatly damage the survival rate of snails and mosquito larvae, and also restrict the habitat of snakes and rats.

Ease of maintenance

A hard lining can result in easier maintenance and cleaning. Provided it is properly designed and constructed, its maintenance can be made considerably cheaper. Weed growth and siltation is restricted, and some types of lining can even be made self-cleaning to some extent. However,, inappropriate design or poor construction can lead to expensive maintenance requirements.

Reduction of management losses

Management losses are often the heaviest source of water wastage. Gates left open when they should be closed, canals overflowing because too much water has been let into them, blockages and minor leaks which should be routinely cleared or repaired. Management is made easier with a lined canal. The progress of water down the system is easier to predict and monitor.

Prevention of seepage out

The prevention of seepage losses are the most commonly cited reason in opting for canal lining. Seepage has two effects: wastage of canal water and waterlogging of surrounding land.

With effective lining the conveyance efficiency can be significantly improved and wastage reduced. Waterlogging of surrounding land is often a problem with canals that leak, in that it can encourage weed growth, salinity and crop damage. It can also lead to the encroachment of crop cultivation in areas not intended for it, causing social and administrative problems.

Interceptor drains on steep slopes usually have to be lined in order to prevent slope failures caused by a high phreatic surface. This is a constant hazard in Brunei, where the loose sandy soils and steep slopes collapse as soon as the tree cover is interfered with.

Prevention of seepage in

Incoming seepage is not often recognised as a problem, but in many groundwater schemes it hampers water management and maintenance, in both unlined canals and others that are lined badly.

It is sometimes impossible to shut down sections of canal due to inflow from the surrounding paddy fields. Seepage flows can often be equal to the full design discharge of the canal, and because they are uncontrollable any attempts at water management scheduling are made extremely difficult.

Small seepage flows can lead to increased weed growth and sediment deposition even in lined canals. A small flow in the bottom of trapezoidal channels will meander from side to side and tend to accumulate sediment in miniature sandbanks on alternate sides of the channel. Under tropical conditions these rapidly attract weed growth and further sediment. Thus it is important that any lining should be properly designed and built, to avoid these self-imposed problems.

Reduction of siltation

Run-of-river schemes are particularly prone to periodic heavy sediment loads. The siltation problem is made worse by the fluctuating nature of flow, particularly on the smaller schemes. This has the effect that for much of the time canals are flowing at well below their full design capacity, and velocities are therefore often less than required to maintain adequate scouring.

Unlined canals tend to have an uneven cross-sectional distribution of flow with areas of low velocity where sediment is deposited. Weed and vegetation growth acts as a sediment trap.

Not all lining is effective in reducing silt deposition. Stone masonry often has a rough surface with protruding joints, which can attract a significant thickness of fine sediment. Trapezoidal sections with shallow gradient and hence low velocity of flow tend to deposit sediment at the junction of sides and base, where there is a reduced velocity. Parabolic or semi-circular sections can be self-cleaning over a wide range of discharge, but heavy bed loads (usually medium sand upwards) are best carried by a rectangular section.

figure 6.10   ***Brick side lining stabilises the outside of a bend on the Maira Branch Canal in***
Pakistan

Erosion prevention

Erosion is a potential problem in many areas of steep topography or light soils. It is avoided either by lining or by drop structures, which can be expensive. A direct cost comparison between concrete lining and drops usually suggest lining to be the cheaper option when longitudinal ground slopes exceed about 1 in 50, but the numerous other advantages of lining may considerably outweigh the cost-only criteria. Figure 6.10 shows brick side lining still functioning well after 50 years.

Farmer damage reduced

Damage caused by farmers to canals and structures is generally a sign of inadequate water management or poor design. When water supply schedules are not met, farmers often make their own arrangements for abstraction by cutting holes in the banks or constructing temporary check structures of rocks or earth. These measures are common in Java, in both lined and unlined canals. However,, they should be seen as a symptom of a more fundamental problem in management which may be attributable to a number of separate factors.

Lining can lead to reduced deliberate damage by virtue of an implied improvement in water management, and can physically discourage the more serious interference such as canal breaching.

Structures simplified

Structures such as turnouts, cross regulators, division boxes, drops and measurement devices can be made considerably cheaper or even dispensed with if lining is appropriately designed. If canals are unlined, all structures require cutoff walls to prevent seepage and undermining. On small canals this can double the amount of masonry or concrete required. Lined canals do not usually require such heavy substructures.

Undershot steel sluice gates are normally used in Java for both turnouts and regulators. They require a substantial supporting structure of masonry or concrete and are therefore expensive. They always leak. Although originally designed for unlined canals they are used as standard structures for lined canals too. The use of parabolic concrete lining allows a complete redesign of all structures with substantial cost savings and important improvements in operation.

Discharge increased

Where canal gradients are very slight, the capacity of a canal can be increased by lining in order to reduce surface hydraulic roughness. Sometimes it can be useful in order to bring extra land under command. In Sri Lanka the 65 cumec Trans-Basin Canal from the Mahaweli River to the Ulhitya Oya was designed to be lined only to overcome the tight restrictions imposed by its flat gradient. Because it was cut through hard rock, excavation and blasting had to be minimised to optimise the cost, and a lined section was significantly smaller than an unlined one.

In situations like this including gravity tunnels, it can be economic to line only the bed. The bed roughness has a greater influence on the overall conveyance than the sides.

Command level increased

The 500 km long Merowe Left Bank Conveyor, recently designed in Sudan, will be lined partly in order to maintain scouring velocities whilst maintaining gravity command over its main irrigation area in the Goled Plains. An unlined canal would have required a much larger cross section with lower velocity in order to command the same land, and consequently would have led to greatly increased excavation costs in construction and also invite siltation problems in operation.

Bank stability

On old unlined canals carrying more than about 1 cumec it is common for the banks to be eroded by people, animals, rainfall and flowing water. The banks may require strengthening by shoring them with side lining. This is one of the few instances where stone masonry may be of useful application, but usually concrete is better, cheaper and quicker to construct.

6.5   Reasons for Not Lining

Cost is usually the prime reason for not lining. But if cost is a constraint then look for partial lining solutions, such as lining only the embankment sections of a canal system, or only the reaches that pass through permeable or unstable soil, or through villages.

Look for other potential problems, such as a faulty management set-up, which may be more insurmountable than the relatively simple engineering recourse of lining. Lining may not cure ills that are created elsewhere. There have been many instances of expensive lining being employed on schemes that had an impossibly unwieldy management organisation, and that continued to malfunction as a result.

Consider the speed and the capability of local contractors to carry out the job of lining. If it can't be done properly, then don't do it at all.

It may be that the scheme works adequately without lining. A scheme such as the Gezira would not benefit much from canal lining, as it is already solidly built in heavy clay soils which don't give rise to much seepage. Its massive siltation problems would never be solved by lining, and although schistosomiasis is a big problem there, the cost of lining as a contribution to improved health standards would probably be out of all proportion to the benefits.

Beware of potential geotechnical problems with soil and rock conditions, particularly with large canals. Gypsiferous and dispersive soils can literally dissolve away if small leaks occur through cracks and joints in a rigid lining, resulting in total undermining and collapse of the lining. This is one reason for the use of an impermeable membrane.

External hydrostatic pressure can develop either from inflowing groundwater or seepage through cracks and joints in the lining, followed by rapid drawdown of the water level in the canal. This can then cause piping and migration of fine soils through the joints, leading to undermining and partial collapse of rigid lining. This is one reason for the use of a filter membrane beneath a rigid lining.

6.6   Strategies for Lining

Durability

Is it permanent or temporary? Must it be designed to last 50 years or just a couple of seasons? If theft is likely to be a problem, then a heavy or contiguous in situ form of construction is called for.

> On large irrigation projects, the technical reasons for lining may well be augmented by strategic arguments which could apply on a national scale, or merely for the project in question.

Hydraulic performance

If it is important to distribute water quickly on demand, an hydraulically efficient canal profile must be used. If the water carries sediment, the design for a non-silting velocity may affect the choice of lining.

Water management systems

A demand-scheduled system may rely on a rapid response time, which in turn will dictate a smooth, hard lining with fast water velocity rather than an earth lining. At the other extreme, a continuous flow schedule with constant flow will be little affected by a slow response time, so lining with a view to improving performance will be wasted.

Construction requirements

What technologies are available in the country? Is it important not to damage existing crops? Is there a problem with land acquisition? If the project is being rehabilitated then is there a need for rapid installation to fit in with water delivery schedules? Is it practical to consider intensive construction methods such as slip-forming?

Labour resources

Is it intended to benefit small local contractors or major contractors having better facilities? It may be a national strategy to utilise local contractors where possible, but this may conflict with the need for quality control. There may be a labour shortage, in which case more intensive use of machinery may be called for, and precast units for example might be made larger than they would be for manual handling.

Material resources

From an economic viewpoint it is usually desirable to utilise construction materials that are locally available, rather than imported. Cement may be in shortage or even in glut, as it was in the 1980s. The use of local materials such as stone masonry needs to be considered with a view to the canal's performance requirements and the near-impossibility of achieving a good standard of workmanship.

Maintenance

Maintenance can be greatly facilitated by a hard smooth lining. Weed growth can be inhibited and desilting if necessary can be done more rapidly. However, some types of lining can lead to excessive maintenance, especially if they are incorrectly built. One argument sometimes used by supporters of stone masonry when confronted with the inevitable spectre of the lining starting to self-destruct is that its continuing need for maintenance guarantees local employment. This is a fallacy because what happens in practice is that the maintenance gets forgotten about anyway. And what corrupt engineers always fail to see is that there will always be new things to build, rather than keep rebuilding the same ones at catastrophic cost to the country.

Levels of technology

Aid agencies are often tempted to consider low level technology as being the only thing appropriate for developing countries. This is a mistake. Many third-world countries have quite sophisticated facilities such as concrete precasting and pre-stressing yards, which can be utilised for canal lining manufacture. It is churlish to infer that only village level technology is appropriate. Beware the oft-cited argument that local materials and labour resources must be used to stimulate the local economy. The prime objective of lining to improve irrigation is forgotten, and dismal quality is the result.

Quality control and supervision

Whatever the type of lining chosen, some degree of supervision will be required. The standard of engineering expertise available must be objectively assessed and if it is not adequate, it must be supplemented from outside.

Reducing wastage

Water wasted at the head of the system will invoke a shortage at the tail. When the system is hundreds of kilometres long, as in Pakistan, India, China and some of the Central Asian Republics, the social, political and economic repercussions of wastage can be enormous. Many a battle has been fought over water rights.

Reducing salinity

The 16 million hectares of Pakistan that make up the largest contiguous area of irrigation in the world is gradually turning into a disaster area through waterlogging and salinisation. Plans are afoot to implement drainage and canal lining in order to prevent seepage. On this scale, decisions based on technical perceptions can have consequences of global importance. The situation is many times worse on 7 million ha in the Aral Sea Basin, in Kazakstan, Uzbekistan and Turkmenistan. Here the problem comes from saline groundwater, poor quality water drained from the upper reaches of the Syr Darya and Amu Darya rivers, wind-blown salt from the dry bed of the former Aral Sea, and seepage from unlined canals contributing to rising groundwater. Rice is grown as a leaching crop in many places just to flush out the salts from the upper soil horizons. Canal lining has been installed in many places to reduce seepage but has mostly failed. Now the time is ripe for rehabilitation but there is no money for it.

6.7   Canal Geometry

Slope

The hydraulic slope, or hydraulic gradient, is analogous to pressure head in a pipe. It is the difference in total energy between two points along the canal, usually expressed as a vertical fall in metres per horizontal kilometre, or more usefully for analysis, as a decimal value.

For uniform flow, the hydraulic slope is the same as the slope of the water surface, which in turn is parallel to the bed slope of the canal. With non-uniform flow the hydraulic slope is the gradient between two known points along the water surface plus the difference in velocity head, $v^2/2g$.

The slope of a canal is normally decided at an early stage of design. It can be made as flat as possible to keep command, or steeper to reduce canal size or increase velocity.

Best hydraulic section

For any shape of channel profile there is a 'best' hydraulic section, meaning the *theoretical* most economical combination of depth and width for a given discharge. It has the shortest wetted perimeter, the least cross-sectional area, the least frictional resistance to flow and the smallest surface area of lining material. The ideal hydraulic section is a semi-circle, and for any other profile the best section is the one which most closely approximates to a semi-circle.

However, this is never the most appropriate section in practice. Theoretically the best trapezoidal section will have a side slope of 60 degrees, which is too steep for easy construction of earth or in situ concrete. Similarly in large parabolic canals the theoretical best section has sides that are too steep to be easily made in in situ concrete. Steep sides are a safety hazard for both people and animals that may accidentally fall into the canal. The need for freeboard and the possibility of a fluctuating discharge, both above and below the design level, will further modify the optimum economic section away from the best hydraulic section. In practice, the cost implications of deviating even quite substantially from the best hydraulic section are often not great, and should always be over-ridden by considerations of safety, structural strength and ease or practicality of construction.

There are numerous academic papers which pontificate over the mathematical intricacies of the most efficient section. Most of these have no place in the real world of irrigation, and are no substitute for practical engineering judgement.

Freeboard

Freeboard is the height from the top of design water level to the top of the lining, the *lining freeboard*, or from water level to the top of the canal bank, the *bank freeboard*. It is a safety reservation which can accommodate waves, a flood surcharge or a surge flow caused by faulty operation of the canal. It can be regarded as an ignorance factor to allow for inaccurate estimates in roughness or slope, or to accommodate the effects of poor construction tolerances on the same parameters. It can be deliberately built-in to accommodate wave action caused by wind or the operation of canal control structures. Or it can be deliberately utilised for outlet structures such as the side-spill turnouts shown in figure 7.11. Small earth canals might be designed with a high freeboard in the knowledge that the banks will get eroded or compacted by the passage of people or animals.

There are many old rules-of-thumb for determining the height of freeboard. In America, USBR practice recommends a bank freeboard varying from 1.5 feet for discharge of 20 cusecs up to 4 feet for 3000 cusecs, and a lining freeboard of 0.6 feet up to 200 cusecs rising to 2 feet at 3000 cusecs. Indian practice is slightly less conservative.

Modern trends in design however, tend to restrict the lining freeboard to the barest minimum on grounds of cost. Canals rehabilitated with a ferrocement overlay have sometimes used zero or even negative lining freeboard, but in these cases the existing lining was sound enough to prevent appreciable leakage. Table 6.3 gives guidelines for bank and lining freeboard which have been found workable in practice.

capacity	freeboard, mm	
cumec	lining	bank
0 - 0.5	50	150
0.5 - 1	100	300
1 - 10	200	500
10 - 50	300	1000
50 - 100	500	1500
100 - 200	700	1500
200 - 500	800	1700

table 6.3   Freeboard

Wind waves

On large canals with long straight reaches it may be useful as a cross-check to estimate the height of gravity waves due to wind. A formula for this is given in the box below. It is an adaptation of one developed for deep open water, and assumes that wave heights in a canal are less than in the sea or in a lake due to the friction effects of the canal bed and sides, and the shallow water along the banks. The Merowe Project in Sudan will have a main canal system 500 km long. During the design phase, the hostile climate, zero rainfall and almost continuous winds from the north, fed some concerns that excessive wave heights might be generated in long canal reaches. The longest reaches oriented roughly in the prevailing wind direction are about 10 km long, and the estimated wave height above mean water level using this formula for the maximum sustained wind speed of 105 km/hr is half a metre, comfortably within the lining freeboard of 800 mm. However, this formula is un-tested in the face of hard evidence from the field, and I would welcome any data from an enthusiastic research student that either confirms or refutes it!

Wind-induced gravity waves in a canal :

$$H = (U.F/1740)^{0.5} + 0.61 - (F/525)^{0.25}$$

Where H = wave height (trough to crest) in m
U = wind velocity km/h
F = fetch in km

This is based on the Molitor formula (Ref. 7) with a 20 per cent reduction for confined waters of a canal.

Waves generated from gate operation

Cross-regulator structures and head regulators which have control gates can generate waves if improperly designed or operated. Sudden closure or opening of a gate will create a single wave that will move upstream or downstream for a considerable distance before diminishing through side friction. Such transient waves can be calculated by unsteady flow models, and can be prevented by ensuring that gates and valves are made to operate slowly. Gate opening rates should be specified in contract documents. Thus a large radial might be specified to lift at a maximum rate of 300 mm per minute. Large gates should be dogged, that is, be equipped with a ratchet device that prevents them from closing suddenly in the event of a power failure or mechanical failure of the lifting gear.

Waves caused by channel slope change

The standing wave of figure 6.11 is a good example of a weak hydraulic jump created at the end of a steep reach of canal in which the Froude Number is close to 1, although still subcritical (see Chapter 10.) The flow in the steep section is undulating, with prominent waves due to the near-critical state of flow. The jump forms at the transition into deeper, slower-moving water, and is also induced by the bridge abutments and the start of a horizontal bend. The height of the splash zone can be clearly seen, and the lining freeboard needs to contain this.

figure 6.11   Standing wave at the end of a steep canal reach

Bends

Sharp bends in a canal can cause head loss or erosion. On a bend, centripetal force creates a dominant secondary current which tends to erode the outer bank and leads to head loss. With supercritical flow a super-elevation, or rising of the water surface on the outside of the bend, and strong sequence of surface waves are formed as well. Conventional wisdom holds that the tighter the curve and the faster the velocity, the greater the head loss. However,, some recent analysis[5] suggests that in a lined canal a tight curve results in less head loss than a long sweeping curve because the effect of the secondary currents is manifested over a shorter distance. So with a subcritical lined canal the aim in designing a bend is merely to avoid creating a disturbance which may be transmitted downstream as surface waves, any head loss is likely to be negligible in practical terms. With an unlined canal the aim is to avoid erosion damage, but if the route dictates a tighter bend then the outside of the bend can be lined at minimal cost. Changes in direction of supercritical channels such as spillways are usually achieved

---

5 J. C. Ackers, personal communication.

with an intermediate pool or hydraulic jump stilling basin, since in a concrete chute neither head loss nor erosion becomes an issue.

Design rules have been worked out in various countries, usually giving a minimum radius of bend for the discharge capacity of the canal. For small canals the following guide[6] can be used for the minimum radius of curvature measured to the canal centreline, with the radius expressed in terms of the canal water surface top width T:

canal capacity cumec		bend centreline radius	
		concrete lined	unlined
0	- 1	$3T$	$5T$
1	- 10	$5T$	$7T$
10	- 100	$5T$	$10T$

table 6.4   Bends

These radii should be increased for velocity higher than 1 m/s, and for erodible soil or lining subject to abrasion damage. In general, the faster the velocity, or the greater the capacity, or the greater the canal width, or the greater the depth, or the more erodible the canal material, then the greater must be the radius of the bend. It is common practice to partially line the sides of earth canals at bends in order to permit a tighter radius.

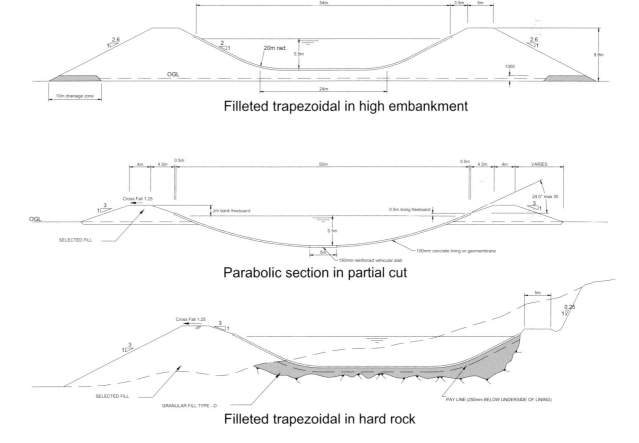

Filleted trapezoidal in high embankment

Parabolic section in partial cut

Filleted trapezoidal in hard rock

figure 6.12   Merowe Left Bank Canal profiles, capacity 230 cumecs

---

6 *These figures are broadly in line with USBR guidelines. Pakistan and the Punjab, where soils are erodible silts, work on considerably greater radii for unlined canals.*

Bank width and slope

Bank widths are also a subject which is often determined by local design rules. In areas of restricted land take, it may be preferable to adopt a parabolic or rectangular flume with zero bank width. In other cases the bank may perform one or more of several functions:

- Supports the canal lining. Masonry, plaster and in situ concrete all rely on firm banks and subgrade for support. Full compaction is essential especially where the canal is in fill.

- Restricts seepage, in which case the width of bank must allow a long enough seepage path that stays below the surface of the soil. South Asian practice has a rule of thumb for canal banks in well compacted fill, in that a notional seepage line drawn at a slope of 1 in 7 from the top of the water surface should always be below the ground surface. Weak or permeable soils need a flatter seepage line. In most soils that are properly compacted this rule is too conservative, and if possible a full geotechnical design should be carried out using flownets or other techniques. In large canals the earthworks can be a high proportion of the cost and it is important to carry out detailed geotechnical studies to develop the most appropriate design.

- Serves as an access road, maintenance track or footpath.

- Contains temporary flood surcharges, in which case the banks must be stable enough to handle the internal pore water pressures induced by a rapid rise and fall in water level.

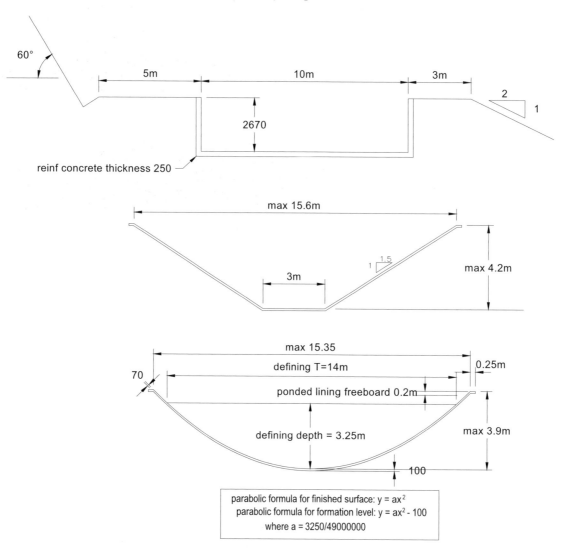

figure 6.13   Alternative profiles for Pehur High Level Canal, capacity 30 cumecs

A reasonable minimum bank top width for canals of about 1 cumec capacity is 1 m. If a small vibrating roller is used for compaction, this is the minimum width that can be handled. However,, this can be reduced if the canal has a self-supporting lining (such as precast parabolic segments) and soil compaction is not critical.

Large canals may have a minimum bank width of 5 m, being the least width that can be easily worked using heavy earthmoving machinery. High embankments may have intermediate side berms also, for seepage control, surface drainage, or ease of maintenance.

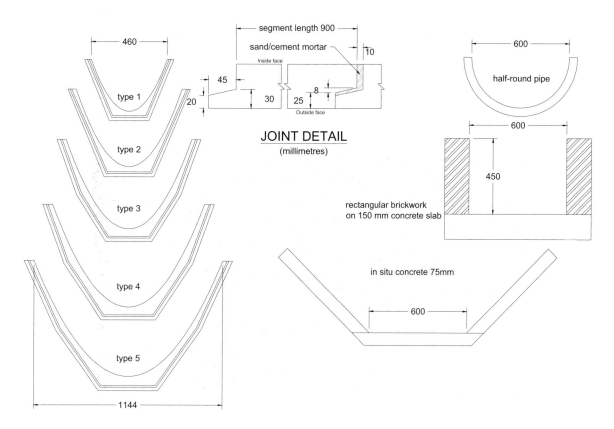

figure 6.14   ***Precast parabolic segments, and the traditional profiles they replaced in Pakistan***

References and further reading for chapter 6

1) Hydraulics Research, Wallingford, DACSE software for vortex tube design, 1993.
2) Plusquellec, H., Burt, C., Wolter, H.W., 'Modern Water Control in Irrigation', World Bank Technical Paper 246, 1994.
3) Withers, B. & Vipond, S., 'Irrigation Design & Practice', Batsford 1974.
4) Greg Rae, http://www3.hmc.edu/~grae/chaos/chaos.html
5) http://cosmos.kriss.re.kr/chaos/
6) http://johnbanks.maths.latrobe.edu.au/chaos/animated/index.html
7) US Army Corps of Engineers, Shore Protection Manual, 1991.
8) Justo et al., 'A Finite Element Model for Lined Canals on Expansive Collapsing Soil', proc.11th conf.soil mech foundation eng, San Francisco, 1985.
9) Barragan, M., 'Circular Canals', ICID Bulletin, January 1976.
10) Chow, V. T., 'Open Channel Hydraulics', McGraw-Hill, 1959.

CHAPTER 7   CANAL CONTROL STRUCTURES

Water flowing in the canal is controlled in several ways by structures of various kinds which perform one or more specific functions. This section was inspired during a farmer training project in India by the realisation that it was the trainers, not the farmers, who needed the training. Most agricultural extension staff and a good many of the engineers had little idea of the real purpose of many canal structures. Farmers may not understand the technical principles, and may cause damage and disrupt the water management as a result. It is useful for extension and management staff to understand these basic ideas even if not the engineering principles behind them.

7.1   Access and Safety

Canal inspection roads

Inspection roads should be incorporated into the design of all main and secondary canals. The road may not necessarily be on the crest of the bank. On gravel or unsurfaced roads, the minimum road width to permit easy passing of large vehicles is 6 metres, which is wider than most canal banks need to be.

> Access, safety and health are normally considered, if at all, as an afterthought in design, but the design and construction effort needed to ensure them are minimal. The best approach in design is to bear these in mind from the start.

Bridges

Bridges for vehicles, people and animals should be provided at locations of existing roads, near villages, and in general at intervals not greater than 5 km along main and secondary canals. Bridges should be incorporated into the design of cross regulators and head regulators. The cost of doing so is small. An architectural feature of bridges worthy of note is to ensure an odd number of spans thus avoiding a central pier. The appearance is greatly enhanced.

Access ramps

Entry points into large canals for maintenance should be provided near cross regulators. If it is envisaged that heavy vehicles will be required in regular maintenance in the canals bed, then any canal lining should be designed for vehicle loading, as a reinforced strip in the bed.

figure 7.1   *An access ramp for canal maintenance, Colorado*

Escape steps

In lined canals, steps should be provided at 500 m intervals to facilitate exit from the canal. It is likely that such steps will also be utilised for washing and bathing, and where required for this purpose wider steps should be provided.

Limited side slopes

In the design of lined parabolic or trapezoidal canals of overall depth greater than 2 m, the side slope should be limited to a maximum of 30 degrees, in order to facilitate escape by people and animals. This slope is flatter than theoretical optima, but the extra cost of materials is negligible and flatter slopes are generally easier to construct.

Trash racks

Trash racks are a cheap and simple addition to the inlet of any structure that could suck in people, other animals, or debris that could cause damage or blockage. They should be installed at the entrance of tunnels and inverted siphons as a matter of course.

7.2   Health

Designers should be constantly aware of designing-out potential hazards and designing-in facilities which will be of social benefit.

Irrigation schemes are notorious as harbingers of water-related disease. Malaria, rift valley fever and schistosomiasis (or bilharzia), are the best known amongst dozens. Engineers and politicians are also notorious in their ambivalence to the social hazards and sometimes tragic consequences of thoughtless development of irrigation canal systems.

Yet there are several ways in which vector breeding grounds can and should be designed out of a scheme. Mosquitoes are vectors of malaria, rift valley fever and many other diseases. Their larvae inhabit stagnant water. Eliminate the stagnant water and you eliminate many of the mosquitoes. Bilharzia is carried by snails which inhabit weed-infested waterways.

There are several strategies in design and planning which can reduce health risks:

- Include a drain outlet in all structures to avoid collecting stagnant water.
- Rigid canal lining to reduce habitats for weeds.
- Periodic drying-out of the canals to kill weed and snails.
- Village settlement planning to provide sanitation and clean drinking water.
- Educational programmes to provide awareness of hazards and their avoidance.
- Piped distribution systems instead of open channels.
- Automation of canals.
- Sprinklers in preference to surface irrigation methods.
- Biological control such as stocking canals with fish which eat weed, larvae or snails has been tried with varying degrees of success but should always be considered.

Although all these measures cost money, a smart socio-economist could easily show the extra cost to be justified in terms of improved health.

7.3   Water Discharge Control

The discharge or amount of water flowing in the canal has to be controlled at certain points in the system, the canal headworks being the first. If the canal is fed from a dam it will usually have a set of sluice gates which are opened or closed to control the flow.

A canal which offtakes from a river diversion weir may also have intake sluice gates. On large canals intake gates will be mandatory to enable shut down or regulation of canal flow for all river stages. Smaller schemes may often have an ungated sill and breast wall intake arrangement which does not regulate flow, but restricts the maximum discharge that can enter the canal when the river is in flood.

A canal which is fed by a pump station is easy to control, by switching on or off one or more pumps, but discharges are then limited to a discrete combination of the available pump duties. Control of flow can also be effected by throttling the discharge valves, although this is rarely done on small schemes because it complicates the operating procedures and reduces efficiency. A better arrangement is to design for downstream control, with the pumps switching on and off automatically within a range of water levels in the delivery canal.

There are three principal families of structure, which have different hydraulic characteristics and as a result are used in different circumstances. They are weirs, orifices and flumes.

The discharge Q over an overshot structure or **weir** is proportional to the head over the crest raised to the power of 1.5:

$$Q = cLH^{1.5}$$

The discharge is therefore sensitive to small fluctuations in head or water level. Long-crested weirs are ideal for controlling upstream water levels, but weirs are unsuitable for maintaining a stable outlet flow if water levels in the parent canal fluctuate.

The discharge through an undershot structure or **orifice** is related to the square root of the head:

$$Q = cA(gH)^{0.5}$$

It is therefore much less sensitive to fluctuations in parent canal water level than a weir.

The discharge through a constricted throat or **flume** is governed by formulae similar to the weir, but with exponent and discharge coefficient varying with throat width, since the side wall friction has an influence over the discharge. Typically:

$$Q = CH^{1.55}$$

Flumes can freely pass sediment and debris that would cause problems in other types of structure. They also operate under conditions of up to 80 per cent submergence and incur only a small head loss.

Discharge is controlled either at the head of a supply-scheduled system, or at the outlets of a demand-scheduled system. The control structures at these locations can be based on undershot or overshot sluice gates, throated flumes, weirs or sills, some of which may have the dual role of controlling both discharge and water level.

Head regulators - sluice gate type

Head regulators are located at the offtake from a conveyor or main canal into a smaller canal and will generally incorporate the following components:

- The main regulator structure housing vertical lift undershot gates. A breast wall will be included if the parent canal is significantly larger and deeper than the offtaking canal.
- Measuring device such as Crump weir or cut-throat flume.
- Stilling basin and transition into the downstream canal.
- Road bridge for canal embankment service road.

figure 7.2   Headworks of Dundian Distributary, Upper Swat Canal, Pakistan. A Crump weir creates a stilling section downstream of the sluice gate and also allows flow measurement

Head regulators – moveable weir type

In some schemes built in ex-British colonies such as Sudan and Iraq, Butcher weirs were commonly used (figure A7.1). They comprise a curved profile steel plate, which can be raised or lowered. However, although accurate as a measurement device, they are sensitive to small fluctuations in upstream water level, and are best used in a supply-scheduled system with near-constant flow.

The equivalent used in Dutch colonies was the Romijn gate (figure A7.2). This was a feature of some Indonesian schemes built in the mid-20th century. It is rarely found now, since the management has mostly degenerated into laissez-faire or continuous flow.

Details of both these can be found in Ref. 2.

Turnouts - undershot gate type

Turnouts are structures for diverting flows from secondary canals into tertiaries, or as outlets into fields. They are generally not regulated, but used either open or closed with no intermediate settings.

Undershot slide gates are a common form of turnout. The gates should be fixed in a concrete structure which can be precast to ensure high quality. They do not normally have a regulating function other than diverting the flow in an on/off situation. Indeed, there is little point in using them as variable-flow structures, because farmers will generally use them either fully open in order to maximise the flow, or else shut when they do not require irrigation water.

A typical undershot turnout is shown in figure 7.3. This is not a good design, as the steel handle can be easily damaged or cut off and stolen.

figure 7.3   *Simple lift gates used for check and turnout, India. Note the cut-throat flumes, one in precast concrete and the other in steel. The crop is ginger*

Turnouts - overshot gate type

This is a less common form of turnout (fig A7.3) which is sensitive to small fluctuations in water level in the parent canal. Its only advantage is that it has less tendency to leak due to the low operating head acting on the gate seals. It also functions as a measuring weir, although with uncertain accuracy due to the small head over the crest.

Turnouts - drop inlet type

Commonly used in Sudan, this is an old design, which is effective if properly constructed and head is available or excess head needs to be dissipated. It is used at main-to-secondary and also at secondary-to-tertiary levels. It comprises a gated inlet box, the invert of which is set at or close to the bed level of the offtaking canal. A steel pipe connects the drop box to the offtaking canal, and is buried beneath the access road and canal bank. In steep terrain this type of structure is convenient for dissipating excess head. The example shown in figure 7.4 is poorly constructed, with inaccurate steelwork and weak masonry leading to excessive leakage.

figure 7.4   *A drop inlet turnout in Sudan – inlet gate and outlet pipe. Note the natural 'onion' shape of the outlet pool*

Modular control gates

The French company Neyrtec developed a series of slots closed by lockable steel flashboards, to be used in conjunction with a constant downstream level control gate as described in section 7.5. The slots are of varying widths, corresponding to a per centage of the total width of the rectangular offtaking channel. A typical bank of these would have widths of 60 per cent, 30 per cent, 20 per cent, and 10 per cent of the total width. Then by opening an appropriate combination of slots, any proportion of the maximum flow (to the nearest 10 per cent in this case) can be attained. The water level upstream of the slots is kept constant at all discharges by the self-regulating gate, and hence the discharge through each slot is always the same. Management effort is limited to opening or closing a combination of slots to give any desired discharge. These are shown in figure 7.5.

figure 7.5   ***Modular control gates (courtesy Wolfgang Stenzel)***

Constant discharge modules

The same company developed a modular flume to give a constant discharge for varying water depth on its upstream side. This design consists of a bank of slots of varying width, with inclined baffle plates directed downwards and upstream. The baffles constrict the flow streamlines in a proportion to the approach velocity, which is in proportion to the upstream head. The effect is to give a constant discharge over a limited range of heads. There is a high-head model, with double baffles, which operates over a different range of heads.

This principle has been used in a variety of designs, but the Neyrtec modules remain the most well-researched. They require a minimum operating head however, and if adequate head is not available an alternative such as proportional dividers may be preferable.

7.4   Flow division

Flow division is a variation on the idea of discharge control. Dividing the flow can be done at any sluiced regulation structure, and also by means of proportional division. A proportional divider splits the available flow into pre-set amounts, which are a fixed proportion of the total. Thus if the total discharge fluctuates, a proportional divider will ensure constant proportions but variable amounts of water flowing into the separate channels downstream. It may or may not be gated. The use of this type of structure is entirely dependent on the mode of operation of the scheme. It is relevant for a continuous-flow management system, which is discussed in Chapter 3.

Dividing the flow can be done at any sluiced regulation structure, and also passively automated by means of a proportional division structure. On larger canals, vertical slot outlets can discharge a calculated proportion of the parent canal flow.

Proportional dividers – bifurcators and trifurcators

A proportional divider splits the available flow into pre-set amounts, which are a fixed proportion of the total. Thus if the total discharge fluctuates, a proportional divider will ensure constant proportions but variable amounts of water flowing into the separate channels downstream. In this respect it is a passive automatic structure. An example of proportional dividers is shown in figures 7.7 and 4.14. The use of this type of structure is appropriate for a continuous-flow, supply-scheduled management system.

When used as a turnout device, it may be fitted with a refusal gate, to enable farmers to shut down unwanted discharge, which is then passed on to the downstream users. A refusal gate is shown in figures 4.8 and A4.6.

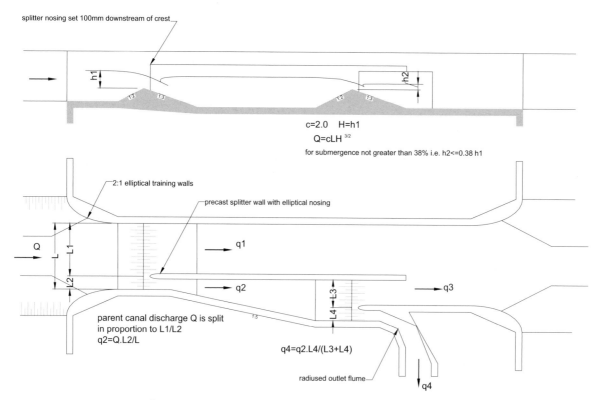

$$c=2.0 \quad H=h1$$
$$Q=cLH^{3/2}$$

for submergence not greater than 38% i.e. h2<=0.38 h1

parent canal discharge Q is split in proportion to L1/L2
q2=Q.L2/L

q4=q2.L4/(L3+L4)

figure 7.6   ***Design for a Crump weir bifurcator***

Proportional dividers may split the flow into a single side channel (bifurcator), or to channels on both sides of the parent canal (trifurcator). Variants have also been developed in which up to four offtakes are served from a single structure, and others in which the flow is split twice (double bifurcator) in order to divert only a small proportion of the parent flow. In practice the width of the offtaking throat should be not smaller than about 1/6th of the parent throat, otherwise the overall width of the structure becomes large, wall friction constricts flow in the throat, and inaccuracy results. They can be combined with a drop, but if head is at a premium then the rules of allowable weir submergence govern. That is, for the 1:3 Crump profile shown in figure 7.6, maximum 70 per cent submergence at each split.

The design is based on a triangular or Crump weir with a splitter wall which is fixed parallel to the flow with its leading edge in the supercritical flow region downstream of the crest. The splitter wall is positioned on the crest such that the ratio of offtaking throat width to the remaining crest width is identical to the ratio of offtaking discharge to the continuing discharge downstream in the parent canal.

In order to operate most effectively, an elliptical transition is required at the inlet, to ensure smooth approach conditions. Critical dimensions are the throat widths and crest levels. Precasting is therefore a preferred form of construction.

figure 7.7   **Double trifurcator. The flow to each outlet is divided twice, to outlets on both sides**
of the canal

Open flumes and proportional modules

Open flumes (figure 7.8) are popular in northern India and Pakistan and have been used there on many of the big proportional-flow systems for a century or more. They comprise a narrow ungated weir with a restricted discharge, designed to take comparatively small discharges from a much larger parent canal. They are intended to give a discharge which varies in proportion to the discharge of the parent channel. But true proportionality only occurs for a single design depth, and their discharge varies in proportion to the depth of flow, rather than the discharge, in the parent channel. They have the advantage that they have no moving parts and cannot be easily tampered with.

Open flume discharge:

$$Q = c.b. H^{3/2}$$

Where

Q = discharge in m^3/s or cfs.
b = width of throat in m or ft.
H = head above crest in m or ft.
c = 1.66 for $b < 0.1$m.
     1.68 for $b < 0.13$ m.
     1.7 for $b > = 0.13$ m.

Or c = 2.9 for $b < 0.3$ ft.
           2.95 for $b < 0.4$ ft.
           3.00 for $b > = 0.4$ ft.

The minimum practical throat width b is normally set at 0.05 m (0.2 ft). The outlet is proportional with setting the crest level at 0.90 of the depth of channel. If the crest is kept higher, it becomes hyper-proportional. The minimum modular head required is 0.2 H.

figure 7.8　　　Open flume, Pakistan

A variant of the open flume is the adjustable proportional module, or APM, which together with its close relation the adjustable orifice semi module (AOSM) was developed in the Punjab in the first half of the 20th century to give a finer degree of control over the outflow from ungated outlets. It is a flume with a cast iron curved roof block, which restricts flow in the same way as a breast wall when the parent water level rises. It is thus less sensitive to fluctuations in parent canal water level than an open flume, and behaves as an orifice. It is not actually adjustable, except at the time of installation, when the height of opening is set to whatever level is required. Refusal gates can be added in conjunction with flume outlets, in order to allow farmers to close down the watercourse during harvesting or heavy rainfall. They do not regulate the flow, only close it off completely or leave it running proportionally as designed.

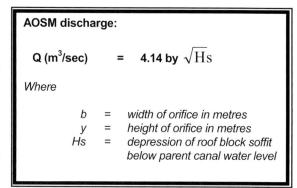

AOSM discharge:

$$Q \text{ (m}^3\text{/sec)} \quad = \quad 4.14 \; by \; \sqrt{Hs}$$

Where

b = width of orifice in metres
y = height of orifice in metres
Hs = depression of roof block soffit below parent canal water level

Neither of these types of outlet gives truly proportional outflow over a variable range of flow in the parent canal, but they give a reasonable approximation when designed for the dominant discharge.

When the Upper Swat Canal in Pakistan was rehabilitated in the 1990s, a precast concrete version of the AOSM was designed. These proved effective in those parts of the scheme where the original continuous flow scheduling was retained.

7.5　Water Level Control - Cross Regulators and Checks

Cross regulators for upstream control are gates or weirs in the parent canal that control water level, upstream of the structure. On upstream-controlled canals, regulators are located downstream of important outlets and branches, where they ensure an adequate level of water to operate the outlets. In the case of downstream control, the gates are located on the upstream side of outlets, since they control the water level immediately downstream.

Cross regulators are essential for the effective water management of a canal system. They are also used to close off sections of the canal during rotational scheduling (see Chapter 3), and thus prevent wastage. Without adequate cross regulators, the management loss of water can be very high. Many of the large protective schemes built before the second half of the 20th century were designed without any main system cross regulation. When they come up for rehabilitation, cross regulators should be the first structures to go in.

In 1980 the Uda Walawe Left Bank Canal in Sri Lanka had a management loss of over 90 per cent. The reason was a complete lack of cross regulators which meant that water offtakes could not be controlled through the turnouts, and most of the water was running to waste. A similar situation occurred on the Kinda Main Canal in Burma, where cross regulators had been omitted from the design to reduce costs (figure A7.7). When the water flow in the main canal was below its design normal depth, which was most of the time because the flow was affected by hydro-electric releases upstream, there was not enough depth of water to get any flow through the turnouts into the field canals. The farmers resorted to building their own.

Cross regulators can take the form of overshot or undershot structures, sometimes combining the two flow principles in the same structure. Both types can have moveable gates. Undershot gates can range from simple slide gates to self-regulating radial gates such as those manufactured by the French company Neyrtec,[1] which open and close automatically in response to changes in water level upstream or downstream. Overshot regulators are sometimes adjustable with a lifting mechanism to raise or lower the crest level (for example Romijn gates, movable weirs), but more often solid fixed weirs.

The advantage of undershot regulators is that one reach of the canal need not be full before water is passed down to the next reach. This is important where a short response time is required in order that downstream reaches can be managed effectively. And in canals that flow intermittently or in rotation, the most rapid filling can be done from the tail-end upward, using undershot regulators.

The advantage of overshot fixed weirs is that they require no operation and are therefore passively automatic for upstream control. Long-crested weirs can maintain water levels within a narrow range even if the discharge fluctuates. For this reason they are most effective in combination with undershot gates, in which the gates are only adjusted for large variations and the weir copes with minor variations in discharge.

Cross regulators are gates or weirs in the canal that control water level, upstream or downstream of the structure. Cross regulators are essential for the effective water management of a canal system. They are also used to close off sections of the canal for maintenance, in emergencies, or during rotational scheduling, and thus prevent wastage. Without adequate cross regulators, the management loss of water can be very high.

These structures are normally gated, with radial gates for larger discharges and vertical lift gates for small discharges. Rarely, flap gates hinged at bed level and hydraulically operated, are used where the maximum head is less than 3 m.

They can also be fixed weirs, submerged as a bed sill or as a free overfall. Cross regulators often serve the dual purpose of velocity control in acting as a drop structure also.

Gated cross regulators for upstream control

In a canal system which is upstream-controlled, the regulators control water levels on their upstream side. Regulators are therefore located downstream of important outlets and branches, where they maintain an adequate level of water to operate the outlets.

Although a simple weir can perform this function, on larger canals it is normal to provide gates so that the system can be operated in a more flexible way with faster response times. The fastest way of filling a canal after a shutdown is to open all cross regulators and fill from the tail-end upstream. With fixed weirs only, each reach has to fill to ponded level before the following reach can start to fill. The process is slowed if outlets are kept open en route.

Gated regulators should always be provided with a bypass weir, having its crest set at or slightly above the design maximum regulated water level. The weir acts partly as a safety feature in the event of failure or incorrect operation of the gates. However, unless the gates are fully automated and computer-controlled, its prime purpose is to facilitate fine-tuning the gate operation. With water level at weir crest level, the gates are correctly positioned. When the weir spills, the gates should be opened,

---

1 Now part of GEC-Alsthom

and if water level falls noticeably below the crest level, then the gates should be closed; unless, of course, the canal is being filled, as in figure 7.9. Minor fluctuations in discharge are accommodated by the weir without recourse to operating the gates. Major fluctuations are handled by opening or closing the gates.

Radial gate discharge: $Q = cA(2gH)^{0.5}$

Where the coefficient c varies with gate opening and submergence conditions. Refs 8, 17.
A is the area of gate opening
H is the head to the centre of the opening
g is gravitational acceleration

Figure 7.9 A small radial gate used as an upstream-control cross regulator, Lower Swat Canal, Pakistan. Note the bypass weir at either side of the gate. It is combined with a fall and was originally designed to be used in a demand-scheduled mode, with water ponded to the level of the bypass weir crest

Choice of gate

Several types of gate are in common use for irrigation cross regulators:

- Radial.
- Vertical lift slide gates.
- Vertical lift wheel or roller gates.
- Bottom-hinged leaf gates.

Radial gates require substantially less operating effort than vertical lift gates, since the hydraulic pressure is directed through the trunnion axis and not against the gate sealing surfaces as in a vertical slide gate, and in a lifting operation the trunnion supports half the weight. They can also be counter-weighted downstream of the trunnion, in order to further reduce the lifting force required. Critical components requiring accurate positioning during construction are the trunnion bearings, and the cheek plates and side seals. Cheek plates are steel plates set into the side walls, against which the seals run. The seals are typically a rubber J-seal, bolted to the gate face. Operation may be through cables onto a manual or electrically powered winch, or through a hydraulic ram. In developing countries with uncertain power supplies, it is common to install a manual winch, which can be readily adapted to electrical operation when a power supply becomes available. Hydraulic components may be susceptible to damage from hostile climatic conditions such as dust storms or high temperatures, and are not often used. Radials are typically used in sizes up to 5 m wide with a capacity of up to 40 cumecs each.

Vertical lift slide gates are commonly used for small gates up to a capacity of about 10 cumecs. They are simple in concept, being a rectangular leaf moving in slots at either side of the opening. The mating surfaces are critical components, which seal the gate against leakage and permit the movement of the gate. In larger gates, the seals may be phosphor-bronze or stainless steel, which are accurately machined to minimise friction. Operation can be through a rack-and-pinion mechanism as in figure A7.4 or more commonly by a screw-threaded spindle and a gearbox with a manual crank or electrically powered mechanism (figure A7.5). They require more lifting effort than radials, since to the dead weight of the gate must be added a hydraulic downpull force of water jetting beneath the gate leaf, and the hydraulic pressure forcing the gate against its seals.

Wheel gates are used in larger sizes, and offer a reduced lifting effort over slide gates. The mating surfaces are wheels or rollers fixed to the gate leaf, running against a track or rail fixed to the structure. Seals are separate and, as in a radial gate, comprise a rubber moulding fixed to the moving gate leaf, running against a corresponding plate on the structure. These gates are usually provided with a counter weight in order to minimise lifting effort, which results in a heavy gantry structure. Lifting mechanism may be geared spindles or chain or cable winches.

Roller gates follow the operating principles of wheel gates, but are usually used in wider spans of 10 m or more. They are used more often in river barrages which demand wide spans. The Stoney roller gates of figure A7.6 are at the headworks of the Upper Swat Canal.

Undershot gate discharge:

$$Q = C\,A\,(2gh_1)^{0.5}$$

Where A = area of opening
h1 = upstream water depth
C varies from 0.52 to 0.58 for free discharge,
and reduces under submerged conditions.
(Ref. 8, fig 17-38)

Bottom-hinged leaf gates are infrequently used in developing countries. They have the advantage of easily passing floating debris. They offer sensitive operation, since the flow passes over the crest as a weir, and a small movement of the gate alters the discharge less than would the same movement in a bottom-opening gate. They are sometimes used in conjunction with radial gates to give fine tuning of both discharge and water level. They are usually hydraulically operated, and due to the torsion stresses induced by water pressure on the bottom hinge, their height is normally restricted to less than 3 metres.

Self-regulating cross regulators for upstream control

There are several incarnations of automatic gates for upstream control. The best-known are the French Amil gates originally marketed by Neyrpic. In this type of radial gate, the leaf incorporates a hollow float, which starts to raise the gate when the design upstream water level is attained. The gate is counter-weighted on either side of the trunnion, with weights being adjusted during gate commissioning.

Self-regulating cross regulators for downstream control

In open canal systems with downstream control, self-regulating gates can be used for control of water levels in the reach downstream of the structure. The best-known are float-operated gates usually referred to by their original trade names 'Avis' or 'Avio'. The free-surface Avis gates operate at heads of up to 1.5 m. The orifice Avio gates can operate at heads of up to 6 m, and seal (imperfectly) against a breast wall.

These gates offer the advantage of being automatic with no moving parts other than the gate itself. They respond immediately to fluctuations in downstream water use.

The original design of these gates was by the French company Neyrtec in the 1950s, since when they have been successfully deployed in many countries. Their movement is controlled by a float which rises or falls with downstream water level. They are radial gates with a trapezoidal leaf, which is not prone to jamming at the cheeks. In recent projects in Pakistan and Sudan, small but significant improvements have been made to the design, based on previous experience with large gates of this type. In particular, the operation is made smoother by enlarging the float chamber, improving the hydraulic flow conditions immediately downstream of the gate, adjusting the size of damping vents, and linking the float chamber to a zone of smooth flow in the canal downstream. During detailed design these gates should be analysed for possible hunting and transient wave generation

> **Downstream control self-regulating gates (Avio and Avis). Minimum reach length for prevention of hunting transients:**
>
> *min wedge storage = QT/2*
> $T = \text{wave travel time} = L/((gh)^{0.5} + V) + 1/((gh)^{0.5} - V)$
>
> *where*   *Q = maximum discharge in cumecs,*
>   *h = cross sectional area / water level width*
>   *V = mean velocity*
>   *L = length of reach*

by using an unsteady flow simulation. During commissioning they should be adjusted so that adjacent gates do not operate in perfect unison, as this would encourage hunting between regulators.

The gates are not watertight and the orifice (Avio) structures are normally provided with additional vertical lift isolating gates. When used as cross regulators it is good practice to arrange the gates in banks of two or more, to enable smooth operation without hunting and to permit partial decommissioning for maintenance. It is also useful to provide a manual override system to facilitate emergency closure. This can take the form of an overhead beam to which can be attached a chain block connected to the float, forcing the gate closed in the event of uncontrolled falling water level downstream. A more sophisticated method is to provide a valved inlet and outlet pipe into the float chamber. When the chamber is full of water, the gate is closed. In the event of emergency such as a canal breach downstream, the connecting valve from float chamber to downstream water level can be shut down and the float chamber filled either manually or through a pipe connected to the upstream reach.

When a series of these gates are deployed as cross regulators in a canal, there is a minimum distance which must be maintained between each structure, in order to avoid excessive hunting. The hunting is caused by transient waves or surges, created when a gate opens or closes rapidly. The transient moves up or down the reach and is reflected back from the next regulator. On its return to the original regulator, its influence on local water level may create a compensating gate movement which serves to reinforce the original transient effect. In a badly-designed canal, the gates may continue to operate in response to each other, rather than to the actual downstream demand. The minimum separation distance depends on the wedge storage volume, discharge, length of reach and travel time.

Fixed weirs as cross regulators

In an upstream-controlled system, these structures will normally be located in secondary and tertiary canals immediately downstream of turnouts, for the purpose of stabilising water level at, and hence discharge through, the turnout. They may be combined with a vertical drop in canal bed level, and will always have a drop in water level across them. They are sometimes known as check structures, but here we refer to checks as temporary structures only.

Cross regulators will normally comprise a free overfall weir, with a low-level drainage outlet which may be gated, in order to permit draining down of the upstream canal reach. In order to stabilise the upstream water level as much as possible, the weir crest can be lengthened and the water level afflux reduced by adopting a folded oblique or duckbill configuration, as in figure 7.10. In small canals where the offtaking flow is greater than 20 per cent of the parent flow, the configuration can be a straight weir perpendicular to the flow.

A folded weir (figure A4.20) is an advantage when a flood surcharge or fluctuating flow has to be passed whilst maintaining a narrow range of water level. A concave duckbill weir has the added advantage of an in-built stilling basin. A convex duckbill (figure 7.10) also gives good energy dissipation. However, beware the dangers of a convex arrangement in an unlined and erodible channel. A recent embarrassment for designers in UK was the adoption of a convex orientation for weirs in the new Jubilee River, a flood relief channel of the River Thames. Erosion of the banks at either side of the structure was severe enough to start undermining the foundations. A labyrinth weir (figure 4.10) is a compact combination of several duckbills.

Weir discharge coefficients and flow formulae are given in figure 7.25.

figure 7.10   *A long-crested weir with convex duckbill configuration, Sri Lanka System H. The crest is round-nosed. This arrangement gives excellent energy dissipation. Note the drainage orifice in the centre*

Notched falls

At the end of the 19th century in northern India it was common practice to design cross regulators as narrow-throated trapezoidal notched weirs, having an operating head equal to the normal depth of flow in the canal upstream. The advantage was a stable channel not prone to siltation, and in the case of variable discharge there was no interference with the operation of proportional open flume outlets, which were

> **Notch weir discharge:**
> $$Q = 2/3 \, (2g)^{0.5} * (l \, d^{1.5} + 0.4nd^{2.5})$$
>
> *Where   l = width of sill*
> *n = 2 tan α*
> *d = normal upstream depth*

intended to give an outflow in a fixed proportion to the flow in the parent canal. In Burma they are still called 'Indian Weirs'. On large canals a series of adjacent notches were arranged across the full width of the canal to spread the turbulence caused downstream. The lower lip of the notch had a semi-circular steel plate, which imparts the characteristic and beautiful flow pattern in figure 7.13.

It would be unusual nowadays to design a new scheme with this type of structure, since proportional flow is becoming increasingly uncommon as a basic design philosophy. However, in the rehabilitation of old existing schemes it may be of interest to retain these beautiful structures, and some design details are therefore given here.

The design of notch falls is more art than science, and draws on practical experience from the Punjab in the 1890s (Ref. 11). The aim in design is to match the head at the notch to the normal depth in the channel upstream for all predicted discharges. Typical geometry is shown in figure A7.10. The side angle to the vertical, α, is a variable and shown here as 15°. They can also be used partly submerged.

Check structures

Checks are small cross regulators in minor canals that increase the water depth immediately upstream so that water can be abstracted by siphons or turnouts. The gated parabolic checks in figure 7.11 serve this purpose, allowing water to overflow the turnout slot on the upstream side. Temporary checks can be made with earth or plastic sheets, which are often used in conjunction with plastic siphon outlets. They are designed as a straight overfall weir.

On a small canal the purpose of a check may be to divert all the flow through a turnout. In this case the notch of figure A7.12 will be closed with a wooden flashboard. The *pucca nucca* of figure 6.4 is a fully gated check, with water being either fully diverted into the field or else flowing on to the next nucca.

figure 7.11   Parabolic check plate, with field outlet closed at left, open at right

7.6   Velocity control

If the water flow is too fast it may cause erosion, even in concrete-lined canals. And if the flow is too fast it can be difficult to get water out of the canal, either through turnouts or siphons. Velocity can be controlled in two ways: either by adjusting the canal slope by means of drop or fall structures, or by increasing the canal's roughness, by using a specific type of lining or flow arresters.

Drop structures

The purpose of drop structures or falls is to control the velocity of water by limiting the canal longitudinal slope to one that generates an acceptable velocity.

Drops are provided where the terrain slope is steeper than that of the designed canal. They consist of a vertical or sloping step or series of steps with a means of dissipating the energy of falling water. Stepped drops, in which water cascades down a flight of steps and expends its energy on the way, are efficient at dissipating energy. Chutes and overfalls concentrate the jet of water, which then has to be controlled in a stilling basin. A glacis drop has a sloping downstream face. If the slope is gentle, say 1 in 5 or flatter, a hydraulic jump forms without the need for a deep stilling basin. In situations which incorporate an offtake or turnout, the long crested weirs may be appropriate as dual function structures with efficient energy dissipation. Baffled chutes (figure 7.12) are also highly efficient provided they are not used in an environment subject to damage from rolling boulders or blockage by floating timber.

figure 7.12   ***Baffled chute drop***

figure 7.13   ***Notch falls, Upper Swat Canal, Pakistan***

Baffled chute drops

Baffled chutes are highly effective for vertical falls greater than 2 metres. They consist of a rectangular concrete chute sloping at 1V:3H, with precast or in situ concrete teeth set into the glacis. Energy dissipation takes place on the chute, and the flow is further smoothed by a simple stilling basin with onion transition. A typical structure is shown in figures 7.12 and A7.17.

figure 7.14   Glacis fall

figure 7.15   Stepped Drop cascade in a parabolic watercourse, and a glacis drop in its parent distributary.

Overfall and chute drops

Where the vertical fall is 2 metres or less, a straight overfall drop structure can be used as shown in figure A7.16

Inclined chute or glacis drops can be used for vertical falls greater than 2 metres, and design discharge less than 1 cumec. A typical structure is shown in figure 7.14.

Pipe drops

Pipe drops are commonly used in Sudan (figure A7.18), using a brick drop well with steel pipe outlet, frequently gated at the inlet and serving as a head or cross regulator. They are less safe than open falls, and children or animals can be trapped in the inlet well. Energy dissipation takes place in the well, but the pipe outlet also concentrates the flow and a substantial stilling basin is required.

Rough channels

Channel roughness can be deliberately increased in order to reduce velocity. The pitched channel on the Upper Swat Canal System (figure A7.61 and 10.2) was built with un-pointed stone masonry with the express purpose of generating a high roughness coefficient and thereby restricting the velocity to a non-damaging value. Another example of deliberately increased roughness is the spillway of Gopishetta dam in Karnataka, which was built cheaply of hand-placed riprap. This had the disadvantage of providing an ideal habitat for crayfish, in the hunt for which the local people would deliberately dislodge the stones.

The same principle can be applied to drop structures, such as the rapid fall of figure A7.13, which is an example of a channel structure deliberately designed with a high roughness coefficient. These were built around 1910, without the benefit of research into stilling basins. They have now been replaced by the chute falls of figure 7.12.

Flow arresters

The steps inserted into the channel of figure A7.15 reduce the velocity by effectively increasing the roughness[2] of the channel. This particular channel was part of an incorrectly designed pumped irrigation scheme, the water is pumped far too high and has to be run down this channel immediately afterward, thus wasting power and pump installed capacity. The concrete quality was so poor that erosion of the canal lining started the same day as the project was inaugurated.

7.7 Turbulence Control

Turbulence arises at any sudden change of flow conditions, such as will occur at a structure, especially beneath sluice gates, over weirs or drops, and at sudden changes in channel geometry such as at the outlet of a pipe or aqueduct. Any turbulence is manifestation of a loss of head, or a waste of energy, which may or may not be desirable at any particular structure. It can generate highly erosive forces, which can be damaging to canal banks and structures. The most damaging conditions arise in the case of diverging flow, when supercritical flow changes to subcritical. Converging flow is less problematical, provided the geometry is smooth. Turbulence is best described by the new science of chaos theory rather than precise mathematical formulae. Since chaos theory[3] is still in its infancy, all the tools we have for designing structures to control turbulence are based on experimental and empirical evidence.

Control measures for turbulence rely either on containment or avoidance. Containment methods, by which excess energy is ***dissipated***, induce turbulence and confine it to an area which is designed to resist damage, such as stilling basins and plunge pools below a drop structure. Avoidance methods, in which energy is ***conserved***, rely on smooth transitions. Transitions are especially important when head is at a premium.

Stilling basins

Stilling basins are an integral part of drops and regulators. Their function is to dissipate excess energy created by the sudden drop in water level across the structure. In most cases the state of flow across the structure will change from subcritical, through supercritical, and back to subcritical by means of a hydraulic jump. The geometry of the stilling basin will depend on the efficiency of energy dissipation within the main structure. Strong hydraulic jumps are created by undershot gates, smooth chutes, and pipe outlets, in which high velocity flow is concentrated. The position of the jump can vary under

---

2 *Roughness is discussed more fully in Chapter 10.*
3 *The Internet is probably the best source for the latest ideas on Chaos.*

changing discharge, and one function of the stilling basin is to contain the jump in one place, either by means of a deep plunge pool or by blocks or sills in the bed which induce the jump to form. Weak or undulating jumps are formed when the change in specific energy is not large, such as with low falls or gates which are large in relation to the channel cross section. Perversely, weak jumps can be more problematic to contain, as they create waves which travel downstream without dissipating much energy. Submerged jumps which form below open sluice gates may not create much surface turbulence, but can create extremely erosive conditions at bed level, which can travel large distances downstream unless they are constrained within a stilling basin.

In determining the size of stilling basins within structures, USBR guidelines can be followed as appropriate. These recommend lengths and depths of basin, based on calculated specific energy, velocity and Froude number. They are appropriate for concrete chute and fall structures, but in my experience tend to underestimate the strength of erosive flow under some conditions. In new projects which have no constraints on space around structures they can be used in conjunction with 'onion' type transitions.

'Onion' type stilling basins have been used successfully in many projects in Malaysia, Burma, Indonesia and elsewhere. They mimic the shape of bank and bed erosion which would form naturally if no stilling basin were provided, as is the case in figure 7.4; see also figure A7.19.

Transitions and fluming

Smooth transitions are to be preferred at inlet and outlet of all control structures, particularly division structures and measuring structures, which depend on smooth flow conditions for correct operation. Elliptical transitions are superior to other shapes, and are not difficult to design or construct. Fluming is a term used for a transition between two channel profiles of the same profile but different dimensions.

The canal transition in figure A7.20 would have been better designed as a reverse ellipse, as shown in figure 7.17. When designing such a transition the first step is to fit a curve in plan between the outer edges (i.e. bank tops) of the two canal sections. The method shown in figure 7.17 is appropriate for most irrigation canals having a velocity of less than 2 m/s. It is based on elliptical segments of eccentricity 2.5 to 1, and a tangential joining line oriented in plan at 1 on 2. If the two canal profiles are trapezoidal or rectangular then the base of the sides are joined with the same curves, except that the straight joining line will be shortened. If the transition is from parabolic to trapezoidal or rectangular, then the lines of the edge of the bed will converge to a point on the centreline of the parabolic. This can be either a straight line or an inverted reverse ellipse, as in figure A7.21.

This procedure is trivial when using Computer-Aided Design, in which this reverse curve can be created as a drawing block and the eccentricity can be increased to any ratio by stretching. An increased eccentricity may be desirable for higher velocities and for supercritical flow. Chow (Ref. 8) suggests the optimum angle of divergence over the length of the structure should be 12.5 degrees, which would be achieved in figure 7.17 by using an ellipse eccentricity of 3.5.

Hydraulic design of transitions is aimed at investigating the change in water level, velocity and energy loss using the specific energy method described in Chapter 10. Some useful detailed examples can be found in Chow (Ref. 8). However, in practice for most irrigation design, it is hardly necessary to carry out a detailed hydraulic analysis, provided these guidelines are followed and the transition is reasonably streamlined.

Construction of warped or flumed transitions is not difficult given the right tools. Concrete of the right consistency can be easily moulded to steep side slopes and top shuttering should not normally be necessary.

figure 7.16 **Inlet transitions to siphon aqueducts on the Shamozai Distributary, Upper Swat Canal. Compare the smooth flow in the elliptical transition on the right with the turbulence in the straight walled one, and the sudden head loss in the square-edged one**

7.8 Conveyance and Cross Drainage

Conveyance structures are used to route canals past obstacles such as depressions, valleys and rivers, and through or around steep terrain. In mountainous terrain, tunnels may be necessary, and in locations having a steep cross slope, rectangular flumes may be required.

There are several basic options for constructing cross-drainage structures:

- Culverts passing drainage flows under the canal;
- Superpassages passing drainage flows over the canal;
- Inverted siphons passing the flow in the canal under the river;
- Inverted siphons passing storm flows beneath the canal;
- Aqueducts passing the canal flow over the river;
- Drainage inlets.

Choice of structure

The choice of structure depends on the relative discharges of the canal and cross-drainage channel, and the local topography. Where canal discharge is much smaller than the cross-drainage flow, an inverted siphon or aqueduct is normally appropriate, with the choice between them being decided on the basis of available head, of which more head loss is required in an inverted siphon, and cost, which is specific to the particular site. Inverted siphons would be preferred where the cross-drainage discharge is unknown or cannot be reliably estimated.

When the canal discharge is greater than or comparable to the cross-drainage discharge, then the cheapest solution is likely to be a culvert or superpassage. The choice between these is dependent on the local site topography. Where the canal bed level is above the level of the cross-drainage channel, a culvert is used. Where the cross-drainage channel bed is near to the canal design water level, a culvert cannot be used if the outlet would be liable to sedimentation, and a superpassage is to be preferred.

Inverted siphons taking cross-drainage flows beneath the canal are rarely used unless the cross-drainage flow is known to be free of sediment. Figure A7.22 shows one that wasn't!

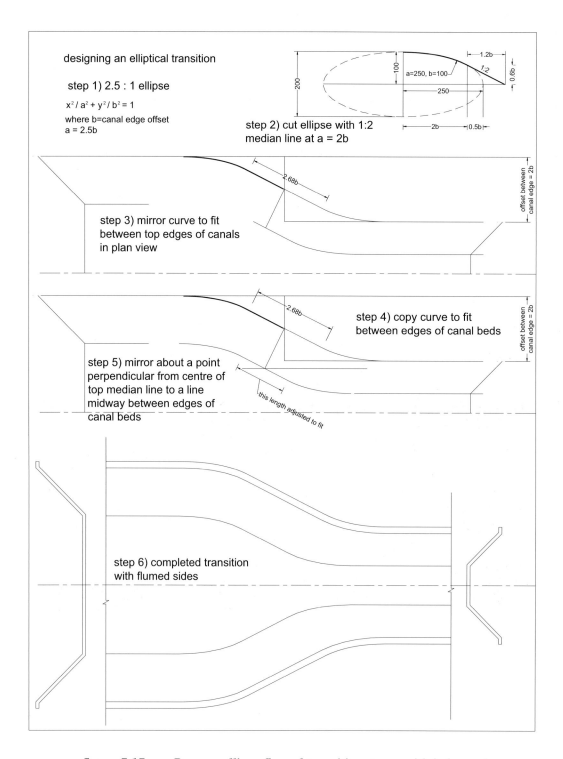

figure 7.17   Reverse ellipse flumed transition, trapezoidal channel

When cross-drainage discharge is small in comparison to the flow in the canal, it may be practicable to use a drainage inlet, thereby directing the drainage flow into the canal.

Culverts

Box culverts or circular pipes can be used. The former are normally used for design discharges of 5 cumecs or greater. Pipes are simpler to construct. They are designed with the aim of passing drainage flows whilst causing the minimum of disturbance to the flow regime that might promote unwanted degradation or aggradation of the stream bed either upstream or downstream of the crossing.

Culverts should be constructed with their exit inverts close to the natural bed level in the stream so as to minimise the risk of their blocking with sediment carried down by the spate flows. The longitudinal slope of the barrel should be high enough to avoid sediment accumulation, and a sloping drop inlet is frequently used to ensure this. Scour protection is provided around the exit of the culvert structures.

The drop-inlet culvert shown in figure 7.18 was designed specifically to encourage the aggradation and stabilisation of deep gullies in an erodible loess soil. The upstream section is sloped steeply at 1:3, so that the inlet was built at an elevation about 10 m above the natural bed level of the gully. After the first rainy season the bed of the gully was completely filled with silt and the gully was effectively stabilised.

figure 7.18   *Drop inlet culvert under construction*

Superpassages

Where natural bed level of the cross-drainage channel is close to the canal design water level, a superpassage may be more practical than an underpass or culvert. This is especially so if there is a steep cross-slope.

The large superpassage shown in figure A7.23 spans the canal with a series of pre-cast and pre-stressed concrete U sections. It is a modular system which can be adapted to different sites by varying the number of channels in parallel. In situations where the occurrence of floods is infrequent they can be designed to carry vehicular traffic also.

Upstream guide bunds are provided to divert the wadi flows into the structure. The arrangement of these will be site-specific. At the downstream end, a stilling basin and scour protection is provided with a combination of gabion mattresses and tipped rock.

figure 7.19   Small superpassage on the Pehur High Level Canal, Pakistan

Aqueducts

Aqueducts have been in use since Roman times as bridges for carrying water. The commonest form of aqueduct is a rectangular channel carried on piers, such as the arch structure in figure 1.1, or the conventional straight beam type of figure A7.24. A variety of construction materials can be used, including masonry, concrete, steel pipe and even timber, but reinforced concrete is now normal for all but the smallest. A rectangular cross section results in an even weight distribution across the piers, is the most economical shape, and is easiest to construct. The top of the channel can be strutted to reduce the wall thickness or closed by a cover slab, to serve as a bridge for people, animals or vehicles.

Siphonic aqueducts are sometimes built. These are a compromise between an inverted siphon, in which the operating head and hence the standard of watertightness is reduced, and an aqueduct, in which the height of piers is reduced. Those shown in figures 7.20 and A7.26 were built in the 1930s in extremely rough terrain in the north-west frontier of present-day Pakistan. They leak.

Conveyance flumes

Conveyance flumes may be simply a change in canal cross-sectional profile from trapezoidal to rectangular in order to traverse a rocky area, or they may be more appropriately described as an aqueduct in timber, plastic, steel, masonry or concrete. Figure A7.28 shows a typical situation requiring a rectangular flume traversing an area restricted between mountain and road.

Canalettes

Canalettes are a continuous form of aqueduct in which a small irrigation channel runs at an elevated level above ground for considerable distances. They are commonly half-round or parabolic in section, and used at tertiary level in the canal system. Typical ones are shown in figures 6.5 and 9.13.

figure 7.20   Siphonic aqueduct, Shamozai Distributary of Upper Swat Canal, Pakistan

Inverted siphons

Inverted siphons are pipelines which carry canal water beneath a river bed or across a depression or valley. They are applicable where the flood regime of the river is not known with certainty, and an above-ground structure may be susceptible to foundation erosion or unquantifiable flood damage. The site conditions and topography will also dictate whether an inverted siphon is cheaper than the alternatives of an aqueduct or a filled embankment with culverts.

The hydraulic requirement for an inverted siphon is an available head, which will determine the siphon diameter or cross-sectional area. Hence the elevation difference between inlet and outlet must be enough to overcome friction loss in the pipe and head loss at inlet and outlet. In this respect an aqueduct may be superior, as its head loss will generally be much less than an equivalent inverted siphon.

Some pitfalls in the design of inverted siphons are:

- Barrel leakage and uncontrolled loss of water.
- Barrel blockage by sediment or debris
- Possible non-acceptance by farmers.
- Surge pressures must be considered in large inverted siphons.

Pipe materials and joints must be watertight at all operating pressures. Concrete pipes may appear to be cheap, but often it is difficult to achieve a sound joint. Small leaks may not appear to be serious, but have a tendency to increase in severity due to the erosive action of water under constant pressure. Unlike a low-pressure distribution pipeline system, which can accept some joint leakage, inverted siphons can fail completely if the joints leak badly.

Welded steel pipe encased in concrete is usually a better alternative. In smaller diameters, fusion-jointed high-density polyethylene (HDPE) is a fail-safe option. In India we recently used HDPE pipe very successfully to rehabilitate concrete inverted siphon pipes up to 300 metres long by internal sleeving.

Reinforced concrete rectangular box culverts can also be used as inverted siphons, but their construction joints will require waterstops and careful construction to avoid leakage.

Sedimentation and consequent blockage of the pipe can be a problem where the discharge is intermittent, and coarse sediment or heavy bed load is present in the water. The design should ensure that under all flow conditions the velocity increases from the canal to the pipe, on the premise that if sediment is carried in the canal either as suspended or bed load, it will also be carried through the pipe. If the slope of the exit end of the pipe is steep, then the velocity will need to be high enough to sweep out any bed load material against the effects of gravity.

As a precaution, a flushing valve can be included at a low point in the pipe, as close as possible to the start of the upward-sloping outlet length. However, for effective flushing this will need to be of a diameter similar to that of the main pipe, in order to generate a velocity in the pipe greater than that which failed to flush through the sediment in the first place. In pipe diameters over about 500 mm, a cheaper stub pipe with a bolted flange can be used instead of a valve, although there are obvious hazards involved in opening it.

Where the outlet end of an inverted siphon is controlled by a gate, **surge pressures** due to gate closure must be taken into account. The effect of surge can be particularly severe if the inverted siphon is long. The 2000 m long Kundal Siphon in figure 7.21 passes 30 cumecs under a static head of 3 metres, but the surge pressures can be more than double this, due to the intermittent closing of the Avio gate at its outlet. Surge was addressed in the design by incorporating a surge tower at the outlet (figure A7.31), with a free surface open to the atmosphere and a weir to discharge surge flows back into the canal.

figure 7.21   Kundal Khwar siphon under construction on the Pehur High Level Canal, Pakistan. At 3.2 m dia and 2 km long, one of the largest ever

On small schemes, farmers tend to be sceptical about inverted siphons, because the water is hidden from view and the pipe always looks too small for the flow. It often occurs in situations where failed structures require rehabilitation. However, once in operation these fears usually evaporate.

Drainage inlets

Where the cross-drainage discharge is small in comparison to the canal flow, a cheap option is to use a drain inlet, which discharges drainage water direct into the canal. The only drawback to this is the likelihood of sediment entering the canal. However, a drain inlet is much cheaper than a superpassage or culvert, and the extra cost of occasional maintenance may be offset by the capital cost saving. Figure A7.38 shows a field drain outlet feeding back into the canal system in Eastern Java.

In the hill country of Sri Lanka it is standard practice to build single-bank canals, in which all cross-drainage flows are directed into the canal. The single bank is on the downslope side, and the upslope side is an irregular contour along the adjacent hillside. Here, however, the rocky terrain and thick jungle mean that sediment influx is not a high risk.

7.9   Overflow Control

Escape structures are necessary in open canal systems in the event of incorrect operation, gate failure, or other emergency. Either because of gates being wrongly operated upstream or downstream, too much water coming in at the headworks, a blockage downstream, or excess rainwater flooding in during the rainy season, the canal will overflow. On a small canal it may not matter where the overflow takes place. In most cases however it is desirable to control it so that it can be safely channelled away into the drainage system without damaging crops or canal banks.

Overflows are controlled by a spillway or waste-weir, often called escapes when used in a canal. Side escapes are located near to or integral with cross-drainage structures, and as close as practicable to potential points of control such as cross regulators or siphon intakes which are liable to cause a backwater effect. Tail escapes are naturally enough located at the tail-end of canals, usually integral with a flushing sluice gate for draining the canal for maintenance.

Escapes are usually some form of weir, which comes into action when the water level exceeds a certain height. This is normally established as the maximum regulated water level, plus a certain amount, typically 150 mm, to accommodate waves and minor fluctuations in level caused by gate operation. Occasionally a siphon spillway might be used. These give a greater discharge per unit width than an uncontrolled weir crest.

Side escapes

Side escapes comprise a long-crested side weir discharging via a channel or chute into a natural drainage channel (figures A7.39 - 41) They are located near to or integral with cross-drainage structures, and as close as practicable to potential points of downstream control liable to cause a backwater effect, such as cross regulators. On upstream-controlled canals they would be located on the upstream side of cross regulators. In a downstream-controlled system, they would be on the downstream side.

Flushing sluices

These are escapes with vertical lift undershot gates, which can be used to drain certain reaches of the canal in an emergency such as a breach, or for maintenance. They can be combined with an overspill weir, as in figure A7.43.

Tail escapes

In an upstream-controlled canal system, tail escapes are required because there is no capability to contain the rejection flow when outlets are closed. They are not required in a canal operated in a mode of downstream control.

Tail escapes are normally an overfall weir incorporating a flushing sluice gate for draining the canal for maintenance.

Siphon spillways

Siphon spillways give a significantly greater discharge per unit width than an uncontrolled weir crest. They are more commonly used on dams, which may have restricted space in which to locate an overfall weir crest. They are automatic in operation, in that the siphonic action is activated when the canal water level rises to a level such that the overfalling jet creates an air seal above the nappe.

figure 7.22   A siphon side escape structure on the Genil-Cabra Canal, Spain

Automatic outlet plugs

There is a Victorian engineering curiosity on the Crinan Canal in Scotland, which can best be described as a giant float-operated bath plug. The Crinan is a navigation canal in a high rainfall area, and is subject to frequent flood inflows which need to be spilled. The outlet is a pipe about 1.5 metre diameter, arranged vertically to drain an inlet chamber which is closed by a plug. The plug is hung from one end of a cast iron beam, pivoted in the centre, at the other end of which hangs a bucket with a hole in it. The bucket holds about four tonnes of water when full, and is fed by a small pipe coming from a small and separate inlet weir on the canal. When the inflow to the bucket exceeds the outflow from the hole in the bucket, the bucket fills, causing the plug to lift, and the water to flow out through the main pipe; see figure A7.45.

7.10 Field Outlets

Outlets into smallholder farms or individual fields can take many forms, depending on the mode of operation of the scheme and the form of construction of the tertiary canal. There are several ways of getting water out of a canal, including siphons, checks, sluice turnouts, constant discharge modules and moveable weirs. Except in the case of continuous flow schemes an effective outlet needs to be easily opened and easily closed, without leakage.

Open cuts

In its simplest form, with tertiary canals constructed of unlined earth, the outlet is merely a cut in the canal bank, which is re-sealed by the farmer after each irrigation (e.g. figure A1.7.) Where labour is cheap, this is standard practice on heavy clay and silt soils, but is not appropriate for sandy soils as it leads to rapid deterioration of the canal banks.

Spiles

Spiles are small pipes through the bank which are closed by clods of earth and do not cause repeated damage to the canal banks. They are appropriate for unlined tertiary canals. They are often used as unauthorised outlets when farmers wish to steal water from larger canals. Figure 3.2 shows them employed in Javanese paddy terraces.

Precast nuccas

'Pucca Nuccas' are precast concrete outlets closed by a circular disc, used on the warabandi schemes of Pakistan and northern India. They are applicable where all the flow is to be diverted through the outlet on a rotation schedule. They are cheap and effective, and although not completely watertight they can be made so by sealing the periphery with mud. These are no more than a formalised version of the age-old breach in the bank to temporarily divert all the flow from a watercourse into a field or farm channel. A typical one is shown in figure 6.4.

Undershot gates

The turnouts shown in figure A7.46 are cheap undershot steel slide gates which are ubiquitous in many countries. However, unless fabricated from heavy (i.e. 4 mm thick) steel plate they have numerous problems in operation, not least because they always leak when closed. They are easily damaged and difficult to seal, especially when debris gets stuck in the grooves. In some countries they are an easy target for thieves who steal the steel.

Overflow slots

A less common variant of the sluice gate is the moveable overshot weir, two versions of which are shown in figures A7.1 and A7.2. These have some advantages over the undershot type, in that the weir can more easily be used for measuring the discharge, and leakage is less likely because water pressure on the gate seals is less. However, being a weir its discharge is more susceptible to water-level fluctuations in the parent canal than is an orifice.

A better turnout system for small discharges is installed in the parabolic canals at Banavasi, Karnataka State, India (figure A1.4.) Here, the turnout is a wide slot that is above the normal water level in the canal, and therefore cannot leak under normal flow conditions. To get water out, a parabolic check gate in the canal is closed, thus raising the canal water level above the sill level of the slot.

Plastic siphons

This under-rated means of getting water out of a small canal is very effective because it does not leak when not in use, and it gets water directly where it is needed in the furrow. It is commonly used on large farms in the USA and on large plantations and sugar estates all over the world. The siphons in figure 7.23 are in the Columbia Basin, USA. They are 1.5 inch diameter polythene tubes and feed water into furrows about a kilometre long. Their capital and operating costs are significantly cheaper than sprinklers.

It is unfortunate that siphons seem to be unpopular with small farmers in the developing world. Attempts to introduce them into India, Sri Lanka and Indonesia all met with limited success. The reasons are rooted in lethargy and ignorance. Lethargy, because it is much easier to open a sluice gate and leave it alone all day, whereas siphons have to be looked after all the time. Ignorance, because the principles of siphon operation are not obvious at first sight, and both farmers and extension workers tend to be wary of unfamiliar concepts. Canal designers intending siphons to be used need to build-in an adequate operating head to the canal.

Designers intending siphons to be used not only need to build in an adequate operating head to the canal, but also to plan a demonstration programme so that all farmers understand their principles of use.

figure 7.23   ***Plastic siphons with long-line furrows, Columbia Basin, USA***

Discharge	Internal diameter mm			
l/s	25	40	50	75
0.25	0.05			
0.50	0.19	0.02		
1.00	0.73	0.09	0.04	
1.50	1.58	0.20	0.08	
2.00		0.36	0.14	0.02
2.50		0.55	0.21	0.04
3.00		0.78	0.30	0.05
3.50		1.06	0.41	0.07
4.00		1.37	0.53	0.10
4.50			0.67	0.12
5.00			0.82	0.15
5.50			0.99	0.17

table 7.1   ***Head loss (m) in plastic siphons***

7.11 Measurement Structures

Flow measurement is frequently considered crucial to the operation of an irrigation scheme. A favoured theme of expatriate consultants is the presupposition that, in order to successfully operate and manage the scheme, it is necessary to know how much water is flowing in the canals at any time. However, in the context of smallholder schemes in the developing world, I have severe doubts about this, for several reasons:

- Firstly, in practice, few schemes have the management resources to maintain flow records or carry out regular measurements.

- Secondly, with proportional flow or downstream control, there is no need for measurement as an *operational* tool. It is far better to design a semi-automatic system which requires no measurement, rather than a supply-scheduled system which is reliant on measurement procedures that cannot be met in practice.

- Thirdly, a preoccupation with constructing and then calibrating measurement structures detracts from more important aspects of the scheme design, such as ensuring that the farmers can utilise the scheme in an effective way (automation being one such recourse.)

Nevertheless, there are clear advantages in designing-in some measurement capability into structures such as head regulators, in which it may be necessary to know the discharge at some point in the project's life. A measuring structure (or more precisely, a control structure which can also be used for measurement when necessary) should be installed at the head of the canal system and often at several points downstream, such as at the offtake of branch and secondary canals.

There are several types of measurement device:

- Weirs;
- Flumes;
- Meters;
- Stage-discharge measurement with rating curves;
- Ultrasonics.

Weirs

Most weirs need more operating head than flumes. However, they might also double as cross regulators, turnouts or as an integral part of a stilling basin. Discharge is determined by measuring the depth of water upstream (and in the case of submerged weirs, downstream also) above the weir crest. There are numerous shapes and configurations in common use.

Sharp-edged weirs are used for accurate flow measurements, either rectangular or trapezoidal (Cipoletti) in profile. Vee-notch weirs (figure A7.47) are useful for measuring small discharges. Broad-crested and Crump weirs are less sensitive but more robust and resistant to damage. Replogle weirs are a version of broad-crested that are especially cheap and simple to make in existing canals. Round-nosed and ogee weirs have a high discharge coefficient, so are not as useful for measurement. Moveable weirs are accurate, but their adjustment varies the flow conditions on their upstream side, so they are best used to measure small offtaking flows from a much larger parent channel.

Crump weirs

Flow measurement structures can be incorporated in the design of head regulators in the form of Crump weirs. This is a triangular section structure with slopes of 1V:2H upstream and 1V:5H downstream. It is a low-head structure, which can operate at up to 80 per cent submergence without affecting its accuracy. It is

| **Crump weir discharge:-** |
| $Q = 1.98\,L\,H^{1.5}$ |
| For downstream head over crest h_2 not greater than 80 per cent of upstream head H |

not susceptible to blockage by floating debris or bed material, and does not require a large stilling basin as the hydraulic jump is formed on the glacis in a fixed location. Crump weirs are also incorporated into the design of proportional dividers described in section 7.4.

Replogle weirs

This is a trapezoidal section broad-crested weir, which is simple to construct, and not susceptible to obstruction by floating or bed debris.

figure 7.24   *A Replogle Weir at Patterson Irrigation Scheme, California*

Vee-notch weirs

The vee-notch is the standard structure for small discharges of a few litres/second, as used in research or for situations requiring accurate measurement of small flows, such as tubewells (figure A7.47). The crest is a sharp bevelled-edged steel plate with a 90 degree notch.

Vee-notch Weir discharge $Q = 0.0147 \, H^{2.48}$ litres/sec

H = upstream head in cm above centre of vee

Flumes

Flumes are a good type of measuring structure because they are not easily blocked with debris and they do not need much head to operate. They are a constriction in the canal, either at the sides (a throated flume) or in the bed (a sill) that causes the water to accelerate through its critical flow state. The height of water level upstream is measured, and this is related to the discharge which can be read off rating curves. Flumes which are drowned or in which flow does not pass through critical need a downstream water level to be measured also.

There are several common types including cut-throat, Parshall, venturi, H flumes and WSC flumes.

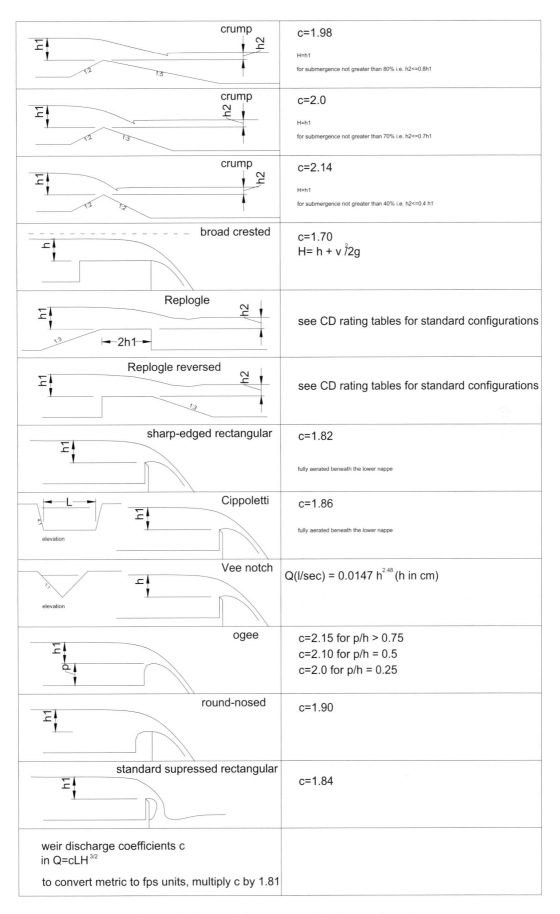

crump	c=1.98 H=h1 for submergence not greater than 80% i.e. h2<=0.8h1
crump	c=2.0 H=h1 for submergence not greater than 70% i.e. h2<=0.7h1
crump	c=2.14 H=h1 for submergence not greater than 40% i.e. h2<=0.4 h1
broad crested	c=1.70 $H = h + v^2/2g$
Replogle	see CD rating tables for standard configurations
Replogle reversed	see CD rating tables for standard configurations
sharp-edged rectangular	c=1.82 fully aerated beneath the lower nappe
Cippoletti	c=1.86 fully aerated beneath the lower nappe
Vee notch	$Q(l/sec) = 0.0147\, h^{2.48}$ (h in cm)
ogee	c=2.15 for p/h > 0.75 c=2.10 for p/h = 0.5 c=2.0 for p/h = 0.25
round-nosed	c=1.90
standard supressed rectangular	c=1.84
weir discharge coefficients c in $Q=cLH^{3/2}$ to convert metric to fps units, multiply c by 1.81	

figure 7.25   Weir types and discharge formulae

Cut-throat flumes

This is a simple but effective structure which is useful for indicating discharge in small canals. It is a low-head, critical flow structure and is not susceptible to blockage by debris. It is suitable for precasting in concrete and can also be prefabricated in steel and backfilled with concrete, as in figure 7.3 and A7.48. It is less complicated than a Parshall flume and easier and cheaper to construct.

The discharge coefficient varies with throat width, roughness of construction material and overall length of flume. Since this design has been the subject of considerably less research than the Parshall flume, the figures given below are open to some debate, and are gleaned from several sources based on a limited amount of laboratory research. Thus the accuracy of the inferred flow rate from these figures may be questionable. However, if accurate and repeated flow measurement is required for a particular project, then a small number of standard flumes should be designed and then calibrated.

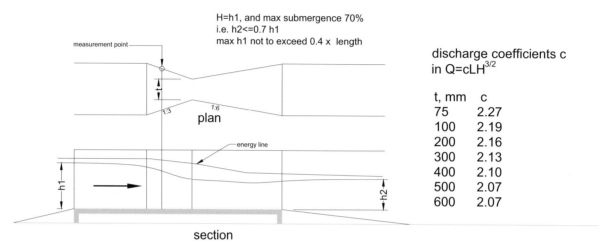

discharge coefficients c
in $Q=cLH^{3/2}$

t, mm	c
75	2.27
100	2.19
200	2.16
300	2.13
400	2.10
500	2.07
600	2.07

figure 7.26   design for small cut-throat flumes

Parshall flumes

Parshalls are a common research tool but less often used in projects as they are more complicated than other structures. Prefabrication of essential dimensions in steel, with infill concrete, is the best way of constructing a small permanent flume.

Meters

Meters are measuring instruments that determine flow velocity or volume over time, usually by counting the revolutions of a rotating propeller. The Australian Dethridge Wheel is an example of a fixed meter that measures volume. The Italian Acquacard device described in Chapter 3 uses a small vaned meter with an electronic counter. And there are numerous portable meters that can be used anywhere. But the appropriateness of a meter has to assessed in conjunction with the management system and policy of water charging in operation. Usually for smallholder farmers on an open channel distribution system, metering is completely impractical.

Dall tubes are used in larger pipelines and pump stations, and measure the drop in pressure across a slight constriction in the pipe. Magnetic flow (Magflow) meters measure the voltage generated between two sensors at opposite ends of a diameter in a pipe, by the fluid passing through a magnetic field. These devices are more expensive than simple propeller meters, but are unaffected by sediment or debris in the water.

Ultrasonics

Where there is no scope for head loss, alternative means of measurement include ultrasonic sensors and stage-discharge calibration, in which water level is measured and the discharge inferred from it. Ultrasonic devices rely on a bank of pulsed beams of ultrasound directed diagonally across a channel, detected at the other side by transducers, and the data processed to integrate the velocity and area of flow to give the total discharge. They do not require a regular channel section, although the integration computations are more accurate if the channel is a regular profile. They do not involve any head loss. They are sometimes installed at the head of canal systems which take water from a multipurpose reservoir, for which accurate estimates of water diversions are required.

Stage-discharge measurement

Correlating observed water level to volume rate of flow is a common method of gauging rivers, and can be used in canals where suitable structures do not exist. However, the process of establishing a rating curve is dependent on repeated measurements of discharge using a flowmeter, and accuracy is usually no better than 10 per cent. Local changes in canal conditions, such as siltation or weed growth, can further disrupt measurements.

7.12 Sediment control

Sediment in the form of silt and sand can be present in large quantities in some canals, leading to partial blockage of canals and hindering the operation of structures. Silt is usually carried in suspension and can be difficult to remove from canal water. Sand is normally transported as bed load in a canal, even though it may have been in suspension in the faster-flowing river from which the canal offtakes. The Gezira scheme has been crippled by excessive silt deposits, thanks partly to deforestation in Ethiopia a thousand miles upriver. In the Gezira, so much silt has accumulated in the canals that there is nowhere to put it even when it is dug out, so the banks get built up as in figure 4.6. The land is actually rising with all the silt contained in the irrigation water. The remarkable example of figure 7.29, showing cattle grazing on what once was the water surface of a major canal, indicates the difficulty of dealing with heavy siltation.

Sources of sediment

The sources of sediment in the canal system can be:

- Suspended sediment in the inflowing water from a river source.
- Washload from surface drainage inflow.
- Dispersive soil from the banks of earth canals.
- Wind-blown sand and dust.

Natural sediment concentrations in the Nile and the Indus can be as high as 6000 mg/l for the flood season in July/August and 2000 mg/l as an annual average.[1] These concentrations are responsible for the severe sedimentation problems in the Gezira and Rahad Projects in the Sudan.

Sediment load in the Nile and many other rivers is seasonal, with the heaviest concentrations during the flood season. This may or may not coincide with peak irrigation seasons.

Sediment and debris washed into canals from drainage inlets is usually a localised problem that can be addressed by regular maintenance.

Dispersive soils are a potentially disastrous condition if canals are built without lining. Gypsiferous soils are notorious for this; I was taken to see a canal in such soils in Upper Burma, and would present a photo of it were it for the fact that was literally nothing to see, because the banks had disappeared completely. The Nubian mudstones in northern Sudan give rise to extremely weak and structureless silt soils, which resulted in the collapsed canal of figure A7.51.

---

1 Concentrations in the Yellow River in China can be 100 times this, so much that normal rules of hydraulic analysis do not apply.

Wind-blown sand can be a major hazard in some areas. The canal in figure 7.27 runs through an active dunefield in the Nubian Desert. The only encouraging aspect of dealing with dunes is that their movement is fairly predictable within localised dunefields, and the dunefields themselves do not move. Dunefields are created by the interaction of steady air flows with topographical features such as hills and low escarpments, which may not be obvious from the ground. In fact wind-blown sand behaves in much the same way as water-borne sediment. Where a steady air flow meets an obstruction such as a rock outcrop, macro turbulent vortices are shed on the downstream side, which result in sand dropping out in zones of low air velocity. The creation of a dune in turn creates turbulence downstream of it, which leads to the creation of another dune, and so on. Wind breaks and canal embankments can have the same effect, with obvious consequent risks of canal sedimentation.

figure 7.27   ***A canal blocked by Barchan sand dunes in the Nubian Desert. These dunes move horizontally at a rate of about 20 m per year***

Modes of operation and design options

The first approach in design is to exclude sediment from entering the canals as far as practicable. The next step is to ensure that any sediment that does enter will be carried through to the fields in suspension. Sediment that is neither excluded nor flushed through the system will at some time need to be physically removed. The management philosophy designed into the scheme at the outset will have to consider the risks of sedimentation, and allow for its treatment.

There will be places in the system where even suspended material will settle out. These will normally be in canal reaches which are designed to be ponded for part of the time, such as upstream of cross regulators or in downstream-controlled level-top canals. Some of this will be flushed through when normal canal velocities are resumed, but there will need to be a maintenance programme for desilting in those reaches which are prone to silting.

There are several approaches to dealing with potential siltation problems:

- Improve watershed management to restrict soil erosion.

- Take water from a reservoir in which sediment is already trapped or flushed through.
- Design canals as alluvial regime channels, if the system offtakes from a seasonal river.
- Continuous flushing through the canal system.
- Continuous flushing from vortex tubes or other silt ejector structures.
- Sediment exclusion at the headworks through a skimming weir or bottom vanes.
- Physical excavation of sediment from the canal by dredging.
- Physical excavation of sediment from silt-trapping structures.
- Periodic flushing of sediment from silt-trapping structures.
- Complete prevention of wind-blown sand ingress, by shelterbelts or windbreaks.
- Culverting the canal to avoid wind-blown sand.
- Physical removal of sand dunes where the canal runs through dunefields.

Watershed management

Watershed management is a favourite subject for aid agencies, and avoiding the problem by preventing soil erosion in the upper catchments is a glamorous objective, and on the face of it, an entirely sensible one. However, it is often not practicable. The catchments may be too large, even situated in more than one country. Burgeoning populations with goats and other grazing animals can prove an insurmountable social barrier. Uncontrolled or even selective logging can devastate tropical rainforest soils. There are potential problems enough, before even considering climate change and global warming.

Reservoir trapping

Large reservoirs trap sediment, sometimes enabling offtaking canals to be free of sediment at least during their early life. The trap efficiency is a measure of the volume of the reservoir related to the annual volume of runoff entering the reservoir. Even large dams may not provide much scope for de-silting the water enough to avoid problems in the irrigation system. Pakistan's Tarbela has a trap efficiency of only 18 per cent, a figure that is reducing rapidly with steady deposition of sediment in its reservoir.[2] Warsak, also in Pakistan was completely silted after 25 years of operation. On the Blue Nile, Roseires Dam is operated to flush through the bulk of the sediment load over a short period in the rainy season; Merowe will operate in a similar way when it is completed in 2008. Sennar operates as a diversion dam for the Gezira Scheme and its reservoir is too small to prevent damaging amounts of silt entering the system. On the Nile only Aswan has a trap efficiency in excess of 100 per cent, and the serious side effects of this are known all too well.[3] The Yeleru Dam in Andhra Pradesh traps all sediment, but the canals are choked with weed which thrives in the clear water.

So for reservoirs that will silt up in the foreseeable future, it is necessary to plan ahead for the time when sediment will inevitably enter the canal system. In the design of the Pehur High Level Canal, an area was set aside for desilting basins to be constructed later, when Tarbela ceases to supply crystal clear water.

Regime canals

Some regions such as northern India, Pakistan and much of China have a long history of dealing with silt-laden canals, and silt is not only inevitable, but is to be regarded as a natural environmental phenomenon. Here it is standard practice to design the larger unlined canals as regime channels, in which sediment is carried through the system and deposited on the fields, taking account of seasonal variations in sediment concentration. The basic principle is that over a full year in which natural river flows and their sediment loads will vary with the seasons, a channel may sometimes aggrade and sometimes erode, but will over time remain stable. The design process emulates hydraulic conditions for which channel slope and width will be predictable and stable. The design of regime canals is discussed in Chapter 10.

---

2 *Tarbela Dam was commissioned in 1978 and it is estimated that by 2020 the delta of sediment at the head of the reservoir will have arrived at the turbine intakes.*

3 *Viz.: destruction of the delta prawn fisheries, degradation of the floodplain soils downstream, prevention of irrigation from the lake due to fluctuating water levels, etc.*

Continuous flushing

Running suspended sediment through the canal system is usually possible to some extent. Bedload however will usually settle out and require removal. The delineation between suspended and bedload, broadly between silt and sand, depends on mean and boundary layer velocities, channel roughness, turbulence and sediment concentration.

The Gezira system was originally designed for overnight storage in its minor canals, which were oversized and level for that purpose. Predictably, they have silted up in a process that has contributed greatly to the general demise of the scheme. However, with the general breakdown in control, the whole canal system operates on continuous flow, in which silt is flushed through the system and onto the fields. The net result is that the land is rising with the addition of silt in the water, and extensive remodelling will eventually become necessary in order to continue irrigating.

A criticism frequently levelled against piped distribution systems is that silty water will block the pipes. Bed load might, but suspended silt will not. Even if it settles in a pipe when the flow is shut down, it will re-transport when the flow is resumed.

Even with sprinklers, the trend is toward low operating pressures with large nozzle diameters that will pass sediment without blocking. Only in drip systems is it necessary to remove silt completely.

Silt ejectors and vortex tubes

There are several designs of silt ejector structure, located in the canal headreach such that sediment can be periodically flushed back into the river. Figure A7.52 shows a skimmer weir, designed to abstract surface flows only, whilst ejecting sediment-laden water close to the bed.

Sediment excluders can follow a variety of engineering designs usually aimed at separating sediment laden water near the bed from cleaner layers closer to the surface. River intake works often employ curved vanes to generate secondary cross-currents in the approach flow that divert bed load away from the canal intake. In hill torrents known to have very mobile beds, sloping sill screens are often used to exclude coarse sediment such as gravel and boulders. Tunnel ejectors are common on large Indian canals.

Vortex tubes are a recent development in silt ejection structures. Extensive development work has been carried out by Hydraulics Research in Wallingford, who have released some useful design software (Ref.18.) Where some excess flow can be made available in a canal headreach, a vortex tube structure may offer considerable potential for reduction of bed load. The structure comprises one or several pipes in the canal bed perpendicular[4] to the direction of flow. The pipes have an open longitudinal slot in the crown and flow in each pipe is controlled by a sluice gate at its discharge end. When the pipes are in operation, the boundary layer flowing along the bed is skimmed off into the tube, carrying bedload sediment with it. The flow within the tube takes the form of a helical vortex, which maintains the sediment in suspension until it is discharged at the end of the pipe. The pipes can be closed when bed loads are light. Even if the pipe gets full of sediment it will still operate once the sluice is opened. It takes typically 15-20 per cent of the incoming canal discharge to waste through the pipes. Since in most field situations maximum sediment coincides with floods or high river flows, there is usually excess water when silt ejection is needed most. Vortex tubes are cheaper and less susceptible to blockage than tunnel ejectors.

The structure is neat and simple. It needs an outfall channel that will not get clogged with the ejected silt, but otherwise is simple to operate and maintain. One was included in rehabilitation of the Upper Swat Canal in Pakistan, where two in-line hydropower stations were constantly under repair through sediment damage to the turbines. Although by design only about 50 per cent of the sediment load was

---

4 *Early designs had the tubes oriented at an angle of 45 degrees to the flow, but this arrangement has since been found to be unnecessary.*

removed, this was the coarse fraction comprising fine sand upwards, which was not only the prime cause of damage to the turbine blades but also the principal material causing siltation of the canals.

The individual tubes are arranged so that each slot covers a separate portion of the canal bed width. The length of slot depends on the required pipe discharge and chosen diameter.

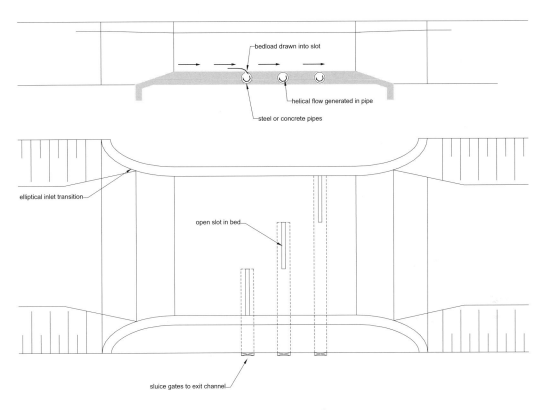

figure 7.28   A vortex tube structure

Settlement reaches and settlement basins

Whilst conventional canals will be designed for a certain sediment transport capability, one approach is to induce siltation by widening a canal reach so that velocity is reduced and sediment settles out in a specific location, which can be flushed out or excavated periodically as required.

On the Merowe Left Bank Conveyor in Sudan, a logical location for a trapping structure is downstream of the Wadi Howar dunefield, which sustains the worst conditions for wind-blown sand. The canal reaches most at risk from sedimentation are those immediately downstream, which are designed to be downstream-controlled and hence susceptible to siltation. Since it is not possible to precisely estimate the amount of sediment to be removed, a trapping structure offers a fail-safe solution, in which the frequency of cleaning can easily be matched to the actual rate of accumulation of sediment. A twin basin will permit one half to be closed off in order to excavate accumulated sediment, whilst maintaining the canal flow in the other half.

An outline design is shown in figure A7.62. This is a twin basin structure with each basin designed to take the full discharge at that location of 120 cumecs. Velocity is designed to reduce from 0.9 m/s to 0.4 m/s over a length of 100 m, and mean depth of 4 m. This is based on the settling velocities for sand sizes given in table 7.2, and is designed to settle out all sand of 0.2 mm diameter and above.

The structure is designed so that either basin can be closed off at both ends by bulkhead gates, for the purpose of excavating accumulated sediment. The floor slopes toward a sluiced drainage culvert which is provided in both basins. The approach and exit geometry is streamlined to minimise turbulence and reduce the velocity steadily. Guide walls are included downstream of the inlet gates for this purpose.

sand size mm	fall velocity *w* cm/s	Fall depth m	Approach velocity m/s	fall length *L*, m
0.10	0.7	4.0	0.40	274
0.20	2.2	4.0	0.40	87
0.30	3.2	4.0	0.40	58
0.40	4.3	4.0	0.40	44
0.50	5.4	4.0	0.40	35
0.60	6.5	4.0	0.40	29
0.70	7.3	4.0	0.40	26
0.80	8.1	4.0	0.40	23
1.00	9.4	4.0	0.40	20

table 7.2   Settling rates of sediment

The sophisticated hopper-bottomed structure of figure 7.30 was designed to be flushed out through bottom sluices. For this method to be effective, the sluice openings need to be large in relation to the channel cross section, in order to generate effective scouring velocities upstream of the sluice. Adequate head is necessary between the water level in the structure and the receiving water, which in this case is the river downstream of the canal offtake.

Dredging

In a running maintenance programme, the aim should be to clean all silt-prone reaches once a year, in order to prevent the establishment of weed and marginal vegetation such as reeds. Vegetation will trap more silt and progressively restrict the canal flow, and the roots of reeds will severely damage canal lining. Conventional earthmoving equipment as in figure A7.53 should not be used if the canal is lined, as it will inevitably damage the lining. For lined canals the most appropriate is a submersible jet pump, operated either from a floating pontoon or from a bankside power unit through an umbilical cable.

Prevention of sand ingress

Wind breaks

Where canals are oriented across the prevailing wind direction (i.e. east-west rather than north-south), wind breaks should be included in the design. These could be fast-growing trees such as Eucalyptus or Neem (*Azadaricta indica*), or Mesquite. The latter requires little water but needs to be tightly controlled as it is invasive of agricultural land and difficult to eradicate from places where it is not wanted. It has been used effectively for stabilisation of mobile sand dunes in Sudan. However, it can also result in a massive and unmanageable build-up of dune sand if the source of the moving sand is not given attention. This has happened on the right bank of the Nile, south of Old Dongola. In places the dunes have completely smothered the riverine terraces. Although it has several other uses such as charcoal and camel fodder, Mesquite has been officially banned in Sudan due to its almost uncontrollable growth along unlined irrigation canals.

Shelter belts have been successfully established in existing projects, but their effectiveness in preventing sand ingress into canals is open to question. They are also difficult to establish in a new project, as they require regular irrigation from the outset, and often before the water distribution system is operational, and can take several years to grow to an effective height. Whilst considered a desirable addition to a scheme, they cannot be regarded as a fail-safe solution.

Culverting

Culverting a large canal is an expensive means of excluding wind-blown sand, even if done only in areas of active dunefields. It may be the only practical recourse in some places. In small canals it is proportionately cheaper. It is good practice to steepen the gradient of a culverted canal reach, to increase the velocity and ensure that sediment does not tend to settle within the culvert. Access points for emergency cleaning should be provided at suitable intervals.

Removal of dunes

Certain areas of the Merowe project contain barchan dunes up to 15 m high, which tend to move south at rates of up to 20 m per year (estimated from air photos taken over a period of 40 years.) In general these areas will be avoided in designing canal layouts, but there are a few locations in which it is not practicable to avert the canal route. Physical removal of sand dunes during construction of the canal is a practical proposition, and was recommended for specific locations. In the Wadi Howar area the dunes are widely spaced, and it is possible to predict where a dune will travel over the next 20-50 years. A typical dune is estimated to have a volume of around 50,000 m^3 of sand, which could be easily transported to the down-wind side of the canal route during construction. It may be necessary to move several such dunes during construction, and the process may need to be repeated after 50 years, so a bridge over the canal should be provided in the vicinity.

figure 7.29   ***Silt accumulation in a Gezira main canal, Managil extension, Sudan***

figure 7.30   ***Desilting basins, Dam Jati, east Java***

References and further reading for chapter 7

(1) Withers, B. and Vipond, S., 'Irrigation Design & Practice', Batsford, 1974.

(2) Bos, M. G., Replogle, J. and Clemmens, A., Flow measuring flumes for open channel systems,Wiley 1984 and ASAE 1991.

(3) USBR, 'Canals & Related Structures', Design Standards # 3, USBR 1967.

(4) USBR, 'Hydraulic Design of Stilling Basins and Energy Dissipators', Engineering monograph #25, USBR 1963.

(5) US Federal Highway Administration, 'Hydraulic Design of Energy Dissipators for Culverts and Channels', Hydraulic engineering circular #14, 1975.

(6) Lewin, J., 'Hydraulic Gates and Valves', Thomas Telford, 1995.

(7) ILRI, Wageningen, 'Discharge Measurement Structures', 1975.

(8) Chow, V. T., 'Open Channel Hydraulics', McGraw-Hill 1959.

(9) FAO Irrigation and Drainage Paper 26, 'Small Hydraulic Structures', FAO 1975.

(10) British Standards Institution, BS 3680:part 4 Weirs and Flumes: 1981.

(11) Buckley, R. B., Irrigation Handbook, Spon, 1928.

(12) Sharma, K. R., Irrigation Engineering', India Printers, 1944.

(13) Sharma, S. K., 'Design of Irrigation Structures', Chand, 1988.

(14) Khushalani, K. B. and M., 'Irrigation Practice and Design', Oxford & IBH, 1990.

(15) Sehgal, P P, 'Design of Irrigation Structures', Khanna, 1982.

(16) Satyanaraya Murty, C., 'Design of Minor Irrigation and Canal Structures', Wiley, 1991.

(17) Henderson F M, 'OpenChannel Flow', Macmillan, 1966.

(18) Hydraulics Research, Wallingford, DACSE software for vortex tube design, 1993.

CHAPTER 8 LOW-PRESSURE PIPELINES

Low-pressure pipelines are beginning to find favour as a means of flexible water delivery in smallholder irrigation schemes. However, when used for irrigation in this way their principles remain largely misunderstood by practising engineers more accustomed to designing water supply schemes in which the optimum pipe diameter is the smallest possible, commensurate with pumping costs or available head. Their acceptance is also restrained by their apparent increased cost in comparison to traditional open canals. This chapter explains the reasoning behind the use of pipelines and seeks to dispel some of the misconceptions surrounding their use.

8.1 Principles of Pipeline Operation

The advantage of low pressure

Flow in a pipeline depends on the available head and the cross-sectional area or diameter of the pipe. The available head at the start of the pipeline is dissipated in wall friction and velocity head, with additional losses due to turbulence at bends and fittings. A downstream-controlled irrigation distribution pipeline has numerous outlets which when operated on a flexible demand schedule will be opened and closed almost at random. The discharge up to any point in the pipe will fluctuate according to the number of outlets which are open at any one time. Therefore the total friction loss up to that point and hence the available head at that point will also vary.

Problems arise with small diameter closed pipelines due to the friction losses along the pipe which vary with discharge. The discharge through an outlet is dependent on the available head at that point in the pipeline, and that in turn depends on whether any other farmers are taking water at the same time. In a small diameter, high-pressure pipe the individual outflow rates are very sensitive to other outflows from the same pipe. The higher the operating pressure, the more sensitive is the outflow rate to fluctuations in friction loss along the pipe.

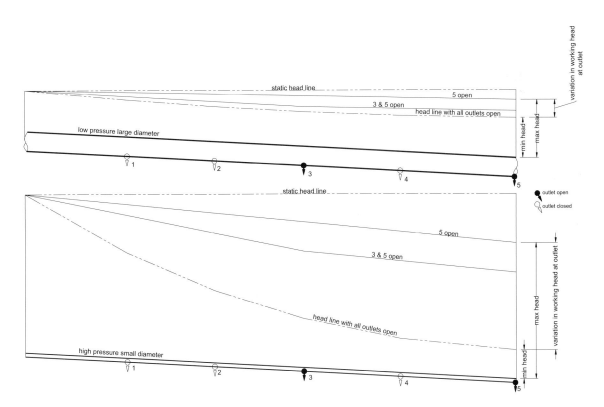

figure 8.1   ***A low-pressure pipeline offers near-constant outlet pressures***

These problems are overcome with a larger diameter pipe operating at low pressure. There is more storage in the pipe, but more importantly, the working pressure and hence flow rate from each outlet is not greatly affected by operation of other outlets. This is shown in figure 8.1. Pressure can be controlled by opening the pipe to the atmosphere, or by pressure-reducing valves. With intermediate float valves the system is made *semi-closed*. Figure 8.2 shows how a pressure-reducing valve simplifies the operation of a pipeline system.

Open and closed pipelines

Pressure in the pipeline is influenced by the arrangement of the pipe system. If the pipeline is *closed*, there are no openings to the atmosphere between the inlet and the outlet. Pressure is the result of only the available head in the reservoir at the pipe inlet, and any friction losses caused by water flowing through the pipe itself. A typical domestic water supply system is closed, between the storage tank in the roof and the taps at the sink. The pressure at the downstairs kitchen sink is greater than that at the upstairs bathroom, because the head difference between storage tank and outlet is greater. The water flow in the system is controlled by opening or closing the valves at the outlets, and is therefore under *downstream control*. However,, if there are many outlets in the same system, the operating pressure can fluctuate a great deal due to varying friction losses under varying discharges. This in turn leads to fluctuating outflows at each outlet as adjacent outlets are turned on and off, which is the main drawback in using closed pipelines for irrigation distribution.[1]

An *open* pipeline has at frequent intervals control points which are open to the atmosphere. It is similar to an open canal with weir-type drop structures at the end of each reach which afford *upstream control*. The pressure in each reach of the pipe is controlled by the open structure, which may be an overflow cistern or weir box, at the downstream end of the reach. When used for irrigation distribution its main advantage over a canal lies, like any pipeline, in the reduction in land take and maintenance; but its operational modes are subject to the same constraints as an upstream-controlled canal. See figure 8.3 for an extreme example of an open pipeline. This is a standpipe near the end of a long pipe run, fed from the California Aqueduct, into which supplementary supplies are pumped from groundwater. The standpipe is a crude means of limiting the internal pressure. In this case we may deduce that the design pressure of the Westlands buried pipe is about 15 m, or 1.5 bars.

A *semi-closed* pipeline system offers distinct advantages in operation. The pipeline is normally closed to the atmosphere but pressure is controlled at intervals so that a predetermined head in any reach is never exceeded, no matter what the discharge. The control structures are normally float operated Harris valves, which give *passive automation* and allow *downstream control*. A Harris valve is shown in figure 8.8, with two Dutch designs which perform a similar function. With this degree of control, the advantages of a low-pressure system can be brought into play. These are:

- A low-pressure and hence cheap specification only is required for pipe and joints;
- Pressure fluctuations and hence outlet discharges can be kept within close limits.

The pipeline needs to be larger in diameter than a closed pipeline delivering the same flow at a remote outlet. However, when applied in the context of multiple outlets required to operate on a *flexible delivery* schedule, the operational advantages of a semi-closed pipeline are clear.

It is possible to achieve the same degree of control using active automation with electrically or hydraulically activated valves, and the trend in California is to replace ageing Harris valves with computer-controlled gate valves.

---

1 *A true story is related by Stuart Styles of California, who on asking a farmer on Pima-Maricopa Project, Arizona, why he kept turning his pipeline outlet valve on and off in rapid sequence, got the reply '....haven't you heard of Surge Irrigation???'*

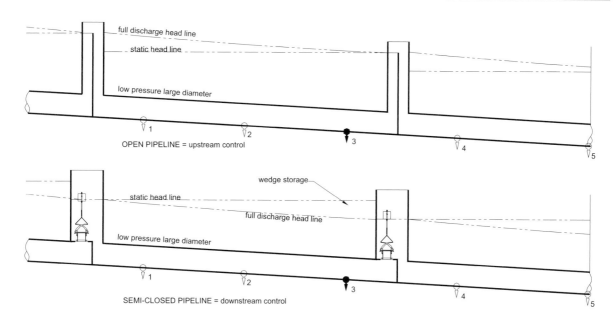

OPEN PIPELINE = upstream control

SEMI-CLOSED PIPELINE = downstream control

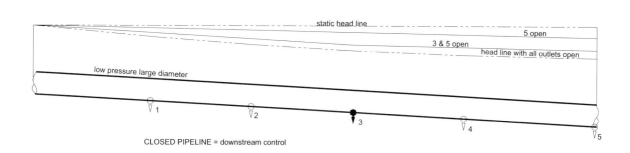

CLOSED PIPELINE = downstream control

figure 8.2   Open, semi-closed and closed pipelines

figure 8.3   An extreme example of an open pipeline, Westlands, California. Maximum static head is at the level of the top of the standpipes (both pictures.) The system is filled by pumping from the California Aqueduct (right)

figure 8.4   Young date palms surface irrigated by orchard valve outlets on a semi-closed pipeline, Coachella, California

Flexibility, and guarantees for success

The guiding principles of flexible scheduling are that:

- Farmers will not use more water than necessary, *provided the supply is fully guaranteed.*
- Farmers will turn off the valves when they have had enough water, as it is not in the farmers' interest to allow waterlogging or wastage.
- Disputes between farmers all wishing to irrigate at the same time will be avoided.
- Disputes over apportionment of water charges within the farm group will be avoided.

Detractors of flexibility invariably cite examples where these conditions have not been met and the scheme has failed or reverted to traditional supply scheduling. However, once the correct infrastructure is in place and an appropriate organisational system is established success should follow. However,, it may take several years of close monitoring and supervision in order to break away from existing traditional concepts of management. The potential problems can be properly assessed and avoided *only* if adequate supervision and assistance is provided from the design stage through to construction and the first years of operation. This must include farmer involvement during in-field design, establishing unit fields, farm groups and the administrative organisation. It must also include training of trainers, farmers and administrators in use of the project.

In-built constraints

Water use is restricted by imposing physical and managerial constraints. Physical restrictions can be applied to limit the discharge through the system and the flow rate that can be taken through any single outlet. These include principally the size and type of outlet valves since that is where regulation of the system takes place. Pressure-reducing valves restrict the operating pressures in each reach of the pipeline and hence limit the possible outflow at any outlet. The outflow rate is physically limited, hence the term *limited rate* scheduling is applicable. Unlike in most other applications of pipelines, the pipe diameter does not present a constraint to the flow since the pipe may be required to deliver the designed outflow at any outlet whether the pipeline is flowing at its maximum or minimum rate.

With limited rate **arranged scheduling** operating procedures must limit the number of outlets taking water at any one time. Managerial constraints must be built into unit field administration procedures. The main constraint is based on each farmer pre-arranging his irrigation days a few days in advance. The farm group administrator will maintain a timetable showing the programme of arranged days. A farm group will have a designed maximum number of arranged days to result in an acceptable level of **congestion**.

8.2   Pipeline Layout

General layout

The pipeline system will take water from reservoirs or level-top canals which contain enough in-built storage to accommodate daily fluctuations in water use. The distribution network comprises main and branch pipelines feeding laterals and group feeders which contain the field outlets. Each unit field (see section 5.4) has at least one valved outlet which can be individually opened and closed. The unit fields are arranged into farm groups, each supplied by one or more group feeder pipelines and amongst which a maximum number of **unit streams** is programmed.

> *Concepts of congestion and system design for flexible delivery scheduling were introduced in section 5.4. Water management of a pipeline system is necessarily different from that of an open canal system, and this has a significant influence on the way in which a pipeline system is planned in the field.*

The canal hierarchy of table 6.1 can be extended to pipelines as in table 8.1.

level	functions	name	target congestion per cent
2	Intermediate conveyance between reservoir and lateral	main, branch	70 - 85
3	Feeds one or more farm groups	lateral	60 - 75
4	Feeds a farm group	group feeder	50 - 70

table 8.1   *Pipeline nomenclature*

Upper levels 0 and 1 of the system will almost always be served by reservoirs or open canals, including level-top canals. The Main will generally be the pipe serving the largest number of farm groups. Branch lines will serve more than one lateral or group feeder. A lateral may serve more than one farm group. A group feeder pipe will serve only one group, although one group may be served by more than one feeder.

The route for the main and branch pipelines should follow ridges, roads and major cadastral boundaries where practicable. Where a command area has a distinct zone of land that is lower than the rest, it is better to serve it by a single pipeline only, in order to minimise the number of pressure-reducing valves.

Field outlets

A typical low-pressure field outlet is shown in figure 8.4. This is controlled by an orchard valve, which is a screw-down disc valve concentric with the top of a riser pipe. Operating head at the riser outlet is enough to run into an open channel system, and typically taken as 150 mm above ground level.

Gated pipes can also be used with a similar valved connection to the riser. Several versions are available, including aluminium and PVC pipe, and lay-flat butyl rubber tubing. Individual outlets are holes closed by sliders, spaced along the pipe such that they can discharge directly into furrows. A higher operating head is needed for gated pipes, although still only about 750 mm above ground level at the riser.

Involvement of farmers

At this stage the existing land tenure patterns need to be reviewed and the implications of blocking out for irrigation assessed. Involving farmers at this stage is the best way to avoid social problems later. If land holdings are very small, some degree of amalgamation into **unit fields** of say about 2 ha will be required for the purpose of practical water delivery, in which case several farmers will be required to share the same outlet. Any land holdings much larger than this will need to be subdivided into **unit fields**. Farmers should be encouraged to put in additional outlets at their own expense to facilitate their own water management.

> *Sprinkler, microjet and drip systems generally require a higher operating pressure and typically run off heads of 3-10 metres. The Aquacard system described in section 3.9 is on a network of HDPE and concrete pipes linked to large open reservoirs. Because the allocation of water is accurately measured and charged by volume used, the need for adjusting all outlets to give a similar discharge rate is removed.*

Blocking out and farm groups

In section 6.2 the process of blocking out for a canal system was described. The process is similar for a pipeline, but less constrained by topography. The upper boundaries of all fields should be angled across the contour to give a manageable slope for surface irrigation (e.g. 0.2 - 1 per cent.) The outlet point from the pipeline into the field should be at the highest point of the field, although some land levelling might be useful in order to attain a practical pipe layout.

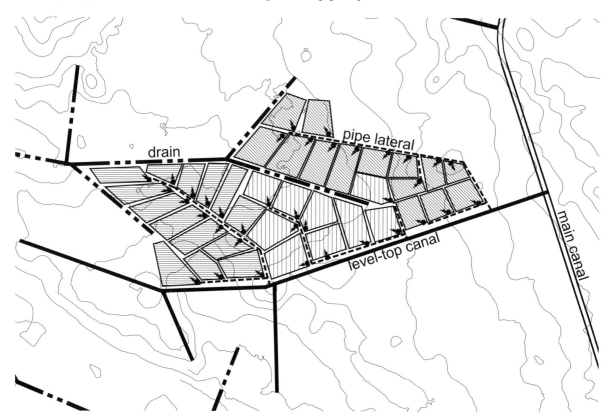

figure 8.5   ***Blocking out for a flexible delivery pipeline system. Three farm groups of 9 – 14 unit farms are shown.***

An irrigation group is typically 10-20 unit fields, amongst which operating limitations will, in the case of arranged scheduling, be imposed by restricting the number of unit fields taking water on any particular day. Normally two or three streams at a time will be permissible within the group, in order to give a congestion factor of 50-60 per cent. The number of streams limits the size of the group; more than three streams at a time may be difficult to organise. A farm group will normally be on a single lateral, but may also include direct outlets on the main or branch pipelines. Draw a boundary in a different colour around each group, and give it a number starting at 1 for the tail-end of the system and working upstream along the main pipeline. Figure 8.5 shows the land of figure 6.1 blocked out for a pipeline system.

8.3   Administrative Organisation

The following is based on institutional arrangements for a pipeline system in Pakistan designed for a limited rate arranged schedule. Other schemes may differ in detail but this is a useful basis for smallholder schemes.

Farm groups and the group irrigator

The *Unit Field* is an area of about 2 ha that can be irrigated in 1 day with the *unit stream* of irrigation water. It may be an amalgamation of several smallholdings or it may be a single field in a large farm. When irrigation is required the representative or owner of the unit field will arrange in advance with the Group Irrigator a suitable day to take water.

The *Farm Group* contains up to 20 unit fields served by one or more group feeder pipelines. Irrigation releases will be restricted within the group to keep congestion to less than 70 per cent. This entails restricting the number of unit fields irrigating on any day to a maximum of between two and four, depending on the number of unit fields in the group. Congestion is described in section 5.4. Each farm group will be regulated by its Group Irrigator.

The *Group Irrigator* will be hired and fired by the farmers, and may or may not be one of them. He maintains a register of all authorised uses and records of irrigation streams requested within each farm group under his control, which could number up to 5. He authorises each application for water and agrees a date on which each unit field may take water. If the farmer requests do not exceed the design number in a group, the Group Irrigator can authorise such uses. If the required number of streams is already allocated, he determines an alternative day with the farmer. No farmer should have to reschedule his irrigation more than 2 days from his preferred date.

Water Users Association

The *Water Users Association* will encompass all groups served by the same pipeline network, and will include several farm groups. Its function is to co-ordinate information on water use and operational performance, to collect water charges, and to carry out routine maintenance on the pipeline system at levels 2 to 4. It should be elected by the farmers and serve as the interface between the farmers and the government organisations concerned with the overall scheme, such as the Departments of Irrigation, Revenue and Agriculture.

Above level 2 the main system will normally be managed, owned and operated by a government-controlled body such as the Irrigation Department.

Each WUA will have a chairman, secretary, accountant, and technical co-ordinator whose job it is to collate records of water use and send them to the Project Hydrographer, and to keep a regular check on maintenance requirements.

For the overall project, the position of Hydrographer would be needed in order to collate records of water use, farmer use, flow rates, for evaluation and information to be applied to future projects.

Ownership

The *outlets* together with their stub pipes should be owned by individual farmers. Where outlets are shared between farms smaller than 2 ha, there should be joint ownership. The farmers' contribution to the cost of the scheme should be in the form of purchasing the outlets. Additional outlets can also be purchased by a farmer as a means of facilitating his own irrigation practices. They would be subjected to the same programming restrictions of maximum streams within the group.

The *group feeder pipelines* should be either commonly owned by the farmers in the group or by the Water Users Association.

The *main and branch pipelines* could be owned either by the WUA or the Irrigation Department. Normally their capital finance and construction would be administered by a government department. Once completed and commissioned the rights and responsibility for operation and maintenance should be transferred to the Water Users Association.

Water charging

As with any demand-scheduled system, charging for water carries the attendant risk of encouraging wastage of water and its consequential problems of waterlogging and disruption of downstream supplies. With a large number of small farms, it is never practical to measure the flows used at every outlet. However,, volumetric measurement is possible on a farm group basis, using flow meters installed in the pipeline at farm group boundaries. Charges can be apportioned to each group on a volumetric basis.

Within the farm group charges can be apportioned according to the areas under each crop type or the number of arranged irrigation days as recorded by the Group Irrigator. To be sustainable, any apportionment system must be equitable, transparent and practicable. It must allocate charges equitably between users, reflecting as closely as possible their actual water use. It must be transparent in operation, to avoid social disputes. It must be technically feasible in its requirements for measurement and record keeping.

8.4 Why and When to Use a Pipeline

The advantages of a buried pipeline system over open canals are:

- No land take required.
- Insensitive to topography.
- Not affected by sedimentation by wind-blown sand.
- Not affected by weed growth.
- Not affected by unstable soils in canal banks.
- Not susceptible to seepage losses.
- Minor structures such as bridges or cross-drainage structures are not necessary.
- Permits great flexibility in delivery scheduling.

With such in-built advantages, a demand-scheduled delivery system is possible. This should lead to improvements in crop production, efficiencies in water use, and more amenable social conditions amongst farmers.

> *Most pipelines that engineers design are for either water supply or for sewerage. In the first case the pipes are under pressure, usually high pressure. In the second case the pipes are often flowing partly full, as gravity channels of circular cross section. The operating requirements of a low-pressure irrigation pipeline for flexible delivery are radically different from either of these.*

The success of demand scheduling is dependent on an assured supply of water, provided by intermediate storage within the delivery system. The response time when an outlet is opened must be immediate.

Storage is provided within the pipeline system itself, and in level-top canals which are regulated by downstream-control gates. Pressure is controlled by Harris valves, and is restricted to a maximum of about 6 m head. Cheap pipes and simple joints and fittings can thus be used.

Hence a pipeline is particularly appropriate in situations with a high density of existing small farms, the boundaries of which would be disrupted by surface canals. It is suitable both for new schemes and rehabilitation of existing schemes which are being converted to flexible demand scheduling in search of better water use efficiency, higher agricultural returns, reduced waterlogging and salinity, and better social cohesion amongst the farmers.

A pipeline may not be practical in very flat lands since it requires somewhat more available head than the equivalent open canal system. However, such situations are relatively uncommon and in most cases if flexible delivery is required and the necessary intermediate storage is available for it, then a pipeline is also possible. In flat terrain the level 1 and 2 distribution can be done by level-top canal which feeds low-pressure pipelines for the lower levels of distribution.

Maintenance requirements of the pipeline system should be minimal. Outlet valves may require occasional repair or replacement, and leakage may develop which requires temporary draining of pipe sections and concrete repairs. It is unlikely that any desilting will be necessary, as silt which is carried into the pipe will also be carried out of it, even if it settles out when the flow shuts down. Weed growth will not arise in the pipeline. The WUA should be responsible for arranging any necessary repairs and for monitoring the condition of outlet valves.

figure 8.6   *Gated pipe system, USA*

figure 8.7 **Installing MDPE pipe, India**

8.4 Designing the pipeline

Unit stream size

The unit stream, as described in section 5.4, is the discharge required to irrigate the unit farm in a single day. It should be based on prevailing conditions of soil, cropping patterns and potential evapotranspiration, and an 8 hour day. (A twelve hour day might be acceptable for a short period in which a heavy water demand occurs, such as planting wheat in November or naize in June.) For the Topi area shown in figure 8.9, a stream size of 60 l/s was adopted. This implied a 12.3 hour peak irrigation day on a 2 ha farm unit, assuming 8 mm/day crop water demand at 60 per cent application efficiency.

Congestion

For each farm group the number of streams must be chosen to satisfy the congestion target of 50-70 per cent. Table 8.2 illustrates the congestion for typical group sizes at 10 day frequency. Congestion is expressed as a per centage and calculated from the formula:

congestion = no. of farms/no. of streams/frequency in days

group	tot.	1	2	3	4	5	6	7	8	9	10	11
area, ha	340	22	19	13	21	17	24	24	17	22	22	22
no. farms	170	11	10	6	10	9	12	12	9	11	11	11
no. streams		2	2	1	2	2	2	2	2	2	2	2
nominal frequency, days		10	10	10	10	10	10	10	10	10	10	10
congestion, 10 day		56	50	60	50	44	60	60	44	56	56	56
optimum streams		2	2	1	2	1	2	2	1	2	2	2
cumulative farms		11	21	27	37	46	58	70	79	90	101	112
cumulative streams		2	4	5	7	9	11	13	15	17	19	21
per cent reduction		0	0	0	0	0	0	0	11	13	14	16
reduced streams		2	4	5	7	9	11	13	14	15	17	18
stream size, l/s		60	60	60	60	60	60	60	60	60	60	60
cum. area served, ha		22	41	54	75	92	116	140	157	180	202	224
pipeline capacity, l/s		120	240	300	420	540	660	780	840	900	1020	1080
duty, l/s/ha		5.39	5.81	5.55	5.62	5.86	5.69	5.58	5.34	5.01	5.06	4.82
length, m		650	600	300	600	600	500	1100	350	600	700	700

table 8.2 **Typical congestion figures**

Calculating the pipeline capacity

The required capacity is the cumulative number of streams multiplied by the stream size at each point in the system. The number of streams may be reduced on a linearly sliding scale by a reduction factor of up to 25 per cent for cumulative farm numbers of 70 or more.

Sizing the pipeline

There are several formulae in common use for calculating flow capacity, velocity and head loss in pipes. These include Hazen-Williams, Colebrook-White, and Manning. Typical friction coefficients are given in table 8.3.

pipe material	Hazen-Williams	Colebrook-White	Manning
	C	Ks, mm	n
Old concrete	120	0.3	0.018
concrete	140	0.15	0.015
steel	145	0.03	0.012
PVC, spigot & socket	150	0.06	0.012
PE fusion jonted	160	0.03	0.011

table 8.3   Typical pipe friction coefficients for diameters 300 - 900 mm

- Establish node points along the main and branch lines. Nodes will be at expected changes of pipe size, normally where branches or laterals join the main. It is not necessary to have a node at every direct outlet or at the junction of every group feeder.

- Establish the hydraulic grade line at zero flow. The start level at the pipe inlet from the reservoir or level-top canal will be the lowest operational water level there. The pressures inside the pipe are restricted to a maximum of 5-6 metres by means of Harris pressure reducing valves. Hence a Harris valve station will be located at every 5 m vertical drop in elevation.

- The procedure is to set a minimum head on usually the most distant outlet but the worst condition can occur at other locations as affected by topography. In operation the hydraulic grade line needs to be at least 0.6 m above ground level at every outlet. A lower head will tend to cause less stable flow at the outlet as pipeline pressure varies. If gated pipe is to be used, 1.2 m should be available.

- For each reach between nodes of the pipeline, establish the hydraulic grade line at maximum flow, starting at the downstream end of the pipe system. The hydraulic grade line is depressed below its static level by friction losses in the pipe, head losses through the Harris valves and form losses at the valve stations and at the pipe inlet. Assume a small pipe diameter (say 200 mm), work out the losses and see if these can be accommodated by the head available. If not, try a larger pipe size (work in standard steps of 100 mm

---

Hazen-Williams formula:

Hydraulic gradient $i = (Q/.2343C/D^{2.63})^{1.85}$

Where   Q = discharge in gpm
(1 l/sec = 13.24 gpm)
C = friction coefficient
D = pipe internal diameter in inches
(1 inch = 25.4 mm)

Colebrook-White formula:

$1/\lambda^{0.5} = 2\log D/2k + 1.74$

Where λ = friction coefficient, $2g\,DS/V^2$
k = effective roughness height, mm
D = pipe internal diameter in mm
S = hydraulic gradient
V = velocity

Manning formula:

Hydraulic gradient $i = Q^2 n^2 / (A^2 R^{4/3})$

Where Q = discharge in cumecs
n = friction coefficient
A = pipe area $\pi D^2/4$
R = hydraulic mean radius = D/4
D = pipe internal diameter, m

internal diameter referring to the inside diameter of commercially available pipe) or a larger size of Harris valve, or two Harris valves in parallel. Harris valve data are given in Table 8.6. The energy line slope is conveniently taken from charts or calculated from the Colebrook-white or Hazen-Williams formulae. It should be a little less than the ground slope.

- Note that the entire pipe in a single group is likely to be of the same diameter, since its full design discharge does not reduce along its length. If the design number of streams for that group is 3, say, then the three lowest outlets or the three highest outlets could be in operation at any one time.

Pipe materials and laying

Buried pipes can be classed into two categories, rigid and non-rigid. The non-rigid pipes are thin-walled and depend for their performance on the surrounding backfill material, which must be carefully placed and effectively forms a component of the pipe by imparting rigidity. Rigid pipes are normally cement-based, such as precast concrete and asbestos cement. Other types such as thick-walled steel or RPM (reinforced plastic matrix) are rarely used in irrigation due to their cost.

Concrete

Precast concrete pipes are universally available, but not always of adequate quality, even for low-pressure applications. The adoption of Harris valves requires a pipe working pressure of 6 m of water. The spun (Hume) pipes, which are common in South Asia, are notorious for poor compaction and blow-holes. Their end joints are often roughly cast and difficult to seal when laying. Ironically, reinforcement in these pipes frequently contributes to their poor performance, by creating voids behind reinforcement junctions as the pipe is spun. The method of manufacture relies on a steel pipe, split into two halves which, bolted together, form the outside mould. This is rotated on a horizontal axis. Concrete is fed into the mould as it rotates and is held against it by centrifugal force.

A better type of pipe is the unreinforced irrigation pipe, which is formed on a vertical axis using an outer mould and internal mandrill. This is common in the USA and can be easily manufactured in developing countries. As for the precast parabolic channel segments described in chapter 9, the important factor is concrete strength, not reinforcement.

Asbestos-cement

Autoclaved asbestos cement was in common use for water pipes until recently, when many developed countries outlawed it on health grounds as a source of carcinogens. However, when used for irrigation purposes it can pose no conceivable health risk, and can be a cheap option in those countries where it is still available. It can withstand high pressures and has accurately machined joints, so is easy to lay.

Steel

Thin-walled steel can be economic in larger diameters over 400 mm, as spiral- or seam-welded pipe. It can be cheaply jointed by site welding, or by mechanical joints such as bolted flanges. More sophisticated bolted flexible couplings such as Victaulic or Viking-Johnsons are rarely used for low-pressure applications due to their cost. As a non-rigid material, backfilling with granular soil needs to be done evenly, although steel will withstand considerable deformation without failing.

Steel pipes may be protected from corrosion with cement, bitumen or epoxy coatings, but these are rarely used in low-pressure irrigation applications, on grounds of cost.

PVC

PVC is notoriously prone to damage from surge pressures, flotation in wet ground, and deterioration in sunlight. It is made as thin-walled low-pressure (15 m head) irrigation pipe in diameters up to 600 mm. As with other plastics, its price tends to track the price of oil, and is currently expensive in relation to concrete.

MDPE

Medium Density Polyethylene can be a cheap option in small diameters less than 250 mm. When properly fusion-jointed it is highly reliable and can be used either for new pipework or as internal sleeving for leaking concrete pipe. It is light and simple to lay (figure 8.7) and will accept considerable deformation without failure.

GRP

Glass-reinforced plastic can be an economic option in larger diameters over 300 mm, especially if a local manufacturing source exists. It is normally jointed with rubber gasketed collars. It requires care in handling and can be susceptible to impact damage during transporting and laying. It can be custom-manufactured to specific wall thickness and pressure ratings.

8.5   Pressure and Flow Control

Harris valves

Harris valves are in-line, float-operated self-regulating devices analogous to downstream control gates in open channels. They close when pressure in the downstream reach attains a predetermined level. They have a scissor linkage which imparts a mechanical advantage of about 6:1 on the valve seal, and this makes them more effective, and able to operate under higher heads (up to 6 m), than the simpler alternatives shown in figure 8.8.

The floats are usually formed from a block of expanded polystyrene, but can also be fabricated as hollow polyethylene boxes. The float chamber is built to a level just above the design maximum head in the downstream reach. With a single valve, the float chamber at its simplest can be a pipe slightly larger in diameter than the float. In multiple valve arrangements such as those in figure 8.9, the floats require a traveller and guide to prevent adjacent floats touching. The turbulence created when the valves are open can be intense.

Harris valves were originally designed and patented in the USA and can be easily and cheaply fabricated. Their main disadvantage in developing countries is a susceptibility to theft and vandalism or incidental damage. However, for low-pressure applications I am aware of no effective alternatives that can be had at a similar cost.

A further advantage of Harris valves is their avoidance of surge pressures. Their rate of closure is slow enough to prevent water hammer.

Pressure-reducing valves

There are several standard designs of in-line pressure-reducing valve that serve a similar function to Harris valves. They consist of a check valve which is closed by a secondary valve activated by the pressure on the downstream side. However, these are all expensive and designed for higher pressures than the 5-6 m of a Harris.

Computerised systems are becoming cheaper and more reliable, and where head is available, a much tighter degree of control can be applied to conventional pipeline systems. In the Coachella valley of California, Harris valves are being superceded by solenoid valves controlled through SCADA networks. In the Ofanto River scheme in Italy, the need for tight discharge control has been superceded by the ability to accurately measure and charge for the water used at each outlet (see figure 3.11).

Float-operated sleeve and disc valves

The self-centering disc valve of figure 8.11 is a downstream-control unit which regulates the flow from a primary pipeline into an intermediate storage reservoir. It is float-operated, and opens in response to falling water level in the discharge chamber. Like the Harris valves, the float leverage arrangement has a mechanical advantage of about 6:1, but this can operate at heads of 50 metres or more. The vertical inlet pipe is covered by a flat disc fixed to the float linkage. The lower surface of the disc is maintained concentric to the pipe under the action of flowing water.

A variation on this is the cylindrical sleeve valve, which operates at low heads only.

Both these valves are Neyrpic designs. Their associated civil works being somewhat cumbersome, they are not suited to in-line applications but are appropriate for controlling the inflow into or, in the case of the cylindrical valve, out of a reservoir.

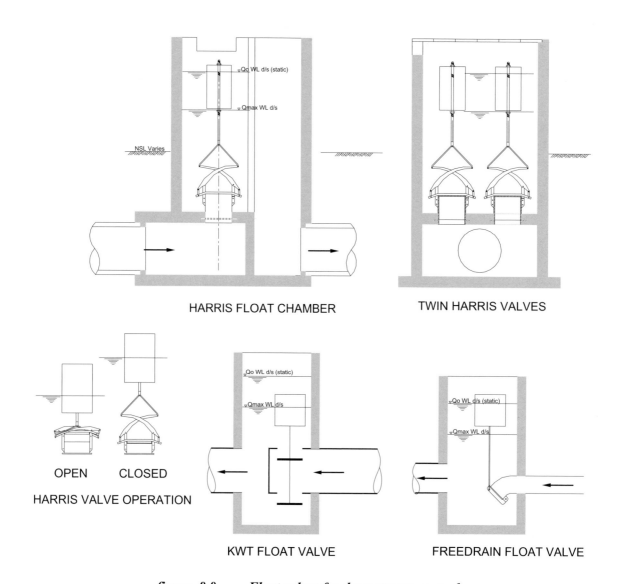

figure 8.8   *Float valves for downstream control*

figure 8.9 **Harris valve stands in Topi, Pakistan. These are very substantial structures, built to deter vandalism. These 24 inch diameter Harris valves are here arranged in pairs to handle the discharge of 1200 l/s**

figure 8.10 **Installing a Harris valve float**

figure 8.11   Float operated disc valve giving downstream control in a pipeline system. The float is in the cylindrical chamber on the far side. The vertical rod in the foreground ends in a horizontal disc which covers the pipe outlet

Outlets

Farm outlets should be on short (3-5 m long) stub pipelines in order to reach property lines from the group feeder lines, and to facilitate setting of the outlet riser at a standard depth below ground level of about 0.6 m.

The cheapest low-pressure valve is an orchard or alfalfa type, which is set into the top of the riser pipe and screws down to close. A typical arrangement is shown in figure 8.12. Flow rates and operating heads are shown in table 8.4.

Gate valves can also be used as in figure 8.13, although they are more expensive as they are designed for higher pressures than orchard or alfalfa valves. As an interim measure on the Topi scheme, they were fitted with orifice plates in order to limit the discharge to a standard rate.

valve size inch	discharge in litre/sec for head loss in metres			
	0.1m	0.25 m	0.5 m	1.0 m
4	10	13	17	24
6	20	31	37	55
8	34	54	64	99
10	57	88	110	147
12	85	120	156	226
14	108	170	215	283
16	133	221	266	368
18	170	269	311	509
20	226	340	453	594
24	325	509	594	934
30	538	736	1047	1274
36	679	1132	1330	1868

table 8.4   Head losses – alfalfa and orchard valves

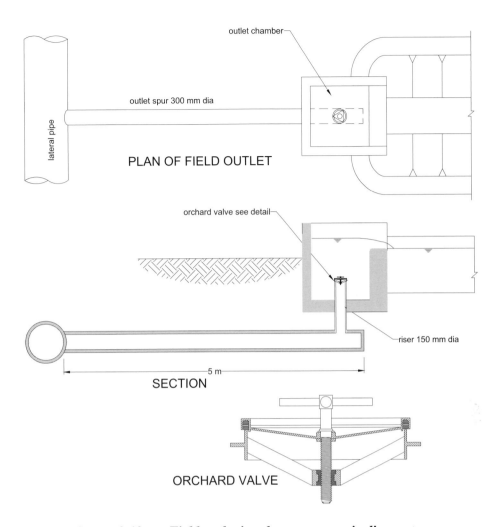

figure 8.12   Field outlet in a low-pressure pipeline system

figure 8.13   A Beaumont circular weir outlet from a buried pipeline, Sulawesi

Pipe inlets

A pipeline system should offtake from a canal or reservoir in which intermediate storage is guaranteed. The pipe inlet will consist of a trash screen and an isolating gate. If required, a flow measurement device such as a volumetric meter can be built in. A typical arrangement is shown in figure 8.14. The screen follows the curved profile of the canal and is located above bed level, in order to reduce turbulence and the consequent likelihood of admitting sediment into the pipe.

figure 8.14   Pipe intake from a level-top canal

8.6   Design Procedures

Design of the main and lateral system needs to be done in conjunction with the farm group feeder network. The main steps in design are:

1) Topographic mapping.
2) Layout of lateral and main pipeline network.
3) Preliminary blocking out.
4) Conferring with farmers.
5) Review of land maps.
6) Cadastral mapping of land ownership boundaries.
7) Operating programme.
8) Final blocking out.
9) Layout of group feeder pipelines and farm grouping.
10) Design of farm outlets and unit stream sizing.
11) Design of pipeline system capacities.
12) Detailed design of pipeline system to meet constraints of pressure, discharge, flow velocity, and topography.

Topographic mapping

Topographic mapping should be plotted at a scale of 1:2000 with 0.5 m contours. Roads, drainage lines, buildings, graveyards and rock outcrops should be clearly indicated. The boundaries of all fields larger than about 0.25 ha should be marked. The corner points of all fields larger than 0.25 ha should be located with x,y,z coordinates. The corner elevations (z) should be determined to the nearest 100 mm.

Cadastral mapping

Cadastral mapping shows farm boundaries that can be superimposed on the topographic maps. Existing maps can be superimposed on the outline layouts, preferably by digitising or scanning into computerised format. However, they may be outdated or inaccurate, in which case field confirmation is required. Land holdings that come under separate family or village groups need to be identified so that Water Users Associations and farm groups can be established effectively taking account of social divisions.

Pipe layout

The broad outlines for pipe layout are described in section 8.2. The group feeder pipeline can take a direct path between points, should not run close to lines of trees and avoid current watercourse ditches. The connecting structure between a group feeder and lateral may be a Harris valve stand on the lateral or a simple branch.

Use of spreadsheets in designing a pipeline

The procedure for congestion and pipe capacity can be entered on a spreadsheet, which iterates to find the optimum streams and required pipe sizes. A sample is shown in table 8.5, and is also on the CD.

distributor sector	1	2	3	4	5	6	7
area, ha	48	70	82	80	72	52	40
no. farms	24	35	41	40	36	26	20
no. streams	6	9	9	9	9	9	5
nominal frequency, days	7	7	7	7	7	7	7
per cent congestion 7 day	57	56	65	63	57	41	57
optimum streams	6	9	9	9	9	8	5
cumulative streams	6	15	24	33	42	51	56
per cent reduction	0	7	11	15	19	23	25
reduced streams	6	14	22	29	35	40	42
unit stream size, l/s	25	25	25	25	25	25	25
cum. area served, ha	48	118	200	280	352	404	444
pipeline capacity, l/s	150	350	550	725	875	1000	1050
duty, l/s/ha	3.13	2.97	2.75	2.59	2.49	2.48	2.36

table 8.5   ***Congestion calculation***

Pipeline design can also be done on a spreadsheet. An interactive procedure is described here, which is also available on the CD version of this book. This uses Colebrook-White to calculate friction gradients.

Sample MS Excel spreadsheets for pipeline and canal design can be obtained on the CD, along with full colour high resolution versions of all photos in the book, together with a large collection of additional related photos, and canal design software.

Survey information along the pipeline route is fed into the spreadsheet and plotted as a graph of the longitudinal section. To this is added an initial pipeline diameter and invert levels. With discharge data, a separate graph of pressure distribution along the pipeline is generated. The requirements and locations for pressure-reducing valves can be seen at a glance. Trial Harris valve data are then inserted into the appropriate line of the spreadsheet, which then carries out a test to check the adequacy of the size of valve chosen.

Figure 8.15 shows the interactive graphics, with the pressure drop caused by the Harris valve clearly apparent at chainage 2400 m. The aim is to ensure that negative pressure does not arise in the pipe, and that there is adequate residual pressure (0.5 m in this case) at every outlet for its correct function.

Separate spreadsheets can be established for branch lines or group feeders, values of operating head and pipeline invert linked to those on the parent pipeline.

This design procedure can also be done graphically using a pencil and paper, but since spreadsheets are now ubiquitous for engineering design, it is not considered further here.

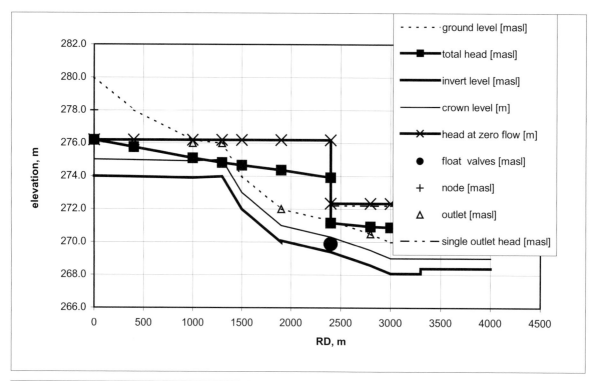

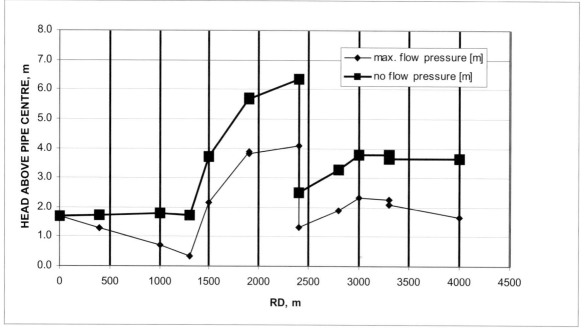

figure 8.15   Interactive pipeline design, longitudinal section and pressure distribution

Choice of Harris valve

The static hydraulic grade line for the structure which must be considered in establishing the height of the stand is the rise above the full flow water level. Rise is the distance the float will move while the Harris valve goes from the open to closed position and with increasing submergence develop enough pull to close the valve against the change in pipeline pressure across the valve and to seal it. Additional stand elevation must be provided for freeboard.

The setting of the float at the start of operation will be with the valve wide open.

This process of selecting pipe sizes for required capacity in a reach to have proper pressure condition at outlet structures and along the pipelines, continues on down through each reach. The required outlet structures and their head losses are considered. They may be in-line structures controlling the lateral line, or they may be off-line structures to group feeders. The float valve provides the two functions of limiting head on the pipelines and providing stable head for variable offtake flows.

| valve size | int. dia | discharge | discharge in litres/sec for head loss in m | | | | | | | float size | float travel |
inches	mm	coeff*	0.5	1.0	2.0	3.0	4.0	5.0	6.0	mm square	mm
4	102	2.19	17	24	35	42	49	55	60	250	152
5	127	2.5	25	35	50	61	71	79	87	250	203
6	149	2.3	36	51	72	88	102	114	125	250	229
8	200	2.2	66	94	133	162	188	210	230	400	330
12	305	3.2	128	181	256	313	362	405	443	400	483
16	381	2.15	244	344	487	597	689	770	844	600	584
20	483	2.3	378	535	757	927	1070	1197	1311	600	584
24	584	2.8	501	709	1003	1228	1418	1586	1737	600	813

Multiply the velocity head ($v^2/2g$) through the valve orifice by the discharge coefficient to get the head loss.

table 8.6   **Harris valve sizes and head loss coefficients**

If this method seems a little vague, it is important to bear in mind the practical implications of an error in estimating the number of unit streams and hence the pipeline capacity. They are not very great. Suppose we change the reduction factor so that the 25 per cent is knocked off only when the number of unit fields exceeds 100. In this example the theoretical pipeline diameter serving group 5 would change only from 484 to 497 mm. In practice it would remain unchanged at 500 mm.

References and further reading for chapter 8

(1) Merriam, John L., 'Design of Semi-closed Pipeline Systems. Symposium Proceedings, Planning, Operation, Rehabilitation and Automation of Irrigation Water Delivery Systems', American Society of Civil Engineering, Irrigation Drainage Division, Darell D. Zimbelman, Ed, pg. 224 – 236, July 1987.

(2) van Bentum, R., and Smout, I., 'Buried Pipelines for Surface Irrigation', IT/WEDC, 1994.

CHAPTER 9   CANAL LINING

Most engineers have their own prejudices about which type of lining to employ in a particular situation. Unfortunately, although prejudice may assist in arriving at a decision, it is usually founded not on technical reasoning, but on what went before. So the end result is a reflection on several generations of engineering which have ignored recent advances in materials and construction methods in the interests of maintaining the status quo. Bureaucracy does not help. Government irrigation departments and many large consulting companies tend to have unwieldy management structures that deter innovation and free thinking in design. It is easier to avoid the need for decision making and argument by using standard designs that have been in use for centuries. This is a sad reflection on the state of the engineering profession.

9.1   Physical Influences on Lining

Chapter 6 explored the reasons and strategic arguments for and against lining. Now, before embarking on any detailed design, we consider the forces of nature and of humankind that will be exerted upon it.

Structural stress

Structural stresses on lining need to be considered, although not necessarily quantitatively assessed, at the outset, and either avoided or resisted by designing adequate strength into the lining. In small canals, in situ lining is rarely subject to substantial stresses other than that arising from soil settlement or movement. In large canals, hydrostatic forces can also be substantial. Rigid in situ lining such as concrete or masonry is almost always designed to be continuously supported on its earth backing and as such it behaves as a beam on elastic foundations. Finite element analysis can be used for larger canals, and can for example indicate the drawbacks of adopting a trapezoidal profile when a curved section would generate far less stress (figure 9.1.)

Shear stresses may be induced at the junction between lining and structures due to differential settlement. Joint design needs to make allowance for movement in these situations. For irrigation canals some weeping of joints is usually admissible and joints can be kept simple.

The stresses induced by the lining's own dead weight can be destructive in the case of stone masonry, which has little or no inherent structural strength.

Hydrostatic pressure

In a large canal, unbalanced water pressure on either side of the lining can cause damage. Concrete lining 100 mm thick can withstand a hydrostatic pressure of about 1 m head. There are several potential modes of failure which need to be considered. Apart from the stress caused by a simple build-up of water on one side of the lining, the secondary effects of washing out of soil fines and consequent piping failure have to be guarded against.

There are two situations in which external pore water pressure may develop. The first is from a high groundwater level, which may be the case for a canal in deep cut. This may also be a localised effect due to heavy rainfall, with runoff concentrated at points where water can run down behind the lining. This happened during construction of Pehur High Level Canal, before surface drains had been properly formed, and caused the damage shown in figure 11.6.

The second case is due to rapid drawdown of the canal, after the surrounding soil has been saturated by seepage from the canal. Locations susceptible to rapid drawdown include the downstream side of cross regulators, and this is one reason for spacing these structures not too far apart.

In small canals, lining can often be made heavy enough to resist any uplift, although external hydrostatic pressure due to irrigation or localised canal leakage can blow out lining joints even at low head. Figure A9.46 shows a bitumastic joint blowing at an external pressure of less than half a metre. Often the occurrence will be intermittent, such as in heavy rainfall when the canal is not flowing.

Lined canals in cut or on cross-slopes may need to be designed to cope with possible external water pressure. There are several ways of doing this. Leakage and the effects of rapid drawdown can be restricted with the use of an impermeable membrane beneath the rigid lining.

Pressure-relief valves, which let water into but not out of the canal can be easily damaged during maintenance and are not often used. However,, a cheap method has been used for membrane lining in China (Ref. 3), which comprises a small disc of plastic covering a smaller hole in the membrane, protected by geotextile to prevent migration of fines, and covered by a precast slab, which will permit the release of water under a small differential head.

Under-drainage can be a practical option, with perforated pipes running parallel to the direction of the canal and discharging at cross-drainage structures.

Weep-holes can let water in either direction, and in situations where external pressure is a problem, the potential loss of irrigation water out of the canal is rarely serious. A gravel and geotextile filter must be used to prevent migration of soil particles. No-fines concrete offers a neater alternative to weep holes.

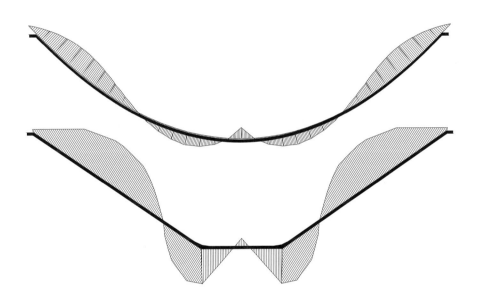

figure 9.1   Finite element analysis shows high stress in a trapezoidal section due to lateral forces from swelling soils. Similar results arise in the case of vertical loading and settlement of the underlying ground

Point loads

Point loads due to the actions of people, vehicles or animals are potentially the most damaging to small canals. The hooves of cattle and buffalo are notoriously destructive to masonry and poor quality concrete. Women washing clothes, men riding bicycles or small boys rolling rocks into the canal are additional factors that must be designed for.

In the case of prefabricated segments, the heaviest stresses derive from handling, either being dropped from the back of a truck or rolled over in the process of being manhandled into their final location. Even careful handling and laying involves point-load stresses due to the whole weight of the unit being concentrated at a single point.

Durability

The durability, or longevity of lining is a subjective measure of its resistance to ageing, without the effects of physical damage. Most materials gradually deteriorate in some way with time, and with exposure to weather or sunlight. Some of the plastic membranes are notorious for physical deterioration; polyethylene and uPVC go brittle when exposed to ultraviolet in sunlight for a few months. Concrete can gradually weaken under the chemical action of sulphates or salinity in the soil. Table 9.3 gives an indication of durability criteria for various types of lining material.

Thermal and shrinkage stresses

Thermal expansion and contraction is an irresistible force which can break concrete if proper allowance is not made in the form of joints. Actually thermal damage is normally unspectacular, in thin concrete appearing only in the form of hairline cracks, but it can be enough to provide a root-hold for weeds. In stone masonry, differential thermal movement is partly responsible for the mortar joint infills parting company with the stonework.

Shrinkage and creep of concrete in the process of curing is a common cause of cracking in thin lining. Properly controlled concrete mixing and curing can avoid serious shrinkage. Joints at adequate intervals won't stop shrinkage but will restrict damage. However,, joints do not work on a thin plaster or concrete lining that is laid on a high-friction surface, such as when overlaying an existing rough lining during rehabilitation. It is often better then to omit the joints, ascertain the position of cracks, and then cut out any damaged zones and refill with concrete later.

Soil movement

Moving soils are potentially the most destructive of all forces acting on a rigid canal lining. There are three main causes of movement: settlement of fill due to poor compaction, consolidation due to the gradual egress of water over a long time period and, by far the most widespread in its effects, swelling and shrinkage of soil with changing moisture content. The most dramatically-swelling soils are usually clays with a high montmorillonite content, but most soils with any appreciable clay content will swell to some extent when wet and shrink on drying, opening up deep cracks in the process. Any canal lining which is supported fully on such a soil will be forced to move with it, and will break immediately unless it is either flexible or self-supporting and strong enough to hold together. A high-friction interface between soil and the underside of lining invites self-destruction, as in the masonry lining of figure 11.5.

Soil dispersion

Dispersive soils are those that either dissolve in water or lose their structure underwater to such an extent that they are moved in fine suspension, and can wash out through joints and cracks in the lining. Weak silts, loess and fine mudstones may all be hard when dry, but can behave quite differently after soaking in water. Some gypsiferous soils can literally dissolve and disappear completely. In these situations, voids can develop beneath the lining, leading to sudden collapse (such as in figure 11.7.)

Scour

Abrasion caused by fast-moving sand and gravel carried as bed-load can erode concrete and masonry. It is wise to limit the design velocity to a maximum of about 5 m/s under these conditions. Erosion of earth canals can occur at velocities of less than 1 m/s, and these need to be designed with an appropriate threshold velocity or considered as alluvial regime canals.

In northern Pakistan it was common practice to partially line large canals with lean concrete. By the time half the cement had been spirited away, and with no proper curing, the residual concrete strength was little more than that of compacted soil. It was so susceptible to erosion in flowing water that the lifetime of this lining was no more than 1 or 2 years.

Vegetation

Aquatic weed growth can reduce canal capacity, but a more disruptive force on lining is posed by perennial plants such as reeds and trees with penetrative root systems which can dislodge heavy concrete slabs as well as small pieces of stone masonry.

Accidental damage

Accidental damage often occurs during maintenance which involves clearing of weeds or silt. Masonry lining is easily disrupted as roots get entrenched in the joints. Unlined canals can rapidly lose their profile shape through unsupervised de-silting operations.

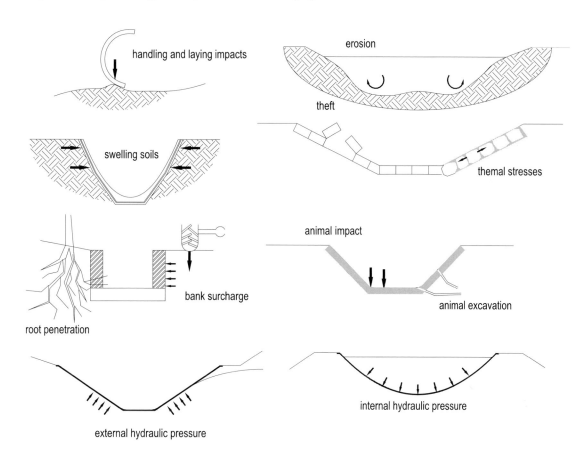

figure 9.2   Canal profiles and forces on them

Animal damage

There are some oft-quoted tales of gophers burrowing into large earth canal banks in the USA and causing massive breaches. Far more pervasive in some countries such as Java are land crabs, which thrive in irrigated conditions and burrow through the joints in stone masonry and weak concrete, and can disrupt rigid canal linings in a short time. Rats and similar rodents tend to burrow into banks above the water line, but cause deposition of soil in the canal and weaken the banks leading to their gradual disintegration.

Deliberate damage

Deliberate damage and theft, often by farmers, is the most disconcerting response to any engineer's end product. But it should also be regarded as an important social indicator which serves the engineer with useful feedback. There is always a reason for deliberate damage, and it usually has little to do with engineering. Conversely, there are plenty of lessons to be learned from the study of deliberate damage, and the good engineer will learn from it and design in such a way that does not invite people to destroy. Chapter 11 considers deliberate damage in some detail.

9.2 Masonry Lining

Included here are all blocks, slabs or bricks of stone, clay or concrete, man-made or natural, that are normally laid by hand with mortar jointing. It is the commonest form of construction in many countries especially for small canals, but it is also one of the worst and its use should be discouraged.

Stone masonry

This is the worst possible method, more suited to building garden walls than hydraulic structures, yet also the one most frequently used. The entirely inappropriate use of stone masonry is usually based on the following faulty reasoning:

- It is labour-intensive, perhaps using local small contractors in line with government policies of distributing the capital expenditure on public works over as wide a sphere of the population as possible.
- It uses low-level technology, and can therefore be carried out by untrained or partially skilled local workers.
- It requires no engineering supervision, since everyone knows how to do it.
- It uses local materials, thereby reducing transport costs within the country and avoiding the need for imports.
- If it uses local resources then it must be cheap.

Most of these criteria can be met by concrete lining, and with a much greater degree of effectiveness than can be achieved by stone masonry. The contention that stone masonry requires little skill to construct is something of a fallacy; in practice it is extremely difficult to make it watertight or crack-free. It always leaks. It is easily disrupted by swelling soils, roots, animal activity and differential thermal expansion. During construction it is difficult to control quality. It is slow to construct. And it is always expensive. Extensive field experience in Indonesia, India and Sri Lanka, in which stone masonry was compared with high-quality concrete and various other types of lining, invariably showed it to be up to twice as expensive as plain concrete.

Stone packing

Many canals are lined with dry stone packing, which is not intended to prevent leakage but merely to maintain the shape of the canal and prevent the sides falling in. It is stone masonry without any mortar in the joints. It can be useful when there is a fluctuating groundwater level as there is no build-up of hydrostatic pressure beneath the lining. It is used successfully in India on small run-of-river diversion schemes only intended to operate during the rainy season when groundwater levels are high. There is in this case no consequent loss of water through seepage because the groundwater balances the canal internal water pressure.

Laterite masonry

This is a common house-building material in south India, where it is often jointed with mud. The durability of mud joints in a fast-flowing canal is unsurprisingly poor. It suffers from the same drawbacks as stone masonry, with the added problems that blocks get stolen for building houses, and it is porous and leaks continually.

figure 9.3   **Stone masonry, India**

Clay bricks

Where locally available, bricks can be cheap and easy to use, but they always leak, are easily damaged by animals and roots, and are easily stolen. Brick lining of small canals and watercourses is still common in Pakistan and India, where it is usually given a coating of cement/sand plaster. However, as a seepage barrier, brickwork is usually a failure. Water seeps through every joint and often the bricks themselves are porous too. This creates ideal conditions in the underlying soil for worms and other burrowing animals which soon construct a honeycomb of drainage paths to further enhance the seepage rate. Often a lined canal of brick is more porous than the unlined canal it replaces.

figure 9.4   **Brick lining, Sri Lanka. Every joint is a potential source of leakage, but at least it's cheap**

Brickwork is commonly used for the vertical walls of rectangular channels and watercourses, sitting on a concrete base slab. Roots easily penetrate the junction of brick and concrete, and the walls are readily pushed over by external root action or animals and vehicles passing along the banks. In Pakistan this is a significant factor contributing to the poor performance of thousands of watercourses.

Many large canals in India were lined with brick or tile in the 19th century, but they had several layers of brick with a waterproofing membrane of mortar or bitumen in between. Today such labour-intensive methods are slow and expensive in comparison with mechanical placing of concrete.

Bricks can be useful for rapid repairs or as a protective covering for plastic membranes, in temporary lining.

Stone slabs

The limestone in parts of East Java and the sedimentary mudstones of southern India are often used for canal lining. But they are easily stolen, easily dislodged by root action and swelling soils, and the joints always leak because it is impossible to get a joint which is smooth, regular and tight-fitting. This does not matter if the function of the lining is merely to retain the canal profile, but leakage is usually excessive.

Precast concrete slabs

Precast concrete slabs are popular because they are cheap and easy to make. However, their performance is no better than stone slabs. Unless the joints are designed to resist movement, shear forces between adjacent slabs quickly lead to their disruption. Rectangular precast slabs are often used, but when a small amount of settlement forces one slab out of plane with its neighbours, the joints crack and the weight of upslope slabs contribute to the mode of failure shown in figure 9.5. They are a prime target for thieves, because their regular size renders them useful for many things. The hexagonal slabs of figure A9.57 did not last long before they were snapped up by local people for making kitchen floors.

figure 9.5   Precast slabs sliding under their own weight, India.

Precast concrete sides with in situ concrete bed

Both potential theft and also difficulties in compaction of earth banks can be avoided by the compound solution shown in figure A9.62. These rectangular channels were constructed very quickly by positioning the vertical precast sides and concreting them into the bed afterwards. However,, they suffer from the problem common to most small rectangular canals, in that they have poor resistance to horizontal loading from the sides through adjacent vehicles or due to expanding soil pressure, and often collapse. Whilst active soil pressure due to dead weight alone may be minimal on a small canal like this, the forces from soils that move through swelling and shrinkage can be extremely great.

9.3　In Situ Concrete

In situ concrete in one form or another is potentially the most durable lining, provided its quality can be assured by means of tight supervision. Where the earth sub-grade is wet or poorly compacted, in situ concrete is likely to fail. All concrete needs proper mixing, curing and placing, and therefore needs competent engineering supervision. In situ concrete comes in many forms, including plaster and ferrocement.

Thin mass concrete

If concrete is of good quality and strong, it need not be thick in order to ensure a watertight lining that will be resistant to damage by cattle and other natural causes. A thickness of 50 mm is enough for small canals, ***provided the concrete is properly vibrated and cured***. As a bed lining it is cheap and effective. Concrete strength should be 25 N/mm^2 (C25, 25 MPa, or 250 kg/cm^2.)

In large canals the maximum thickness required is 125 mm. The Ghazi-Barotha Canal in figure 9.7 has a capacity of 1,600 cumec and a lining thickness of 125 mm. Pehur High Level in figure 9.6 has a capacity of 30 cumec and a lining thickness of 100 mm. The 350 cumec Toshka Canal in Egypt has a lining thickness of 200 mm and a geomembrane, but this is over-conservatively designed.

Figure A9.1 shows a vibrating plate compactor used on rehabilitation of small canals in India. This bed concrete was used in conjunction with ferrocement sides. Getting the right workability of the concrete mix is important. Curing was done by ponding water in the canal for a week after placing. The mix was only a standard 1:2:4, yet its strength was so good that a month later a gang of miscreants intent on doing damage left the site with their crowbars bent and the concrete intact.

In trapezoidal canals compaction of the sloping sides is the biggest problem. A vibrating screeder bar is required, long enough to give control over the concrete thickness and heavy enough to give adequate compaction. The one shown in figure A9.2 is too short and too light to be effective.

The compactive effort discernible in figure A9.3 is even worse. The man standing on the slope is holding a hand rammer to compact the concrete which is then screeded off to the trapezoidal wooden profiles. Doing it this way requires as dry a mix as possible so that it does not slump down the canal sides. The normal result of this method of placement is 75 mm of honeycombing topped with a 5 mm thick smooth plastered surface to camouflage the imperfections.

Figure 9.6 shows a form of placing concrete which was developed specifically for constructing canals in California in wet and unstable ground conditions. It is marketed under the name 'Bunyan Tube' and comprises a hollow steel roller which is winched across the concrete surface as it is placed, rotating against the direction of movement. It thus compacts the concrete up to a limited depth of about 100 mm, and finishes the surface to an accurate profile. If used properly it is extremely effective and quick, and is especially suitable for in situ parabolic lining or other curved profiles. It is important to get the concrete mix right for optimum workability. A mix designed for pumping is usually specified in the UK, where the tube is commonly used for laying flat floors. In the USA it has also been used for curved shell roofs.

The tube was initially introduced into Pakistan on the Swabi SCARP rehabilitation project (figures A9.4 and A9.5) with limited success, mainly because it required a modicum of supervision at a cost perceived as unacceptable by the management. The preferred method of canal lining in Pakistan was to leave semi-skilled concreting gangs with no supervision at all. Under normal conditions of international-standard construction, the time and effort saved by using the tube can outweigh any expense of supervision. Its use was soon to be proven on the stunningly beautiful Pehur High Level Canal.

A full sequence of construction photos are available on the CD or on the web at
  http://www.adrianlaycock.com

When using the striker tube for laying floors, it is usual to run it on precast concrete screed bars which are first adjusted to the correct level and then concreted into the finished floor. Similar arrangements can be applied in a canal, as in figure 9.6 where the templates are cast in situ to the correct profile. This method naturally lends itself to the construction of large in situ parabolic canals.

figure 9.6   Concrete placing on the Pehur High Level Canal

figure 9.7   Slip forming the Ghazi-Barotha power canal, Pakistan

Slip-forming

Slip-forming is a mechanised concrete placement method in which a steel former shaped to the finished canal cross-sectional profile is travelled along the length of the canal and concrete poured in continuously, screeded and compacted in a single pass. Joints are usually cut afterwards in the green concrete. A winched former needs long straight reaches of canal. Larger canals are often slip-formed one side at a time, with travelling forms that are guided by rails or wires along the top of the canal bank. For medium and large canals, a paving train will consist of a trimmer, a concrete placing rig, and a finishing rig for final smoothing and spraying of curing compound. Figure 9.8 shows a design of paving train for a large parabolic canal. Other pictures are shown in figures A9.25-A9.31.

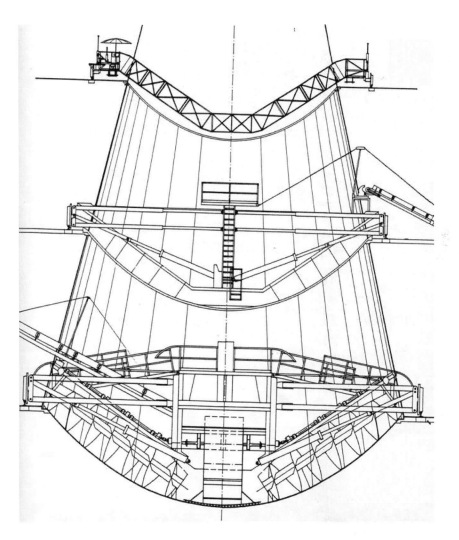

figure 9.8   *A paving train for a parabolic canal*

Joints

Joints in concrete lining can serve several functions:

- Expansion joints, to accommodate thermal expansion.
- Contraction joints, to accommodate shrinkage as the concrete matures or as temperatures fall.
- Construction joints, to provide a seamless junction between concrete pours.
- Movement joints, to accommodate differential settlement at the junction with structures, or local movements induced by swelling or settling soils.

Expansion joints require an insert of compressible material, such as 10 mm thick expanded polystyrene. This allows movement of the concrete without causing high stresses.

Contraction joints are a deliberate, clean-edged discontinuity which induce the controlled formation of cracking due to shrinkage. The first edge to be formed is normally painted with bitumen to enable slight movement without spalling the concrete.

Construction joints are situated to fit the contractor's methods of construction and are normally proposed by the contractor to be approved by the engineer. They are used where the concrete is required to be homogenous but where continuous concrete placement is not practicable. They should normally have a mechanical key and should be scarified while the concrete is still green, within 24 hours of placing. Reinforcement if present should be continuous across the joint.

Movement joints normally have a mechanical key, or ledge, to limit the vertical shear displacement of one component against the other. In canal lining this situation would arise where lining abuts against a heavy structure. If a continuous impermeable membrane is used, it must be folded at the joint in such a way as to allow any movement to be taken up in the folds.

Cold joints are to be avoided. They are unintentional discontinuities in concrete, caused by delayed arrival of consecutive pours. A delay in placement of adjacent pours of more than about 1 hour (less in hot weather) may result in a cold joint, which effectively breaks the concrete structure at that point and creates a permanent crack.

In small canals practical experience suggests that joints are often inappropriate. Friction between the underside of the lining and the subgrade beneath renders it difficult to predict where any cracking will occur unless the thickness of concrete is completely uniform and it is bedded on a smooth membrane or sand. Thin concrete lining tends to crack transversely, due to shrinkage and temperature effects, at intervals of 6 to 7 m. Transverse joints if used should be at this spacing.

In rehabilitating an old masonry canal with a concrete or plaster overlay, it is impossible to predict exactly where cracking will occur. It is then practical to lay continuous concrete, observe any subsequent cracking, and then if necessary cut and fill a joint to coincide with the location of the crack.

There are some potential weak spots where joints will be called for. If the lining is on a convex curve, any expansion will tend to force the lining toward the centre of the canal. Lining adjacent to a heavy structure liable to settlement or in a swelling soil will require expansion joints.

In large canals, transverse joints at 6 m intervals have generally proved satisfactory, with every third joint an expansion joint and the rest as contraction joints. The Pehur High Level Canal (figure 9.6) used a hybrid joint type at every cast-in situ screed bar. The mechanical key effectively prevented any movement of soil through the joints, which were not sealed.

Longitudinal joints are only necessary in large canals wider than about 15 m. Below this width, expansion can take place naturally provided there is no restraint at the top of the canal bank. Longitudinal joints would normally be run at the junction of base and sides, or at the start of a fillet in a curved section.

Joint sealing is not necessary. A bead of polysulphide or bitumastic sealant is often specified to cap contraction and expansion joints. It is standard practice in design of water-retaining structures which must not be allowed to leak, but in canals it is both expensive and ineffective. Joints tend to seal naturally with fine silt, and sealant material is sometimes stolen or dislodged by people or animals.

Reinforced concrete

Although it is sometimes used for canals, reinforcement in concrete is usually not necessary. The danger in using reinforcement is that engineers tend to think that reinforcement automatically confers strength. They then forget to ensure that the concrete is properly mixed, vibrated and cured, and so it leaks. This method is also expensive and wasteful, because in order to get the cover (thickness of concrete over the steel reinforcement) required by most codes of practice, it needs at least 100 mm thickness of concrete, which in a small canal is more than is necessary for a watertight lining. The structural stresses on lining once it is in place are minimal, provided its earth base is solidly compacted; there is usually no need for reinforcement on grounds of structural strength. Even precast segments do not need reinforcement unless they are heavier than about half a tonne.

Jointless lining

Jointless concrete technology utilising steel fibre reinforcement is commonly used in Europe for factory floors. The design philosophy assumes that stresses are distributed evenly throughout the concrete matrix, and that the steel fibres result in an evenly distributed resistance to cracking, so that any cracks that do develop are microscopic. A typical specification would be a C35 concrete containing 40 kg/m^3 of steel fibres. To my knowledge, this method has not yet been used in canal lining, but there is no reason why it should not be, in situations which demand some form of reinforcement such as soft ground or transitions into large structures.

Plaster

Cement plaster is useful for small canals and as an overlay in the rehabilitation of old concrete or masonry lining. It is cheap and easily repaired, but quality control can be difficult. Its thickness is usually specified as 20 to 30 mm but ensuring this thickness in practice is not easy. Normally specified as a cement/sand mix in the ratio of 1:3, 1:4 or 1:5, complete mixing is important and frequently not done properly, leaving weak spots which are easily damaged by animals and people. Curing is rarely done, and vast amounts of tropical plasterwork are wasted because of it.

Ferrocement

Useful in rehabilitation work, as an overlay to stone masonry, and canals in embankment, ferrocement is easy to make and repair, and is very strong. It consists of a fine chicken-wire mesh embedded in mortar, typically 25 mm thick. It needs care in ensuring that the wire is encased in the centre of the mortar (figures 9.9 and A9.33.) It cannot be done with separate layers of mortar as it goes off quickly in hot weather and will de-laminate quickly and crack up. The whole profile of a small canal should be done at the same time, otherwise the construction joints at junction of base and sides will open and cause leakage.

figure 9.9 **Laying ferrocement**

Fibre reinforced concrete

Steel fibre reinforcement is a recently developed method in which short crimped wires are added to the concrete during mixing. Polypropylene and glass fibres are also used in the same way. One advantage over standard steel bar reinforcement is that a thick concrete section is not necessary in order to provide cover to the steel. As in ferrocement or mass concrete, the internal stresses are distributed evenly throughout the concrete profile, and even a thin layer can give adequate strength.

However, some field trials of steel fibre-reinforced precast canal segments showed some interesting side effects. Firstly, the fibres exposed on the surface quickly rusted and invited a slow but steady deterioration of the concrete surface through micro-spalling. Secondly, the units were no more effective than high-strength mass concrete and they were much more expensive.

Sprayed concrete

Sprayed concrete, also known as Gunite or Shotcrete, can be a quick and effective means of lining a new canal or rehabilitating an old one, given the availability of the necessary equipment, including a concrete pump, and vehicle access all alongside the canal. Figure 9.10 shows sprayed concrete applied over a basic former with light mesh reinforcement.

figure 9.10   Shotcreting a small canal, Swaziland (courtesy Chris Cronin)

9.4   Precast Concrete Segments

Precast units made to the full cross-sectional profile of the canal have lots of advantages. They can be accurately made to hydraulically efficient and apparently complex profiles. They can resist lateral forces from swelling soils, and can resist shear forces induced by poorly compacted subgrades. They don't need backfill and can be made free-standing, thereby reducing land-take. They can be installed by semi-skilled local contractors with minimal supervision. They can be installed in wet conditions, as in figure A9.48.

Quality control is the main reason to insist on precasting. Quality can be easily controlled, provided they are made in a properly equipped casting yard with basic facilities for concrete mixing, vibrating and curing. They must be properly designed, with full attention to jointing details, but if these basic rules are followed there is no better lining for small canals. The parabolic units shown in figures 9.11 and A9.34 - A9.51 were considerably cheaper than the stone masonry that would otherwise have been used.

Design

Field trials in Java compared parabolic channels of various strengths and various types of reinforcement, including steel fibres. The best design proved to be mass concrete of high strength (40 N/mm^2, 400 kg/cm^2, or 5000 psi) without reinforcement. Reinforcement only adds to the cost, and induces internal points of weakness in the structure, especially if bonding is not good. As with all precast concrete, the biggest stresses occur in handling and laying, with lateral soil pressure an important factor. Hydrostatic forces from water in the canal are negligible. With this strength of concrete, the thickness can be reduced to 50 mm or even less.

Shapes - parabolic is best

Compare the shapes of figure 6.14. Parabolics are a natural choice in every respect. There are no stress concentration points, so they do not break easily. The potential weak spots at the junction of base and sides are strengthened by the gradual thickening around this point. A flat base aids laying, increases strength at these weak spots, and is a natural result of inverted wet casting (see below.) Very importantly, neither cattle nor bicycles will go near them. They are hydraulically good. They do not trap suspended silt and tend to be self-cleaning over a wide range of discharge.[1] They deflect the forces from swelling soils. And above all, they look beautiful.

figure 9.11   Parabolic lining is so smooth

Other, less useful shapes include half-rounds, which are in many countries available from local concrete pipe factories, and trapezoidals of various configurations.

Half-round pipes of semicircular profile are theoretically the optimum shape for highest carrying capacity versus lowest material cost, but only with zero freeboard. However, in practice this criteria is irrelevant because it is eclipsed by the inherent structural inefficiency of a design with constant thickness, and the increased hydraulic roughness due to the spun-casting methods usually employed. Nevertheless, half-rounds share many of the hydraulic advantages of parabolics, and because they are readily and cheaply available in most countries they can be a useful method of lining.

Trapezoidal precast units are all too frequently designed by less adventurous engineers. There is no excuse for precasting trapezoidal or worse, rectangular canals. It is easier and far better to make parabolics.

In India a precast concrete maker proudly brought us a sample of his new trapezoidal design to test. It was almost rectangular in fact, and made of poor quality concrete but reinforced with mild steel bars. The thickness was the same throughout the profile, including the sharp corners. It had been designed to withstand a supposed internal hydraulic pressure so the reinforcement was close to the internal face. In fact all the main stresses come from the other direction, either during handling and transporting or from swelling soils after laying. And the bond of thick steel bars with the weak concrete was so poor that the bars served as internal stress-concentration points against which the surrounding concrete was easily crushed. Within 30 seconds flat and with our bare hands we had reduced it to a heap of rubble. But the contractor soon realised the folly of his design and switched immediately to parabolic canals of international quality, using exactly the same equipment and facilities as he had before.

---

1 *They don't work too well with a heavy bed load, that is, medium to coarse sand which moves along the canal bed by rolling and saltation. A wide flat bed having a large area of boundary layer is better in this case.*

coordinates in mm																	
Size 1144 x 686			size 920 x 610			size 760 x 530			size 700 x 480			size 640 x 460			size 540 x 305		
x	y	y2	x	y	y2	x	y	y2	x	y	y2	x	y	y2	x	y	y2
635	686	686	521	609	610												
610	686	629	495	609	541												
584	686	572	470	609	473												
572	686	544	460	609	445	450	530	530									
495	515	374	439	556	390	405	530	420									
470	464	317	414	494	322	380	530	359	400	480	480						
445	415	260	391	441	260	355	463	298	350	480	353	370	460	460	310	305	305
381	305	170	366	385	222	340	424	261	315	389	264	345	460	401	290	305	253
340	243	112	340	334	185	315	364	200	290	330	200	320	460	342	270	305	200
315	208	76	315	286	147	290	309	161	265	275	163	300	404	295	245	251	164
290	176	40	290	242	110	265	258	121	240	226	125	275	340	235	220	202	128
264	147	4	264	201	72	240	211	82	215	181	88	250	281	176	195	159	92
239	120	-32	239	164	34	215	170	43	190	141	50	225	227	117	170	121	56
216	98	-64	216	134	0	190	133	4	165	107	13	200	180	58	145	88	20
203	87	-82	191	105	-37	165	100	-36	140	77	-25	175	138	-1	120	60	-16
191	76	-100	165	78	-75	140	72	-75	120	56	-55	150	101	-60	100	42	-45
127	34	-100	124	45	-75	100	37	-75	100	39	-55	125	70	-60	100	42	-45
102	22	-100	99	28	-75	100	37	-75	100	39	-55	100	45	-60	100	42	-45
76	12	-100	76	17	-75	75	21	-75	75	22	-55	75	25	-60	75	24	-45
51	5	-100	51	7	-75	50	9	-75	50	10	-55	50	11	-60	50	10	-45
25	1	-100	25	2	-75	25	2	-75	25	2	-55	25	3	-60	25	3	-45
0	0	-100	0	0	-75	0	0	-75	0	0	-55	0	0	-60	0	0	-45

table 9.1   Profile coordinates for precast parabolic canals

Size

The size of precast units has to be determined by the practicalities of handling, transport and laying. There is no theoretical limit, but if manual labour only is available, then the maximum weight for easy handling is limited to about 200 kg. The units in figure A9.48 weigh about 100 kg. The minimum weight should be great enough to deter theft![2] The 2 m long units in figure A9.44 weigh over 200 kg, and a gantry was used for handling and placement.

Manufacture

Some engineers are too readily deterred from the parabolic shape, which is perceived as being too difficult to make. In practice for small canals it is easier to make a steel mould in the naturally-flexing shape of a parabola than it is to make a trapezoidal one. There are several ways of precasting, including wet casting with or without steam curing, hydraulic pressing, spinning, and plastering onto an inverted former.

Wet casting

The method which gives the best finished product is to cast the segments upside-down as in figure 9.12. If steam-curing facilities are available, 5 hours curing at 120 degrees C is enough to ensure full strength. Otherwise, the moulds can only be stripped after 12 hours and the units kept wet for 4 to 7 days. Wet casting gives a very smooth finish and ensures a densely compacted and high-strength unit. The parabolic segments of figure 9.11 were 15 mm slump concrete, 1:1.5:3 mix, compacted on a vibrating table for 45 seconds. Finished cube strength was 350 kg/cm^2, above the 300 specified.

---

2 *In India we had problems with overnight disappearance of newly-laid precast parabolics. It seems they became a favourite target of thieves who found they could be used as horse-troughs, baby baths, and many other domestic purposes.*

The rules for making segments like this are:
- Steel moulds
- Proper concrete mixing
- Proper vibration on a vibrating table
- Proper curing.

Once these rules are met, precast parabolics can be made anywhere. Figure 9.12 is a field in rural India, 200 km from the nearest large town.

figure 9.12   A field casting yard for parabolics, India

Hydraulic pressing

Modern mass-production of small items such as building blocks or interlocking paving slabs use a hydraulic press to force a cement/sand mixture into a steel mould. The mould is stripped immediately afterward and the curing process begun with immersion or exposure to water and, sometimes, heat in the form of steam. Steam curing hastens the process considerably. Pressed dry-cast concrete is quick and cheap but there is a limit on the size of unit that can be precast. Trials in Java using locally available facilities in which the canal segments were cast on end had a maximum unit length of 1 metre (figures A9.45- A9.48.) Their surface finish was somewhat rough due to the dry concrete mix, and vibration (provided by a horizontal vibrating table) was often uneven and incomplete near the top of the mould.

Pipe spinning

Half-round pipes are usually manufactured by spinning, as for the complete pipes. A tubular steel mould is rotated on a horizontal axis on rollers, and wet concrete is evenly spread inside as it spins round (see figure A8.19). The concrete adheres to the mould with assistance from centrifugal force. To get two half-rounds from one pipe, the pipe moulds are longitudinally split with steel spacers before casting. The main disadvantage with this method of casting is that the internal surface finish is usually rough, and if steel reinforcement is used it is difficult to ensure an even cover. And very often the extra work involved in manufacture means that the cost of a half-round is nearly as much as the cost of a complete pipe.

Building-up

This is a method commonly used with thin ferrocement and glass-reinforced concrete (GRC) and also for GRP (glass-reinforced plastic), although this material is not appropriate for canals at its present cost. Layers of plaster are trowelled onto an inverted former, incorporating reinforcement at the appropriate juncture. It is easy to set up a manufacturing yard in the field. The process is labour-intensive but quick, and the resulting segments are light in weight, although this is not an advantage when designing a canal that will last a hundred years. When GRC units were developed for small watercourses in Egypt they were made light enough to be lifted by two men, but they proved to be collapsible under internal hydrostatic pressure when the earth support of the banks was eroded (see figure A9.63.)

Jointing

The joints in precast segments are crucial:

- They must prevent leakage of water between segments.
- They must restrict vertical or horizontal shearing movement between adjacent segments.
- They must absorb any longitudinal shrinkage or expansion movement of the canal lining.
- They must also act as location lugs for accurate and easy laying.

Best is the lap joint of figure A9.50. The joint gap should be filled with fine mortar. Trials with a flexible bitumastic joint filler which would not crack under stresses from moving soils were not encouraging. Flexible foam rubber joints have also been used as on the Maskane Project in Syria in figure A9.52, but the material does not last long and soon perishes or gets damaged.

In practice the vertical shearing forces even in the heaving montmorillonite soils of east Java were found to be very small in relation to the horizontal forces on the canal sides. Mortar is therefore adequate for these joints, and although it will crack eventually it is easy to repair. High strength and plasticising additives can also enhance the durability of jointing mortar.

The lap joints should be laid in sequence so that the top lip is laid after the bottom lip of the adjacent unit is in place to ensure a clean joint surface. Most canals will be constructed from their head-end down in order to get water flowing quickly or to avoid drainage water and silt being washed down slope and interfering with on-going laying work. Therefore the laps will 'point' up-slope. This may look wrong but in fact the hydraulic force of flowing water tends to suck out jointing material vertically and not in the direction of flow. In practice the orientation of the laps does not matter so long as cleanliness is maintained during jointing.

Laying

Laying precast units is easy because they do not require much preparation or compaction of the underlying soil. Even in wet or loose soil, they merely need a reasonably smooth and level surface to rest on and they can then be manoeuvred into position by tamping soil at the sides. In the wet soils of figure A9.48 it would have been impossible to lay any other type of lining. It is easier but not essential to lay them on a bed of sand, and a flat bottom on the segments assists in initial positioning.

Guide segments should be levelled in at about 20 m intervals, then the intermediate segments boned in or laid to a string line. Backfilling of the sides should be done before jointing, to avoid disturbing newly mortared joints. Compaction does not need to be very thorough, its purpose is only to wedge the segments in place and stop them moving sideways. Unlike in situ trapezoidal lining, the banks do not have to support the weight of the canal sides.

Canalettes

Canalettes are elevated channels carried on piers as in figure 9.13. They can be half-round, parabolic or rectangular in section. They are a useful and simple form of aqueduct for crossing long, shallow depressions that would otherwise require raised embankments. Their main advantage is minimal land take, and any leakage is easy to observe, enabling timely repairs to be carried out.

In the states of the Former Soviet Union parabolics are used almost everywhere in this way. They are typically 4 metres long and hence reinforced in order to support their self-weight over this span. Precasting also fitted in well with the centralised control of production and construction exercised in the Soviet Union, and allowed fabrication to take place during the winter when weather conditions are not conducive to good construction.

figure 9.13   **Parabolic canalettes, Kazakstan. Parabolics are ubiquitous in the Former Soviet Union**

9.5   Flexible Impermeable Membranes

There are three groups of flexible membranes: plastic or polymer sheet, bitumen-based compounds including asphalt, and geosynthetic clay liners. Plastic membranes are hydrocarbon-based and their price is closely linked to the oil price. That is to say, it rises when the international price of oil rises, and it does not rise so quickly when the oil price falls. Using plastic membranes seems a quick and attractive way of building canals, but there are some pitfalls.

In canal works, most membranes need to be used in combination with a protective covering above, and a blinding layer of sand or a geotextile blanket below it.

Impermeable membranes are appropriate in the following situations:
- In water-sensitive rocks and soil which are liable to break down or disperse on contact with water.
- In soils with high permeability.
- In embankments in which a piping failure could cause severe disruption and high cost of repair.
- In sections in cut or fill in which water seepage through cracks or joints in concrete could saturate the adjacent soil, and result in external hydrostatic pressures on the concrete in the event of a sudden drawdown of the canal water level.

Polythene LLDPE

This is obtainable almost everywhere. Linear low-density polyethylene film is extruded in tubular form and in its commonest configuration usually leaves the factory in a long flattened roll up to a metre wide, suitable for making plastic bags. Bigger sheets can be had up to 8 metres wide. For small canals it is quick and convenient to lay out the roll as a single sheet. Thickness needs to be at least 1000 gauge.

The main drawback of low-density polythene sheet is its weakness. It is easily torn during laying, easily pierced by roots and animal hooves, easily ruptured by abrasion against stones or hard earth, and even the black ultra-violet stabilised type eventually turns brittle after exposure to sunlight. Jointing on site can be done by hot air welding, or on small canals simple lap joints may suffice.

In order to last longer than a few weeks, polythene has to be covered with a solid protective layer of concrete, bricks or earth. There are documented cases from USA and China of polythene surviving unscathed for 30 years.

HDPE

High density polyethylene (HDPE) is stronger but less pliable than ordinary polythene, which is low or medium density. It can be made in heavy sheets up to 2.5 mm thick. In this thickness it is a very tough material indeed. It is commonly used in thickness of 1 - 2 mm for lining small intermediate storage reservoirs or lagoons for water or effluent treatment. It can be heat-welded on site.

Extensive use of HDPE in the Middle East has highlighted a common problem, that of expansion in high air temperature which causes rippling and bubbling of the surface and difficulty in placing a protective covering layer without creating voids beneath it. LLDPE is usually preferred, even though its cost is about 10 per cent more, as any wrinkles which form can be more easily smoothed and flattened.

Butyl rubber, EPDM

Butyl rubber was hailed as a wonder lining material when it first appeared 35 years ago. It is tough and very elastic, unlike the polyethylenes. The joints are heat-welded. It is vulnerable to root action, and even small roots can find their way in between the joints. More vigorous roots will penetrate straight through it. On a large canal or reservoir there may not be many roots because the plastic will generally be deep down in the subsoil. But small canals are usually at or close to the surface and surrounded by roots all the time.

PVC

Polyvinyl chloride in its unplasticised form uPVC used to be notorious for deteriorating with age and sunlight. It is made in thicker membranes and is cheaper and tougher than low-density polythene, but it is not elastic, and is easily penetrated by roots.

However,, recent advances in its manufacturing have led to a plasticised PVC which is immensely durable and guaranteed for 15 years against the type of deterioration for which its predecessors were notorious. It has multi-directional strength as opposed to polyethylene, which tends to tear. As a canal or reservoir lining it can be employed in conjunction with a thin covering of lean concrete to protect it from puncturing by people, animals or machinery. Some proprietary membranes are bonded to a geotextile fleece, which protects the membrane from stones and imperfections in the subgrade.

Buried membranes

Trials in Sri Lanka compared various types of lining, including polythene sheet with an earth covering which appeared to be very cheap (figure A9.60). But it required an earth cover of at least 200 mm in order to avoid damage from wandering cattle or children. In a small canal the excavation required is roughly twice that for an unlined canal. There is a large labour effort required in double-handling the spoil and removing sharp stones from the subgrade and from the covering soil.

But more significantly, the plastic membrane when in place and buried acts as a slipping plane for the earth sides to slump on. Side slopes steeper than 1 on 3 will slip.

Membranes with in situ concrete

In the design of the Merowe Left Bank Conveyor, the most appropriate lining was found to be in situ concrete with a layer of geotextile or impermeable plastic. The soils contain zones which are highly dispersive, and it was important to avoid the risk of cracks and joints in the concrete allowing seepage and the migration of fine soil particles. Four cases were considered in arriving at the final design of LLDPE membrane with 100 mm in situ concrete.

Case 1 Rapid drawdown with hydrostatic forces behind the impermeable membrane

Rapid drawdown will occur in some reaches of the canal when gates are closed. The maximum rate of drawdown estimated in a hydraulic model study was 500 mm / hour. Drawdown downstream of some cross regulators could be up to 2.9 m. The impermeable membrane is intended to prevent leakage from the canal into the surrounding soil. There are no locations on the Project where significant external groundwater pressure is expected. Even if small amounts of water leak through the membrane, it will be in a very limited area and pore water pressures will disperse rapidly enough to avoid damage.

Case 2 Rapid drawdown with hydrostatic forces between membrane and concrete

This situation could occur if water seeps through joints and cracks in the concrete, and there is no bond between the concrete and membrane. The effects of this could be to exert a hydrostatic force, but only if the underlying soil were also saturated, since the membrane would tend to deform before the concrete lifted. In practice, any water that entered through joints should also exit by the same route. Some membranes are produced with a textured surface that bonds to in situ concrete overlay. This can be important in that it reduces the risk of water seeping between the membrane and the overlay, and building up enough pressure to burst either the plastic (if there are voids behind it) or the concrete.

Case 3 Rapid drawdown with hydrostatic forces behind geotextile filter membrane

The function of geotextile is to prevent migration of fine soil particles through cracks and joints in the concrete when hydrostatic forces develop behind the lining. During normal canal operation, some seepage may occur into the surrounding soil. Impermeable membrane is used at all potential points of rapid drawdown.

Case 4 Displacement of the soil support

Several options were considered using finite element analysis. Any significant displacement of the soil will result in damage to the concrete, but progressive failure will be avoided by the filter geotextile or membrane.

Membranes with bricks or precast slabs

Unprotected membranes such as LLDPE are weak and easily ruptured by people or animals walking on them. If it is covered with a protective layer of bricks or slabs (figure A9.58, A9.59) then it can be a very useful method of emergency lining for repairing a breach. Precast concrete slabs as thin as 30 mm have been successfully used in many countries, but for permanent lining, in situ concrete is to be preferred. There is a constant risk of slabs loosening and failing as in figure 9.5, or being stolen as in figure A9.57.

Material	Minimum thickness (mm)
PVC - polyvinyl chloride	0.5
EPDM - ethylene propylene diene monomer (synthetic rubber)	1.2
EPDM (reinforced)	1.2
Polyurethane/geotextile composite	1.2
HDPE – high-density polyethylene	1.0
LLDPE - linear low-density polyethylene	1.0
PP - polypropylene (reinforced)	1.0
Bituminous geomembrane	3.0
GCL - geosynthetic clay liner - sodium bentonite	3.7 kg/m^2

table 9.2   Minimum thickness of membranes for use in canal lining

Bitumen and asphalt

Trials in Sri Lanka tested two methods of lining using hot-run bitumen, one as a buried membrane and the other as a surface dressing on bare soil with a blinding layer of granite chippings using road surfacing techniques. As a seepage inhibitor it was cheap, and simple to construct. But it did not last long before the roots had penetrated from the bank top down to the waterline, and after 2 or 3 years there was little sign remaining of the original surface bitumen. The seepage rate had been slightly reduced by the buried membrane which remained in place despite being penetrated by roots.

Asphaltic concrete as used for road wearing courses and increasingly in dams has been used in canal lining in the USA and elsewhere, usually laid with adaptations of road paving techniques. Its advantage over concrete is that it doesn't need joints and will accommodate temperature stresses and soil movement without damage. It is worth considering if suitable placing and compacting equipment can be made available. It is generally more expensive than plain concrete. It is also susceptible to degradation in strong sunlight.

Geosynthetic clay liners (GCL)

GCL is a sandwich of needle-punched and woven geotextile, infilled with bentonite. The bentonite swells in contact with water, by about 35 per cent of its volume, and forms an impermeable layer which is held in place by the geotextile. The membrane needs to be buried to a depth of at least half a metre, or else protected by a concrete overlay.

9.6   Geotextiles, Geogrids and Geocells

Geotextiles

Geotextiles are membranes primarily intended for use as filters, permitting the passage of water but not fine soil particles. There are three types, classified by their method of manufacture: woven, non-woven and needle-punched fleece. Some of the heavier woven polyester types are also used in soil reinforcement.

Woven geotextiles comprise strands of polypropylene interwoven in perpendicular directions. Polyester and nylon are also used in heavier grades which are used for embankment strengthening over weak soils such as peat.

Needle-punched geotextiles are usually composed of polyester filaments. They have an uncompressed thickness of 10 mm or more. They can be used as drainage blankets, often bonded with thicker meshed structures to provide water passages. They can also be used purely as a protection layer against physical damage of buried impermeable membranes.

Non-woven geotextiles are a thermally bonded mixture of polyethylene and polypropylene, formed by spraying filaments of molten plastic. Thermo-bonded is not used in slopes, since it is more slippery than the needle-punched geotextile and the required friction between geotextile and underlying soil will not be achieved. Pouring of concrete directly on thin thermo-bonded geotextile will clog its pores. This is also a problem with needle-punched geotextiles, so that geotextiles directly receiving concrete (i.e. side drains in highway construction) are thick needle-punched textiles with an opening size small enough that the surface tension of the cement wash does not penetrate and clog it.

Impregnated Geotextiles – a mistake!

In Brunei some lined interceptor drains were needed for a new town in a hilly area with sandy erodible soils. The location of the drains halfway up a steep hillside called for a lining material that was easy to transport and lay in this almost-inaccessible position. The initial design was a lightweight flexible lining that was also very cheap; a geotextile filter fabric, laid in the excavated drain, pinned to the earth with plastic nails, and then impregnated in situ with hot-run bitumen. This was a mistake (mine!). Fortunately this absurd design never became reality because the client was shown some of the same geotextile that crumbled in his hands after having been exposed to the sunlight. The lining would not have lasted 3 months, before sunlight and root action destroyed it.

The real mistake had been a strategic one. The lining strategy had not been completely thought through. Searching for a lining that was cheap and transportable had overlooked the facts that Brunei had plenty of money to pay for a permanent solution, and that heavier materials could still be got into position albeit with some extra effort and cost. The correct solution, and the one eventually adopted, was to use precast concrete segments.

Geogrids

Geogrids are mesh-type membranes usually manufactured from high density polyethylene, and primarily used for soil reinforcement. They can also be used for reinforcing concrete slabs and asphalt roads.

Geocells

Geocells filled with concrete are a recent development, which have been used successfully for canals and mine tailings reservoirs in South Africa. Geocells are formed from geotextile or thin plastic in a honeycomb configuration. It is primarily intended as a slope stabilisation method, covering steep earth slopes and permitting vegetative growth through the cells. When laid out it forms a blanket of individual cells from 70 to 200 mm high, which are then filled either with soil or concrete.

An advantage in using concrete-filled geocells as canal lining is the complete avoidance of joints. Any expansion or contraction is taken up in the thin walls of the cells. They are suited to a curved or filleted canal profile.

9.7 Other Prefabricated Materials

Steel

Sheet steel is occasionally used as a lining material, and can be surprisingly cheap. Second-hand material can often be bought cheaply, usually as a stop-gap measure which cannot be expected to last more than a few years. Steel plate aqueducts are sometimes found on old schemes in the mountainous areas of Pakistan, but more modern materials such as HDPE pipe are easier to use.

GRP

Glass reinforced plastic or fibreglass was first developed for medium to large diameter pipes in the early 1970s. It is lighter than steel or concrete, and does not rust like steel or get affected by sulphates like concrete. It can be moulded easily and accurately into complex shapes. It had been used in

Thailand for small irrigation canals and in 1984 I was looking for a quick and easy means of lining small canals on a new sugar project in Sulawesi. I had a sample parabolic section made up at the local GRP pipe factory. It did not work because it was twice the cost of concrete, and it was too light and easily damaged or removed. I do not know what became of the Thailand canals but there can be no great future for GRP as a canal lining.

GRC

Glass-reinforced concrete makes a strong, lightweight lining which is easy to form into precast units of any shape. A suitably equipped factory is required in the vicinity, and for a large amount of lining it is worthwhile considering a customised factory for the project. The process of making GRC entails the mixing of glass-fibre rovings in with the concrete. The mix can then be sprayed or manually placed onto timber, steel or plastic formers. It results in a material that is extremely durable and does not crack. Repairs are simple. The concrete thickness can be as little as 1 centimetre, and individual units several metres long can be handled manually. However, GRC as thin as this can fail under hydrostatic pressure, so it has to be fully earth-supported. The results of failed trials in Egypt are shown in figure A9.63. As with many of these materials, it pays to test out new ideas with some pilot projects, in order to arrive at the most practical size and thickness.

Other similar methods of reinforcing precast concrete include polypropylene fibres and steel fibres. Polypropylene fibres about 5 cm long can be got cheaply from the makers of cheap plastic rope. Steel fibres are typically from 3 to 5 cm long and the thickness of heavy office staples. The trials in Java indicated that fibres exposed at the surface soon rusted. They were also more expensive than the extra cement it took to make a high-strength mass concrete.

Asbestos cement

Asbestos cement has for a long time been a common material for water pipes and roofing materials in Africa. It is now little used in most countries due to health risks of asbestos fibres in domestic water supply, even though the autoclaving process in manufacture alleviates most of the risk. Several manufacturers have produced small channel sections from AC and it is sometimes used for small irrigation canals (figure A9.65.) But it breaks too easily and is therefore of limited value compared with other methods.

Timber

Figure A9.64 shows some beautiful timberwork in Montana, USA, which was constructed in the early 20th century but is still working well. Such workmanship is difficult to find now, and modern methods are bound to be cheaper. Plywood can be useful in rectangular flumes for temporary diversion channels across earth dams, but as a permanent lining material timber is obsolete.

Recycled plastic

A very recent development in small canals is the use of recycled plastic waste. The idea comes from Germany, where the Green party had a big influence in creating laws that oblige producers of plastic packaging to recycle it. The German people have co-operated so well in returning their empty plastic bottles that the producers (and hence the Government who subsidise its collection) have huge amounts of waste plastic they cannot get rid of. It is light, unbreakable, flexible, tough, non-biodegradable, waterproof and very cheap. Therefore it was proposed as suitable for half a million kilometres of tertiary canals in Pakistan, but so far it has not been used in this way.

9.8   Stabilised Earth Lining

In its broader sense, the term earth stabilisation applies to any method of soil treatment which increases its strength, density or impermeability.

Compacted earth

The commonest and best form of earth stabilisation is simple compaction. It is the single most important process in canal construction and is rarely done properly. The canal bank that has not been properly compacted is easily noticed. Breaches, steady seepage and sudden failures can usually be attributed to inadequate compaction.

Compaction is the process of rearranging the particles in the soil matrix to increase the overall soil density. In the process the particles knit together and the soil matrix as a whole becomes stronger. The soil moisture content is critical. A completely dry soil is almost impossible to compact, as is one that is too wet. Water in the pores of the soil acts as a lubricant for compaction, but too much water takes up space which could be used by soil particles after compaction. The optimum moisture content gives the strongest soil matrix for a given soil and compaction effort. This optimum is expressed in engineering terms as the result of a laboratory test, usually the Proctor test or a variation of it. As for any earthworks, when building earth canal banks in fill, water and proper compaction equipment, whether simple hand rammers or a vibrating tamping foot roller, must be available.

Engineering specifications indicate the lowest acceptable density (usually 90 to 95 per cent of optimum) after compaction. This can be easily measured after construction, either accurately using a sand replacement test or more rapidly using a Proctor needle. A well-graded soil with a fair amount of clay in it has a wide range of moisture contents which give it a good strength. On the other hand, a poorly-graded silt has a very narrow range of water contents either side of optimum, so it is difficult to attain the correct water content. The strongest soils, and those which are easiest to compact, usually have a wide range of grain sizes and an even grain size distribution.

Consolidation is the process of squeezing water out of the soil under its own weight. This can take many years for a large embankment, and cannot be achieved in the same way as compaction. It can be encouraged by internal drainage.

Soil cement, and cement-bound fill

Soil of the right consistency can be made into a strong, near-watertight material by mixing it with cement and water and then compacting it. In Africa it is still common to construct very hard-wearing roads this way, in a laterite soil with cement mixed into the top 150 mm with a rotavator. However, it is not so easy to do it on a sloping canal bank, and generally not worthwhile for a small canal. It requires a well-graded soil with not too much clay, and appropriate equipment for mixing and compacting.

Cement-bound granular fill (CBG) is commonly used for filling in canal banks in order to reduce the likelihood of settlement of concrete lining, but its main function is to create an incompressible foundation rather than an impermeable one. A sand/cement mix in the ratio of 24 to 1, or a river gravel/cement mix of 22 to 1, depending on available materials, are used, machine-mixed as for concrete.

Bentonite

Bentonite is a thixotropic[3] clay which swells in volume by 40 per cent when wetted. It is used as a drilling mud and as a support fluid during construction of diaphragm walls, sandwiched in cardboard or geotextile for foundation waterproofing in buildings, and occasionally also in dams in a trench to seal off a plastic impermeable membrane. It has been employed dispersed in canal water as a sealant but any such applications are rare and uncertain. It can be injected as a grout into open-grained soils. This was a solution proposed for the Yeleru Canal, where poorly-compacted lateritic soils were causing huge seepage losses. Bentonite injection holes stitched along the bank would permit sealing of the canal without draining it. It forms the impermeable component of GCL membranes, in which format it is more useful.

---

3 *'Thixotropic' refers to the property of variable viscosity, in which the mixture of clay and water is stiff when left undisturbed but becomes very fluid as soon as it is agitated by pumping or other mechanical disturbance.*

Silty water

If there is fine clay or silt suspended in the water, then as water seeps through the banks the suspended sediment can fill up the pores in the soil matrix, effectively sealing it. Silty water is a good and cheap means of sealing an earth canal against seepage. The flood waters of the Swat river on the north-west frontier of Pakistan are loaded with fine silt, which has done a remarkable job of sealing canals in the loess soils of the Peshawar Vale, but not so effective in sealing the coarse mineral soils of the upper Machai canal on the same canal system. If the soils are too coarse then no amount of silt will seal them.

Oil, molasses

The buried bitumen membrane mentioned in section 9.5 above is a method of soil stabilisation that seals the pores of the soil matrix whilst contributing no inherent strength itself.

Molasses, being a by-product of sugar processing that is sometimes difficult to dispose of, has been used on some sugar estates for stabilising roads and canals in a similar way. Oil, including used engine oil and crude oil in those countries fortunate enough to have it in over-abundance, has also been used. Its environmental effects are not recorded.

9.9   Unlined canals

The correct solution for any particular canal might be no lining at all. This should always be the starting point, and the conclusion to spend money on lining should be the result of a series of objective decisions.

figure 9.14   How not to build a canal. Dozing up the banks without compaction, Sri Lanka

Tight soils

Plenty of soils are so densely structured that they are almost impermeable. Many do not need lining if prevention of seepage is the objective. Some like the black cotton soils (alkaline clays with a high montmorillonite content) of Africa swell and shrink on wetting and drying, and hence develop wide cracks which absorb a lot of water if the canal flows intermittently, and transmit the cracks to any inflexible lining which may be supported on it. Canals in the volcanic *Grumosols* of Java may be impermeable but their banks easily lose their shape and become wide and shallow with high dead storage losses.

Field trials are invaluable in assessing potential seepage losses. Compaction tests will indicate potential problems in forming embankments.

Earth cut

Very often a soil in its natural undisturbed state will be stable and impermeable, but when excavated and formed into an embankment it will be prone to seepage and difficult to stabilise. The compromise solution of partial lining in only the embankment reaches is often economical and effective, leaving the cut sections unlined.

Rock

Hard rock is usually impervious, but not always. The Machai Branch Canal in Pakistan runs partly in several tunnels through hard limestone, but the shear zones where tight folds have been created in the strata contain very weak rock which over the years has developed deep solution cavities which drain away large quantities of water. Some canals in Upper Burma were constructed through gypsiferous rock which just dissolved away.

Usually rock needs no lining, and the canal can be cut with vertical sides to a minimum-area rectangular profile. However, a block-jointed rock such as dissected granite may create a very rough surface when excavated, and it may be worthwhile to smoothen the bed and sides with a concrete overlay or shotcrete in order to reduce the hydraulic roughness.

The Merowe Left Bank Conveyor in Sudan will pass through hard gneiss over the first 40 kilometres, then through Nubian sandstone for the next 250 km. Both these rock types have potential problems which led to the decision to line the entire canal system. The gneiss is heavily fractured in places, and massive in others. It is riven with faults and macro joint planes which have contributed to a rough surface topography with numerous deep gorges and ancient river channels. More than half the length of the canal in gneiss will be in fill or partial fill in crossing these drainage channels. The cross-drainage embankments, some of which will be 15 m high, will require lining as their component materials of crushed rock and gravel will be highly permeable. Transition lengths between embankments and rock-cut sections, and sections partly in cut and part in fill, will also require lining. The length of canal in solid rock which would not require any lining was only 10 per cent of the total length, so it was deemed easier and no more expensive to line continuously.

The Nubian sandstone is hard in its natural dry state, but can disintegrate rapidly in water. It also contains zones of mudstone which are extremely fine-grained and are dispersive in water. The risk of not lining in these materials was clear. Collapse of the canal profile would be inevitable, and this has already happened in other unlined canals in the same area (see figure 5.6.)

References and further reading for chapter 9

(1) Justo *et al*, 'A Finite Element Model for Lined Canals on Expansive Collapsing Soil', proc.11th conf.soil mech foundation eng, San Francisco, 1985.
(2) Kishel, J., 'Seepage and Contraction Joints in Concrete Canal Linings', Proc. ASCE v115 # 3, June 1989.
(3) Chengchun, K.E., Singh, V.P., 'Chinese Experience on plastic membrane concrete thin slab lining for canals', Irrigation & Drainage Systems 10:77-94, 1996.

	suitability for: easy water management	good hydraulics	low health hazard	low land take	seepage prevention	easy maintenance	maintaining canal shape	easy construction	easy supervision	durability	labour intensive	local materials	resistance to: soil swelling/ shrinkage	poor soil compaction	uplift pressures	physical damage	crabs, rats	root penetration	theft	cost: cheap	low foreign exchange
IN SITU CONCRETE																					
mass concrete	o	o			o	o	o		o	o	o				o	o	o	o	o	o	o
reinforced concrete	o	o			o	o	o		o	o	o		o	o	o	o	o	o	o	o	
sand/cement plaster	o	o			o	o	o	o		o	o	o		o	o	o	o	o		o	o
ferrocement	o	o			o	o	o		o	o	o		o		o	o	o	o		o	
sprayed concrete	o	o			o	o		x	x	o	x	x				o		o			
geocells	o	o				o	o		o	o			o	o	o	o		o			
slipformed concrete	o	o			o	o	o	x	x	o	x	x			o	o	o	o	o		x
PRECAST CONCRETE																					
parabolic	o	o	o	o	o	o	o		o	o		o	o	o	o	o	o	o	o	o	o
half-round	o	o	o	o	o	o	o		o	o		o	o	o	o	o	o	o	o	o	o
trapezoidal/rectangular	o		o	o	o	o	o		o	o		o		o	o	o	o	o			
fibre reinforced	o		o	o	o	o	o		o	o			o	o	o	o	o	o			
reinforced concrete	o		o	o	o	o	o		o	o				o	o	o	o	o		x	
prestressed	o		o	o	o	o	o	x	o	o			o	o	o	o	o	o	o	x	
asbestos cement			o		o	o	o	o	x				x	x	x	x	o		x		
GRC			o	o		o	o			x			x	x	x	x	o		x		
SLABS																					
precast concrete	x				x	x		o	o	x	o	o	x	x	x	x	x	x	x	o	
stone	x				x	x		o	o	x	o	o	x	x	x	x	x	x	x	o	o
MASONRY																					
brickwork	x				x	x		o	o		o	o	x	x	x		x			o	o
clay tiles	x				x	x	x	o	o		o	o	x	x	x	x		x		o	o
mortared stone	x				x	x	x				o	o	x	x	x		x	o			o
stone packing	x	x	x	x	x			o	o		o	o		x	o		x	x	x	o	o
PLASTIC MEMBRANES																					
LDPE	x	x	x	x	o	x	x	x	o	o	x		o	o	o	x	x	x			o
HDPE	x	x	x	x	o		x					x	o	o	o						x
butyl rubber	x	x	x	x	o		x	o	o			x	o	o	o		x				x
plasticised pvc	x	x	x	x	o		x	o	o			x	o	o	o						x
OTHER																					
asphalt, bitumen	x	x	x	x		x				o			o				x	o			
soil cement	x	x	x	x	o	x	x			o	o		o			x	x	o		o	o
GCL membrane	x	x	x	x	o	x	x	o	o	o	o	x	o	o	o	o	o	o	o		
clay	x	x	x	x	o	x	x			o	o		o	x	o		x	x	x	o	o
compacted earth	x	x	x	x	o	x	x	o	o		o	o	o	x	o		x	x	o	o	o

o SUITABLE	**x** UNSUITABLE

table 9.3   Durability of canal linings

CHAPTER 10   CANAL HYDRAULIC DESIGN

This chapter deals with several design tools including the Manning formula, which is the most widely used of all engineering design methods and, for all practical purposes, entirely adequate for most irrigation canals. An appraisal of methods for designing mobile-bed alluvial canals is given, with a description of a field method for assessing existing unlined canals which can then be used as a design guide.

10.1 Basic Tools

There are plenty of good books on hydraulics and hence no in-depth theory is repeated here. However, it is useful to recapitulate some of the basic principles upon which the procedures of design in the following chapters are based. A body of water tends to flow when subjected to a differential pressure across it. Its movement is resisted by friction between the water and the sides of the channel or pipe, by physical barriers such as sluice gates, and by internal friction or viscosity. The latter is an unchanging factor in irrigation save for exceptional cases such as the soupy silt-laden waters of the Yellow River.

Stable channels and Manning's formula

There are several formulae for canal design, but the one in most common use is that of Manning, which came into general use in the early years of the 20th century. Precursors of the Manning formula were those of Chezy, Bazin, Kennedy and Thompson. Successors are the Tractive-force method and Colebrook-White. The Strickler formula is an alternative form of Manning, used in mainland Europe. These are all applicable to steady flow in stable channels or pipes with non-changing profiles. All rely on assumptions of roughness, which have to be based on subjective engineering judgement and experience.

Manning's formula is universally known and familiar to almost every engineer involved in irrigation design. This is enough to justify its continued use. There has been plenty of research into ***Mannings n,*** the roughness coefficient that is the crux of the matter. Appropriate figures are given in table 10.2.

Manning gives velocity in terms of channel geometry, slope and roughness:

$$v = \frac{R^{0.67} s^{0.5}}{n}$$

or, in terms of the discharge,

$$Q = Av = \frac{AR^{0.67} s^{0.5}}{n}$$

where v = velocity m/sec
  R = hydraulic mean radius = A/P
  A = cross-sectional water area
  P = wetted perimeter
  s = longitudinal slope, or hydraulic grade line
  n = roughness coefficient, Manning's n

The ***conveyance***, $K = Q/s^{0.5}$, is sometimes used to describe the canal capacity, and as an intermediate step in the iterative calculation required to solve the Manning equation. With computers using numerical methods of calculation, solving Manning is straightforward. A spreadsheet method, and a more comprehensive piece of software which produces charts for specific shapes of channel, are included on the CD of this book.

Alluvial channels and Lacey's regime formula

Many canals are built in environments less conducive to accurate analysis. Canals which offtake from sediment-laden rivers may have special characteristics which are not constant. The alluvial canals of northern India fluctuate in discharge and sediment load over the course of the year, depending on the seasonal flow conditions in the main rivers. Yet over the course of a full year the canals create for themselves a natural cross-sectional profile and slope which pertains to the *dominant discharge* conditions. There is for a given dominant discharge a unique channel profile which depends on the type and size of sediment carried in the water. A canal that has attained this long-term state of maturity, in the same way as a natural river, is said to be *in regime*.

Over a long time span Gerald Lacey carried out many observations on existing canals in northern India, and devised from them a series of mathematical formulae relating the channel profile to the sediment size. The smaller the silt particle size, or the larger the discharge, the flatter the longitudinal slope. And the finer the silt, the deeper and narrower is the channel cross-section. The channel sides form a curved surface underwater approximating to a slope of 1 on 0.5 for all regime canals. If this seems steep for a stable earth slope, consider the flow velocity profiles in figure 10.3, and think of silty water as a two-phase fluid which solidifies at low velocities. There is a threshold velocity below which the silt particles will settle out and be deposited. The finer the silt, the lower the threshold velocity. In a trapezoidal channel profile, this threshold velocity will occur close to the channel bed and sides, and adhere roughly to the squashed circular pattern of figure 10.3. So below this threshold velocity line is deposited the silt which creates the final stable channel section. And the steep sides are supported underwater. They are part of the water itself.

Lacey's regime function can be stated as:

The hydraulic gradient $s = f^{(5/3)} / 3340\ Q^{(1/6)}$ in metric units
$$s = f^{(5/3)} / 1840\ Q^{(1/6)} \text{ in imperial units}$$

cross sectional area $A = 2.283\ Q^{(5/6)} / f^{(1/3)}$ in metric units
$$= 1.26\ Q^{(5/6)} / f^{(1/3)} \text{ in imperial units}$$

wetted perimeter $P = 4.84\ Q^{0.5}$ in metric units
$$= 2.67\ Q^{0.5} \text{ in imperial units}$$

Lacey's silt factor $f = 1.75(d_{50})^{0.5}$

Where d_{50} = mean sediment size in mm.
Typically, f = 1 in the Punjab, and usually in the range 0.6 (very fine silt) to 1.6 (coarse sand and gravel)

Regime design applies an appropriate measure of sediment size, the **silt factor, or Lacey's f**, to the required discharge and delivers a unique combination of channel geometry and slope. These equations can be solved by iteration, for which there is a spreadsheet on the CD of this book.

In the Punjab Lacey is used everywhere, even for small canals. But how relevant is it, and when should it not be used? In Pakistan the Upper Swat Canal rehabilitation project presented some interesting questions. How to double the capacity of canals that had been operating in regime for almost a century? Do we keep the same slope and just increase the cross-sectional area? Do we modify the silt factor? At first the answers were not very obvious.

The dimensions and slope of some existing canals were measured, ranging in capacity from 5 cumecs down to less than half a cumec. Lacey's f was estimated by back-calculating from the present discharge and channel architecture. For the distributary canals with a capacity larger than about 2 cumecs there was a close fit between theoretical longitudinal slope as predicted by Lacey and the actual situation in the field. But on the smaller minors Lacey did not fit. The slope of minor canals was much steeper than predicted by Lacey, even for the original discharge capacity. For the regime formulae to be applicable, the silt factor would have to be very high, approaching 2, whereas it would be expected to decrease down the canal system from about 1.0 in the headreaches to perhaps 0.8 near the tail as the heavier sediment dropped out. In the end, Manning was used successfully for all canals.

The depth/top width ratio

Rehabilitating the Upper Swat Canal system offered an opportunity to investigate the profiles of a great many unlined canals that had been in operation for a century or more and therefore had attained a stable state. It is well known that the cross-sectional profile of a natural channel depends on the type of soil, longitudinal slope and the dominant discharge, but apart from Lacey's theories, which are applicable to capacities greater than about 10 cusecs, useable guidelines for design of small unlined stable channels are hard to come by.

Because the profile of these canals was roughly parabolic, the only practical measurements possible were the top width of the dominant discharge water surface and the depth at the deepest point. The ratio *D/T*, depth to top width, is plotted in figure 10.1. Theoretical values of *D/T* were superimposed as calculated by Lacey's regime formula and Thompson's formula, the latter being an empirical relationship in use in northern India in the early part of the 20th century. Although the scatter of measured points is quite wide, several lines of best fit were tried out, and the semi-logarithmic linear relationship shown fitted best. Its correlation with Lacey's theory for discharges greater than 10 cusecs is remarkable. The derived empirical formula was then used as a design guide to check that all unlined canals could reasonably be expected to remain stable over the long term.

In SI units, *D/T = 0.162 - 0.08 log(Q)*
In imperial units, *D/T = 0.29 - 0.08 log(Q)*

The trapezoidal canals were designed using Manning, adjusting the bed width to give an acceptable *D/T* ratio and adjusting the slope to give a non-scouring velocity. Table 10.1 shows a range of bed widths applicable for various discharges.

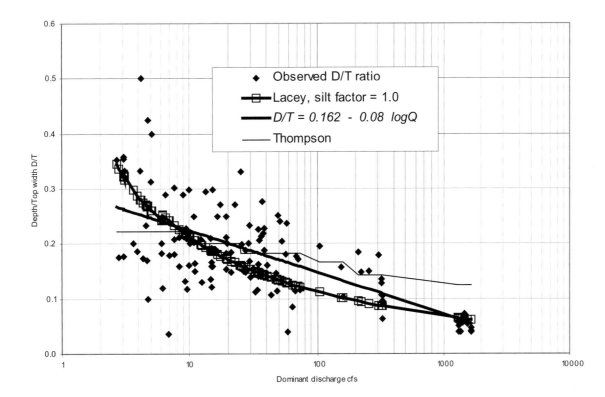

figure 10.1   Depth/top-width ratio for unlined canals

Design Discharge cumecs	Bed width in m for given slope...										
	0.0002	0.0003	0.0004	0.0005	0.0006	0.0007	0.0008	0.0009	0.0010	0.0015	0.0020
0.25									1.00	1.00	1.00	1.00
0.50					1.75	1.50	1.50	1.50	1.50	1.25	1.25
0.75				2.00	2.00	2.00	2.00	2.00	1.75	1.75	1.50
1.00			2.50	2.50	2.50	2.25	2.25	2.25	2.25	2.00	2.00
1.50		3.50	3.50	3.00	3.00	3.00	3.00	3.00	3.00	2.50	2.50
2.00		4.00	4.00	4.00	3.50	3.50	3.50	3.50	3.50	3.00	
2.50		4.50	4.50	4.50	4.00	4.00	4.00	4.00	4.00		
3.00	5.50	5.50	5.00	5.00	4.50	4.50	4.50	4.50	4.00		
3.50	6.00	6.00	5.50	5.50	5.00	5.00	5.00	4.50	4.50		
4.00	6.50	6.50	6.00	5.50	5.50	5.50	5.00	5.00	5.00		
4.50	7.00	7.00	6.50	6.00	6.00	5.50	5.50	5.50	5.50		
5.00	7.50	7.50	7.00	6.50	6.50	6.00	6.00	6.00			
6.00	8.50	8.00	7.50	7.50	7.00	7.00	6.50				
7.00	9.50	9.00	8.50	8.00	8.00	7.50	7.50				
8.00	10.50	10.00	9.00	9.00	8.50	8.50					
9.00	11.50	10.50	10.00	9.50	9.00	9.00					
10.00	12.00	11.50	10.50	10.00	10.00						
11.00	13.00	12.00	11.50	11.00	10.50						
12.00	14.00	13.00	12.00	11.50	11.00						
13.00	14.50	13.50	13.00	12.50	12.00						
14.00	15.50	14.50	13.50	13.00							
15.00	16.00	15.00	14.50	13.50							
16.00	17.00	16.00	15.00	14.50							

table 10.1   Design guidelines for unlined canals in silty soil,
permissible velocity range 0.5 - 1 m/s, side slopes 1:1

Other sediment transport theories

There are a number of alternative methods which have been developed for the analysis of flow in alluvial channels, and are well described in DORC (Ref. 9). These include the Rational Methods and the sediment transport theories of Ackers and White, White Paris and Bettess, Chang, Brownlie, van Rijn, Einstein, Engelund and Hansen, Yang, Colby, Westrich and Juraschek, and others. We do not consider them further here as they are of limited applicability in the context of most irrigation canals. We did test them on the Upper Swat Canals, but found them to give extremely variable results, and none of them was more applicable or relevant to those conditions than Lacey and Manning.

10.2 Canal Architecture

A canal can be described in terms of some physical parameters which enable us to predict its behaviour. These are:
- Roughness;
- Longitudinal slope;
- Cross-sectional geometry.

Channel roughness

Over the past century there has been a good deal of research into roughness, some results of which are given in table 10.2. The roughness coefficient is affected by the state of maintenance of the channel. Weed, silt and physical damage will all increase the roughness, but the extent to which it will alter also depends on the size of the channel. A 10 centimetre growth of grass might completely block a small field canal, whereas it would have very little effect on the carrying capacity of a main canal carrying 20 cumecs. Note that the figures for weeded or damaged canals are all subjective, and are typical values which reflect the likelihood of their being encountered in practice rather than definitive fixed values.

The depth of water in the channel can also affect the roughness. A trickle of water in the bed of the channel will be subject to a higher roughness coefficient than if the channel is flowing full. Experiments with closed conduits (Ref. 1) showed a 30 per cent variation in roughness coefficient with changing depth. And in natural river channels the variation is as much as 50 per cent. If the roughness is visualised as sand grains of particular size which interfere with the water flow, then clearly a channel of large boulders will be effectively very rough (i.e. offer a lot of resistance to the overall flow) if the water is very shallow, but will be effectively less rough if the water is deep. Hence the value of Manning's n varies slightly with the hydraulic radius R.

figure 10.2   A rough canal. The Pitched Channel of the Upper Swat System, Pakistan. This was intentionally designed as rough as possible in order to control the velocity over its 4 mile length. Slope of the channel is 1 in 100, with additional ramp falls shown at intervals of between 10 and 30 metres. Manning's n is estimated at 0.025.

In the case of grassed waterways used in soil conservation and field drainage works, the roughness coefficient varies with the state of flow, because at higher velocities the grass gets flattened and its roughness reduces. Chow (Ref. 1) quotes research evidence relating the roughness to $V * R$, the velocity multiplied by the hydraulic radius.

If the channel is irregular (i.e. non-prismatic) then the roughness coefficient can be effectively increased by up to 0.005 if the changes in cross-section are sudden and frequent. The horizontal alignment or curvature of the channel can also increase the effective roughness, if the curves are tight, by up to 0.002.

Manning's n also varies with size of channel, reducing as the size of channel reduces. Small precast parabolics half a metre wide may have an n as low as 0.012 whereas a channel 5 m wide cast to the same standard of concrete finish would have an n of 0.016. For these reasons, the Colebrook-White formula, in which roughness is allotted a physical dimension is slowly gaining acceptance in some academic circles.

Since the real value of roughness coefficient depends on engineering judgement, in unfamiliar situations it may be necessary to use some lateral thinking. Take a likely range of values, then test out the effects on channel profile, velocity and water depth of adopting either a high or a low value. It may be that there is little difference between them. But avoid the mistake of over-estimating the roughness coefficient just to be on the safe side; this can lead to under-estimating the velocity which can be potentially more troublesome than over-estimating the depth. Feed-back from the field is invaluable; the easiest way of checking the roughness coefficient is in the form of measured canal response times, like those in figure 4.1

type of lining	Manning's *n*		
	clean	weeded	damaged
Smooth glass laboratory flume	0.009		
smooth GRP, HDPE, PVC pipe	0.009		
steel	0.01	-	0.012
slip-formed concrete	0.011	0.013	0.014
precast concrete segments	0.01	0.012	0.014
smooth cement/sand plaster	0.012	0.014	0.016
ferrocement	0.012	0.013	0.013
concrete, tamped finish	0.014	0.015	0.015
sprayed concrete, gunite	0.016	0.018	0.018
asbestos cement, GRC	0.012	0.014	0.014
polythene membrane, unprotected	0.012	-	0.018
precast concrete slabs	0.013	0.015	0.017
bricks, flush pointed	0.014	0.016	0.016
bricks, dry laid	0.018	0.02	0.02
stone masonry, mortar joints	0.018	0.02	0.022
stone packing, dry joints	0.02	0.025	0.025
earth, trimmed & compacted	0.02	0.025	0.025
earth, rough excavation	0.024	0.03	0.03
rock cut	0.025	0.026	-
grassed waterways	0.025	0.04	-
natural channels	0.03	0.1	0.15

table 10.2   **Canal roughness** [1]

Hydraulic geometry

The normal depth, *d*, is the water depth measured perpendicular to the bed of the canal when the flow is uniform.

The wetted perimeter, *P*, is the cross-sectional length of contact between the water and the channel.

The cross-sectional area, *A*, of the water profile is directly related to the discharge, or volume rate of flow:

$$A = Q/V$$

The hydraulic radius, *R*, is the ratio of area to wetted perimeter:

$$R = A/P$$

The hydraulic depth, *D*, is the ratio of area to surface width:

$$D = A/T$$

If a channel reach has the same cross section throughout it is said to be *prismatic*. Design is much more straightforward in a prismatic channel, and it is normal to divide a long canal into prismatic *reaches* for the purpose of design. A reach can be any length, usually designated between cross regulators, offtake structures or changes of cross-section.

---

1 For a graphic and comprehensive description of roughness in natural channels, grassed waterways and canals in varying states of maintenance, see Ven te Chow (Ref. 1), the hydraulic engineer's bible.

10.3 Water flow in canals

The nature of water flow is a subject ripe for the attention of the emerging science of chaos. But until such time as chaos takes over, more familiar concepts are the rule.

Permissible velocity

The velocity of flow is the target parameter for most canal design. Usually the aim in design is for a stable flow having a velocity that is neither too fast to cause erosion, nor too slow to allow silt deposition or weed growth. It also has to be reasonably fast in order to transmit water at an economical rate. The slower the velocity, the larger and more expensive the canal section has to be. Some permissible maximum velocities to avoid erosion or scour are given in table 10.3. Minimum velocities to prevent siltation are shown in table 10.4. Note that these are mean velocities for small canals. Large deep canals can carry a faster mean velocity.

In designing unlined or earth-lined canals, the aim has long been to get a velocity that is both non-silting and non-scouring. The Kennedy and Lacey formulae were developed for the large alluvial canals of Egypt and Punjab with this intent. With small canals, subject to the vagaries of intermittent maintenance and often intermittent use, alluvial channel analysis is not usually appropriate.

With hard linings erodibility depends on the quality of construction. Poor quality concrete can erode as quickly as bare soil. Cement can wash out of the surface of poorly-cured plaster at even quite low velocities.

canal material	maximum velocity, m/s
fine sand	0.45
sandy loam	0.55
silt, silty loam	0.6
loam	0.75
fine gravel	0.75
clay	1
clay loam	1.1
coarse gravel	1.25
densely compacted clay loam	2
low density polyethylene, unprotected	2
grass	2
bricks, dry laid	3
stone masonry, mortared	3.5
cement/sand plaster	4
concrete	>5

table 10.3   *Permissible non-scouring mean velocities*

sediment type	minimum velocity, m/s
fine silt	0.3
fine sand	0.4
sandy loam	0.5

table 10.4   *Non-silting mean velocities*

Velocity distribution

The above velocities are all mean values for the whole channel cross section. In fact the velocity varies widely across the depth and breadth of the canal. Figure 10.3 shows a typical velocity profile in a canal with a rough bed. Close to the bed, the velocity is slowed due to the influence of boundary layer turbulence and friction. The effect of friction diminishes gradually with distance above the bed. The maximum velocity occurs at around two-thirds the water depth, then reduces again toward the surface.

So for bulk discharge calculations only the mean velocity is of interest, but this is greater than the bed velocity, which is the principal factor in causing erosion.

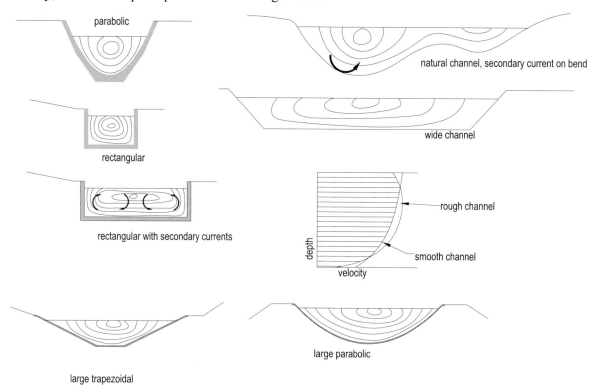

figure 10.3   ***Channel velocity profiles and distribution with depth***

Steady and non-steady flow

If flow is ***steady***, the depth and velocity do not change with time. Non-steady flow exists for example at the advance wave when water is let into a canal, or when a gate is opened and the flow is suddenly changed. Routine canal design always assumes steady flow, but special conditions may create unsteady flow, the consequences of which need to be considered during design. Non-steady flow can be gradually varied, such as in a slowly changing flood wave or recession wave after closing a regulator, or it can be suddenly varied, such as the surge wave caused by a canal bank breach or a regulator gate being suddenly opened. A hydraulic jump or a hydraulic drop are examples of steady flow that is non-uniform.

Uniform and non-uniform flow

When the depth, discharge and velocity are the same[2] along the whole length of the canal reach, the flow is ***uniform***, and the hydraulic slope is the same as the bed slope of the canal. The depth is unique to any given slope and discharge and is called the ***normal depth***.

When the depth varies, due to a changing cross-sectional profile or a control structure downstream, the flow is ***non-uniform***, and the hydraulic slope is the gradient between two known points along the energy line.

Non-uniform flow can exist with both steady and non-steady flow. Uniform flow in practice (though not in theory) is always associated with steady flow. Uniform flow does not apply when the channel is very steep or the velocity very high. Under these conditions, which might be found on a steep spillway, air is entrained and the flow becomes ***ultrarapid***.

---

2 *These definitions are a little ambiguous and are sometimes quoted only in terms of the depth, or on the assumption that velocity does not change with depth for uniform flow. However, for all practical purposes the definitions given here are the most useful and easily understood by practising engineers.*

Energy and head

Standard mathematical analysis is derived from Newtonian mechanics. For the straightforward design of a canal the Manning formula is enough, but in designing structures, transitions, and investigating the flow at changing sections of the canal, it is necessary to turn to the principles of conservation of energy and momentum.

The ***total energy*** of moving water in a canal is comprised of potential and kinetic energy, or potential head and velocity head:

$$E = z + d \cos \vartheta + \alpha V^2/2g$$

where z is the height of the canal bed above datum
d is the depth measured perpendicular to the canal bed
ϑ is the slope of the canal with the horizontal
α is the velocity distribution coefficient, of which more later
V is the velocity
g is gravitational acceleration

For almost-flat prismatic canals (i.e. when ϑ is less than about 10 degrees or a slope of about 1 in 6), this is usually approximated to:

$$E = z + d + V^2/2g$$

The ***velocity head*** $V^2/2g$, is ubiquitous in hydraulics analysis.

There is a convenient physical law of ***conservation of energy***, which can be applied at consecutive points along a flowing canal, for instance at the inlet and outlet of a structure or transition section:

$$E1 = E2 + \text{internal friction losses}$$

where internal losses through turbulence are dissipated in heat.

Another useful concept is ***Specific Energy***, where elevations are related to the channel bed only:

$$E = d + V^2/2g$$

The velocity distribution coefficient α, also known as the Coriolis coefficient or the energy coefficient, is an empirical figure that takes into account the additional energy arising from variable velocity across a channel profile. In complex profiles such as natural streams with flood plains, α can be as high as 2, but for most prismatic channels it is generally between 1.1 and 1.2. (Ref. 1.)

Momentum

The ***momentum*** of moving water is:

$$M = \beta WQV/g$$

where β is the velocity distribution coefficient
W is the unit weight of water
Q is the discharge.

The main difference between this and the energy equation is that momentum theory deals with vector forces or pressures. It is useful for situations such as the flow through radial gates.

The law of ***conservation of momentum*** between two points on the canal is:

$$M1 = M2 + \text{pressure difference} + \text{external friction losses}$$

Another useful concept is ***Specific Force***, where the parameters are related to the channel bed and the hydrostatic pressure is introduced:

$$F = Q^2/gA + ZA$$

where Z is the height of the centroid of the water profile area A.

The velocity distribution coefficient β is an empirical figure that takes into account the additional momentum arising from variable velocity across a channel profile. In complex profiles such as natural streams, β can be as high as 1.33, but for most prismatic channels it is generally between 1.01 and 1.07. (Ref. 1.) It is also known as the Boussinesq coefficient or the momentum coefficient.

Critical flow

Introducing the *Froude number*:

$$F = V/(gD)^{0.5}$$

where D is the Hydraulic depth or hydraulic mean radius A/P

Consider a channel with a certain discharge and an adjustable bed slope. Starting with a flat slope, the flow will be subcritical, deep and slow. Any disturbance of the water surface will produce a wave that travels both upstream and downstream. If the slope is increased by tilting the channel, the velocity will increase and the depth will decrease. Eventually as the slope gets steeper the flow gets so fast that However, much the water is disturbed no waves can move upstream, they all get washed down along with the current. The flow is supercritical. The channel slope at which critical flow begins is termed the *critical slope*.

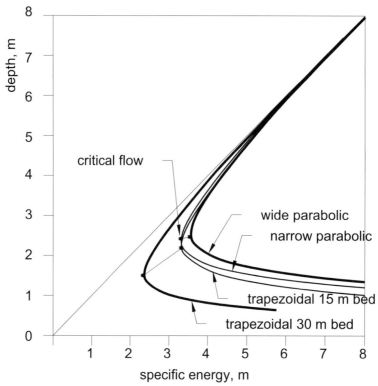

figure 10.4   Specific energy curve for different canals

Considering the specific energy of this same channel, $E = d + V^2/2g$, it has two components based on the depth and the velocity, which go in opposite directions as we vary the slope. As the depth decreases, the velocity increases. A graph of specific energy plotted against depth is in the form of the U-shape of figure 10.4, with the specific energy a minimum at a certain depth. This is the critical depth, where the flow passes from subcritical to supercritical. At critical flow, $V^2 = gD$. In other words the Froude Number F = 1. At subcritical flows, the Froude Number is less than 1, because the velocity is low and the depth is large. At supercritical flows the Froude Number is greater than 1.

The specific energy curve shows that for any energy level there are two possible depths, one subcritical and one supercritical. These are *conjugate depths*, and if the flow state changes from one side of critical to the other it also changes from one depth to its conjugate depth. The most dramatic instance of such a transition is the hydraulic jump.

An appreciation of the flow state is important in practice when designing flow control structures in a canal. Subcritical flow is controllable from upstream or downstream, but supercritical can be controlled only from upstream. Thus a gate or constriction in the canal will not influence the speed or depth of supercritical flow upstream of it.

Secondary flow

Water in a canal does not flow in a straight line. Even in a straight reach, there is a slow three-dimensional spiralling flow aligned along the canal. This flow becomes significant when the canal turns a bend. The secondary current then acts downwards on the outside of the bend, tending to cause scour, and upwards on the inside of the bend, causing deposition.

In designing gravity river offtakes a strong secondary flow is often induced by increasing curvature of the approach channel, in order to deflect sediment away from the canal intake.

The secondary currents in a wide, straight reach periodically invert, creating a meandering *thalweg*, or line of maximum velocity. This is important in river engineering, when a curved channel may be more stable than a straight one. It is also noticeable in flat-bottomed canals under low flow conditions, when sediment is sometimes deposited in a series of miniature sandbanks on alternate sides of the canal.

Turbulence

Laminar flow is not found in irrigation practice outside the laboratory, and all irrigation water flow is ***turbulent***. Turbulence is described by the Reynolds Number:

$$Re = VD / v$$

where V is mean velocity, D is the hydraulic depth,
and v is the kinematic viscosity of water, 0.000001004 m^2/s at $20°$C.

When the Reynolds Number is greater than about 500, the flow becomes turbulent. This has little practical significance in the field of irrigation, but is included here for the sake of completeness.

It is more informative to consider macro-turbulence, which is important in the design of canal structures. Any surface waves, eddies or 'white water' indicate turbulence. The exact nature of turbulence is not fully understood. It can be partially explained from the study of vortices, which are created by shear forces between fluid and a solid boundary, or between two bodies of fluid of different density or temperature or moving at different velocities, or from within the fluid body itself through internal viscous shear. A vortex is a swirling motion with localised high velocities and low internal pressures, which can be intensely erosive. Energy is dissipated as heat and in erosion.

The new and incomplete science of Chaos[3] often cites fluid turbulence as an example of a chaotic system, which breaks down into increasingly smaller and intensive patterns of vortices the more you look into it. But however interesting the concept may be, it is not yet of any practical value in the design of canals.

Design may focus on either one of three objectives:
- To encourage turbulence, in order to dissipate excess energy;
- To control it, in order to prevent scour damage to the canal and structures;
- To minimise it or avoid it completely to reduce energy loss or head loss.

Turbulence can be purposely ***created*** with a sudden change in canal profile or bed shape, such as a step, a vertical sill or a hydraulic brake in the form of concrete blocks in the bed.

Turbulence is ***controlled*** by containing it within a certain area such as a stilling basin which is not liable to erode. A containment structure must be designed with sufficient volume and depth to permit

---

3 *Chaos is the idea that it is possible to get completely random results from normal equations. It also covers the reverse: finding order in what appears to be completely random data.*

the full development of turbulence, and with converging flow conditions at its exit. Converging flow is an effective principle in smoothing out residual turbulence.

To **avoid** turbulence, canal structures are designed with smooth curved transitions, avoiding any sudden changes in geometry.

10.4 The Process of Design

The process of canal design can be oriented in one of several different directions, for example:

1) The discharge or capacity is known, the longitudinal slope is known, and the object is to determine the normal depth and velocity.
2) The discharge is known, and the object is to examine a range of canal slopes to see the effect on depth and velocity.
3) The longitudinal slope is known, the depth is known, and the object is to find the discharge capacity. (This is a common situation in the rehabilitation of existing schemes, or as a rough guide to estimating drainage flows in ditches or natural channels.)
4) The discharge and slope is known but the velocity needs to restricted within the range of non-silting or non-scouring. The object is to test out several different bed widths or profile shapes.
5) The canal slope is known and the object is to know the increase in depth and velocity in the event of a flood surcharge.
6) The object is to compare the cost of several different canal profiles and shapes having the same conveyance.
7) The object is to know the area of critical flow conditions to avoid or accommodate supercritical flow.
8) The object is to test the effects of low flow or flood surcharge on the depth and velocity.

But, as is often the case with engineering design, with hydraulic analysis of channels there is never just a single answer. Even if the canal lining type, the range of discharge and the canal profile shape are known, it is possible to construct an infinite combination of bed widths, water depths, longitudinal slopes and velocities that are possible for any particular case. The computer-generated design charts described in the next chapter can speed up the selection of an appropriate solution by providing individual charts custom made for any range of canal types and shapes and sizes.

10.5 Using Computer Spreadsheets for Canal Design

When in the early 1980s desktop computers became powerful enough to carry out most engineering design, many large design organisations developed their own canal design software using a programming language such as Fortran or Basic. But by the mid 1990s spreadsheets had become so sophisticated that they could handle most situations arising in the design of irrigation projects. The advantage of spreadsheets is their replication capability by copying rows, and *'what-if'* capability. Any variable can be changed and its effect on other parameters can be seen immediately. They also have the added advantage that all data are visible, graphical checking is easy, and data can be imported or exported easily to feed into other software such as CAD or surveying packages. Since they are used almost universally for all manner of calculations, every engineer in a computer-literate environment understands their use. Their drawbacks are that they use a lot of memory, calculation is relatively slow, and they require strict management control amongst multiple users to avoid perpetrating serious errors. Databases, which have also become highly sophisticated, are now often used as the carrier software once development has been completed on a spreadsheet. Databases can operate faster and with more efficient use of memory, but with the latest desktop computers restrictions on memory and processor speed are becoming far less onerous.

The great power of a spreadsheet stems from its *copy* function. Any set of calculations that can be set up in a tabular format of rows and columns can be established on spreadsheet. Once the first row or column is set up with the required formulae in each cell, it can be copied at will to as many rows or columns as required, and indeed to as many other spreadsheets as required. The canal design spreadsheet is a mathematical model of the canal which simulates steady flow. Unsteady flow can also be simulated on spreadsheet, but for the design of most canals, modelling steady uniform flow from reach to reach is sufficient. In the spreadsheets described here, each row describes the canal cross-section at a certain distance from the head of the canal. The columns represent existing and calculated water levels, bed levels, bank levels, and the intermediate design data needed to calculate them, plus the hydraulic characteristics of structures such as falls, weirs, cross regulators and outlets. Additional columns can be added to calculate quantities and abstract data for export to CAD drawing software.

A canal cross section is straightforward enough to design. So in theory is the design of a uniform flow, constant discharge canal, since it is just an extrusion of the standard cross section over the length of the canal. The reality is somewhat different, especially in the case of rehabilitation of existing canals, which is nowadays a more common design case than a brand-new canal in virgin ground. Existing canals often have cross-sectional profiles which vary depending on the lining materials used at various times during their life, the location in open country or village, and the situation whether in embankment or cut. They also might have numerous structures such as bridges, outlets and falls, and frequent changes in bed slope, width and bank level. It will need a new row in the spreadsheet for every change of geometry and every structure. Some structures such as falls or proportional dividers will need two or more rows at the same chainage.

Table 10.5 is an example of a canal model on spreadsheet. This was developed for two canal rehabilitation projects in Pakistan, in which there were over 1000 canals, varying in capacity from 20 l/s to over 90 cumecs, to be designed. The model was varied slightly for parabolic profiles and level-top canals with downstream control, and underwent regular upgrading throughout the 6-year design period but in principle remained as in table 10.5. The complete spreadsheet is not shown here, as it would take 10 pages to show the full 4 km length of this canal, plus another 200 columns which are not shown but include calculation of alternative water levels for low discharge (usually taken as the existing discharge prior to rehabilitation), construction quantities, echoed data for abstraction to ASCII files for structure design and longitudinal or cross-sectional drawings, and macros for inserting structure formulae and creating graphics for checking purposes.[4]

This model is based on energy levels and normal depths for uniform flow. Starting with a known energy level at the head of the canal, the levels in each reach are computed and can be adjusted as required to ensure correct command level at every outlet. The velocity is kept within a target envelope to be non-silting, non-scouring where possible. If the canal is unlined then an appropriate ratio of depth to top-width is maintained, for a stable channel profile. It would also be possible, though less precise, to work off the water level rather than the energy line since with low velocities there is little difference between the two.

The adjustable design variables are:
- Starting energy and water level at head;
- Longitudinal slope;
- Bed width;
- Side slopes;
- Lining type and hence roughness;
- Height of drops and falls.

The design criteria which have to be met are:
- Safe velocity range, non-silting, non-scouring;
- Meet correct command levels at outlets;
- Correct proportions of cross sectional profile for stable section if unlined;
- Economic proportions of cross sectional profile if lined.

---

4 A complete version of the spreadsheet is available on the CD.

The first column A contains the chainage or RD (Reduced Distance, in South Asian terminology) from the head of the canal in metres. This is also the *x* data for all longitudinal section graphics.

Columns B to G are field survey data, most of which can be inserted directly from an ASCII file provided the surveyors are briefed to supply the data in this format in the first place. On a large job it can save a lot of time and avoid potential typing-in errors if survey data is customised to fit into the design calculations.

Columns H to J detail the proposed structures which apart from the head regulator are a Crump weir, an open flume outlet, an AOSM outlet, three falls, two bifurcators and a village road bridge. These structure formulae occupy the cells boxed out on table 10.5. They can be inserted into the spreadsheet by macro, which is a computer routine that with the press of a single button loads all the formulae required for a particular structure into the correct cells.[5] Column J contains a design variable, the fall in water level or height of drop. During the course of design this can be varied to make the canal water levels suit the required command levels.

Columns K to N are the design criteria which have to be met in order for the canal to operate correctly. The function of this canal is to deliver water to its various outlets at a given command level. The discharge required through each outlet in this case is a function of the area commanded. The design discharge in the minor is a summation of the outlet discharges plus an allowance for losses in each reach. This is imported from a separate spreadsheet.

Columns O to T are the geometric properties of each reach of the minor, and are design variables in that they can be altered as necessary during the course of design in order that the canal meets its required command levels. The profile type and its corresponding Manning's *n* are derived from a lookup table by referring to the pointer number in column O. To enter data this way is much easier and less prone to errors. When computerising to this extent, it is important to standardise descriptive data which may be abstracted into derivative spreadsheets for the design of type structures.

The side slope of a trapezoidal canal 1:*m* in column R is chosen on practical grounds. With in situ concrete lining, a side slope of 1:1 would normally be practical only for small canals less than 1 m deep. A flatter slope of say 1:1.5 is necessary in order to place wet concrete on a deeper canal. A slope of 1:0 applies to a rectangular profile. In the case of an unlined canal, the side slope might be constructed as 1:2 but if the canal is in regime with silt deposition likely, a side slope of 1:0.5 might apply due to silt build-up on the sides approximating to a semi-elliptical profile for a stable channel. The bed width might be constrained say to fit into an existing profile under rehabilitation, or it may be chosen for optimal cost of lining. If the canal is unlined it will be proportioned to give an appropriate ratio of depth to top-width.

The longitudinal bed slope in column T has the most profound effect on the velocity of any of the design variables. It may be constrained by topography, in which a flat slope is dictated by a small available head between outlets. In a rehabilitated canal the pre-existing slope will often be a guide to start with, but there may be advantages in making the slope different. The *'what-if'* capability of a spreadsheet permits the testing of a large number of different possible arrangements in a short time.

Columns U to AA are normal depth calculations based on Manning's formula. Solving Manning's equation is an iterative process which works with successive approximations of a normal depth figure. It calculates the parameter $ar^{2/3}$ and the conveyance *z* from two different starting points. If they turn out to be equal then the assumed normal depth is correct. If not, then the assumed value is adjusted in column AA until the answers match. The *delta* in column Z is the difference and should be close to zero. With each iteration the trial depth in column V takes the value of the normal depth in column AA from the previous iteration. You have to experiment a bit to get the most efficient iteration function. We found the best to be making the depth in column AA = column V × (1 + column Z/column Y.)

---

5 *The original version of this spreadsheet, which is still doing the rounds in Pakistan, was developed using Supercalc5, with macros for these operations. The Excel version on the CD does not use macros.*

Different software may treat iteration slightly differently, but we found the relatively unsophisticated SuperCalc 5.1 to be straightforward to use and most of our spreadsheet development was done on this.

Column AB is the normal water level, which is given as a starting point in the first row, either directly or as the sum of a given starting bed level plus the calculated normal depth. Successive rows are calculated from the longitudinal slope times the reach length plus any drop in water level in column J. Design bed level in column AC is the water level minus the normal depth.

Velocity in column AD is calculated as discharge/area, columns N/W. Froude number in column AE is velocity divided by the square root of (g times normal depth.) A graphical check on velocity can be done the graph of figure 10.6. This is useful in checking for non-silting or non-scouring velocity at a glance along the whole canal or a selected length of it. It is very much easier to eyeball a graph than scan a long column of figures. The graph shows velocity for both the design discharge and the low flow situation, the calculation of which is not shown in table 10.5. Earlier versions of the spreadsheet for large unlined canals also included regime theory calculations of depth, slope and velocity, the latter also being shown on the same graph.

The normal water level may be modified by structures in the form of cross regulators, falls, measuring weirs or bifurcators. Operation of outlets such as flumes or orifices is affected by the water level. Columns AF to AI introduce any structures that are dependent on or affect the water level. In column AF is the crest level which is normally automatically inserted as a formula, a calculated depth below water level. It can be adjusted manually in order to meet the checking parameters. Crest width in column AG may initially be calculated as a proportion of the bed width, but may be manually adjusted to give some degree of standardisation in construction, or in the case of proportional dividers, to avoid drowning of the crest or to raise command level. Discharge coefficient and head in columns AH and AI are calculated from standard weir formulae for the structure in question.

The regulated design water level in column AJ may be modified by a manual raising of crest levels. In this case a backwater effect will be created, raising the water levels upstream. The formulae in column AJ therefore compare the normal water level at the RD in question with the water level downstream, and adopt the highest. This gives a horizontal backwater curve, which is not accurate but nevertheless is good enough for most situations in small canals, when the afflux is only a few centimetres and can be accommodated within the lining freeboard. In cases which merit more detailed analysis such as cross regulators on main canals, a more rigorous estimate of the backwater curve can be done on a separate spreadsheet and imported into the working model spreadsheet. In development of this method we found it best not to clutter the spreadsheet with too many subroutines that require iteration; a backwater curve might require 100 iterations and would slow down the calculation of the main spreadsheet which normally only needs about ten iterations.

Not shown in table 10.5 are the hydraulic calculations for the situation of low flow. In this case all crested structures will create a backwater when operating at less than their design discharge.

Columns AK to AQ are design data relating to freeboard, bank levels and widths, and profile geometry. Most of this is abstracted to the construction drawings, preferably automatically using ASCII files direct into CAD software. However, the depth/top-width ratio in column AP is a useful indicator of channel cross-section stability in the case of unlined canals. This is described more fully in section 10.1. If the value is obviously too high or too low for the discharge in question then the bed width in column S can be adjusted.

Columns AR to BD contain formulae for various structures, and checking cells to ensure that the structures will operate correctly. Five different structures are shown here: a Crump weir for measurement, an open flume outlet, an AOSM outlet,[6] and two proportional dividers, one of which is a double bifurcator. These structures are described more fully in Chapter 7.

---

6 *Adjustable Orifice Semi Module, see chapter 7.*

The Crump weir is checked for submergence in column AY. If this is less than 0.7 then an *'OK'* is returned in the next checking cell, column AZ. If it is drowned out, then the crest level would need to be raised or the crest width increased.

The operating head for the open flume outlet is calculated in column AR, and the width of the flume to give the required outlet discharge is in column AS. The relevant discharge coefficient, which varies with flume width, is given in column AT. A check on the minimum practical width for construction is given in column AZ. If the width is less than about 5 cm then construction will be difficult and inaccurate. The head loss through the outlet is calculated in column BA and hence the maximum possible command level is arrived at in column BB. This is compared with the design command level in column K and an *'OK'* or *'LOW COMMAND'* warning is returned in column BC. If the command level is too low then adjustments have to be made in the canal bed levels, falls or slopes.

In the case of the AOSM outlet the operating head is calculated in column AR and a check on required submergence of the roof block is given in column AV. The orifice dimensions *Bt* and *y* are adjusted manually in columns AW and AX, until the required discharge is attained in column AY. If this is not within 5 per cent of the design outlet discharge in column M then further adjustments in orifice size or crest level are made until an *'OK'* is returned in column AZ. The ratio of *Bt/y* is checked in the next row to ensure that the orifice is reasonably square.

The proportional dividers can take the form of a simple bifurcator, a trifurcator if there are two outlets at the same RD, or double bifurcator or trifurcator in which offtake discharges are small and a second split of the flow is required. These are automatic structures, in that they guarantee a proportional split of the flow at all discharges in the parent canal. They are simple and near-foolproof in operation provided the design avoids problems of submergence. Hence the formulae comprise mainly checks on submergence below each separate crest of the structure. The proportional divider formulae occupy four rows of the spreadsheet, representing the left and right offtaking weirs (either of which may be of zero width in the case of a bifurcator), the parent channel weir, and conditions downstream which depend on the fall across the structure inserted in column J. The overall crest width in column AG will need manual adjustment if the offtaking crest widths are too narrow. We took a minimum width of 150 mm to avoid possible choking of the offtake by interference of the splitter wall. Column BD contains a check on submergence of the second crest in the case of a double divider.

Figure 10.5 is a typical output from the graphical checking routines, which give a rapid check on bed, bank and water levels, and velocity along the reach as in figure 10.6. Figure 10.7 is a construction drawing generated automatically in Autocad from data abstracted directly from the spreadsheet using a Lisp routine. It is a more detailed version of the longitudinal section of figure 10.5.

Table 10.5 DESIGN SPREADSHEET FOR TORU MINOR

RD (m)	exstg bank Right	exstg bank Left	exstg bed Centre (M)	exstg water (M)	bed width (m)	exstg Structure Type	proposed Structure Type	proposed arrgt	DROP IN W.L.	command level (m)	offtaking area (Ha)	offtake q (m3/s)	design disch (m3/s)	profile type no.	profile type	manning n	side slope 1:m	bed width (m)	slope	z=Qn/√s	trial depth (m)	a (m2)	r	ar^2/3	delta	normal depth (m)
0	356.24	356.24	355.64	356.29		Head Regulator	Head Regulator						0.91	7	Rectangular lined	0.016	0.00	2.00	0.0008	0.52	0.53	1.05	0.35	0.52	0.00	0.53
20	356.24	356.24	355.39	356.08			Crump Weir						0.91	7	Rectangular lined	0.016	0.00	2.00	0.0008	0.52	0.53	1.05	0.35	0.52	0.00	0.53
20	356.24	356.24	355.39	356.08					0.15				0.91	7	Rectangular lined	0.016	0.00	2.00	0.0008	0.52	0.53	1.05	0.35	0.52	0.00	0.53
30	356.24	356.24	355.39	356.08		Start of proposed lining							0.91	0	Unlined	0.023	1.00	1.10	0.0008	0.74	0.73	1.33	0.42	0.74	0.00	0.73
50	356.95	356.61	355.55	356.01	1.34								0.91	0	Unlined	0.023	1.00	1.10	0.0008	0.74	0.73	1.33	0.42	0.74	0.00	0.73
100	356.66	356.49	355.53	355.96	2.01								0.91	0	Unlined	0.023	1.00	1.10	0.0008	0.74	0.73	1.33	0.42	0.74	0.00	0.73
150	356.52	356.11	355.49	355.94	1.07								0.91	0	Unlined	0.023	1.00	1.10	0.0008	0.74	0.73	1.33	0.42	0.74	0.00	0.73
200	356.49	356.36	355.43	355.93	2.01								0.91	0	Unlined	0.023	1.00	1.10	0.0008	0.74	0.73	1.33	0.42	0.74	0.00	0.73
250	355.48	355.38	355.31	355.91	1.42	Outlet/L	OF			355.14	22	0.016	0.91	0	Unlined	0.023	1.00	1.10	0.0008	0.74	0.73	1.33	0.42	0.74	0.00	0.73
250	355.48	355.38	355.31	355.91	1.42								0.91	0	Unlined	0.023	1.00	1.10	0.0008	0.74	0.73	1.33	0.42	0.74	0.00	0.73
250	355.48	355.38	355.31	355.91	1.42	Outlet/R	AOSM			355.25	39	0.028	0.91	0	Unlined	0.023	1.00	1.10	0.0008	0.74	0.73	1.33	0.42	0.74	0.00	0.73
250	355.48	355.38	355.31	355.91									0.87	0	Unlined	0.023	1.00	1.10	0.0008	0.74	0.73	1.33	0.42	0.74	0.00	0.73
260	355.48	355.38	355.22	355.67		Fall	Fall	replace Glacis	1.25				0.87	0	Unlined	0.023	1.00	1.10	0.0008	0.70	0.71	1.28	0.41	0.70	0.00	0.71
260	355.48	355.38	353.52	353.75		Fall	Fall						0.87	0	Unlined	0.023	1.00	1.10	0.0018	0.47	0.57	0.95	0.35	0.47	0.00	0.57
300	354.48	354.41	353.51	353.96	1.44								0.87	9	Trap lining	0.016	1.00	1.10	0.0018	0.33	0.47	0.73	0.30	0.33	0.00	0.47
350	354.24	354.15	353.37	353.85	1.04								0.87	9	Trap lining	0.016	1.00	1.10	0.0018	0.33	0.47	0.73	0.30	0.33	0.00	0.47
400	354.29	353.99	353.38	353.85	0.86								0.87	9	Trap lining	0.016	1.00	1.10	0.0018	0.33	0.47	0.73	0.30	0.33	0.00	0.47
410			353.25	353.68		Fall	Fall	replace Glacis					0.87	9	Trap lining	0.016	1.00	1.10	0.0018	0.33	0.47	0.73	0.30	0.33	0.00	0.47
410			351.30	351.89		Fall	Fall		2.00				0.87	9	Trap lining	0.016	1.00	1.10	0.0010	0.44	0.55	0.90	0.34	0.44	0.00	0.55
430						Footbridge wooden	ignore						0.87	9	Trap lining	0.016	1.00	1.10	0.0010	0.44	0.55	0.90	0.34	0.44	0.00	0.55
450	353.23	353.10	351.21		0.91								0.87	9	Trap lining	0.016	1.00	1.10	0.0010	0.44	0.55	0.90	0.34	0.44	0.00	0.55
500	352.18	352.21	351.25	351.80	1.05								0.87	9	Trap lining	0.016	1.00	1.10	0.0010	0.44	0.55	0.90	0.34	0.44	0.00	0.55
550	352.04	352.18	351.26	351.49	0.97								0.87	9	Trap lining	0.016	1.00	1.10	0.0010	0.44	0.55	0.90	0.34	0.44	0.00	0.55
590	351.91	352.15	351.30	351.70	1.75	Footbridge wooden	ignore						0.87	9	Trap lining	0.016	1.00	1.10	0.0010	0.44	0.55	0.90	0.34	0.44	0.00	0.55
600	351.91	352.15	351.30	351.70	1.75	End of proposed lining							0.87	9	Trap lining	0.016	1.00	1.10	0.0010	0.44	0.55	0.90	0.34	0.44	0.00	0.55
610	351.91	352.15	351.30	351.70	1.75	Start of right side lining/S							0.87	10	Trap lining with dem 1 side only	0.016	1.00	1.10	0.0010	0.44	0.55	0.90	0.34	0.44	0.00	0.55
635			351.00	351.49		Outlet/R	Bifurcator					0.000	0.87	10	Trap lining with dem 1 side only	0.016	1.00	1.10	0.0010	0.44	0.55	0.90	0.34	0.44	0.00	0.55
635			351.00	351.49			Bifurcator	, right		351.14	239	0.172	0.87	10	Trap lining with dem 1 side only	0.016	1.00	1.10	0.0010	0.44	0.55	0.90	0.34	0.44	0.00	0.55
635			351.00	351.49			Bifurcator						0.69	10	Trap lining with dem 1 side only	0.016	1.00	1.10	0.0010	0.35	0.48	0.77	0.31	0.35	0.00	0.48
635			351.00	351.49			Bifurcator		0.15				0.69	10	Trap lining with dem 1 side only	0.016	1.00	1.10	0.0010	0.35	0.48	0.77	0.31	0.35	0.00	0.48
645	351.53	351.36	350.92	351.70	0.89	End of right side lining/S							0.69	10	Trap lining with dem 1 side only	0.016	1.00	1.10	0.0010	0.35	0.48	0.77	0.31	0.35	0.00	0.48
655			351.00	351.49		Fall	Fall						0.69	9	Trap lining	0.016	1.00	1.10	0.0013	0.31	0.45	0.69	0.29	0.31	0.00	0.45
655			350.07	350.52		Fall	Fall	replace vertical	1.40				0.69	9	Trap lining	0.016	1.00	1.10	0.0013	0.31	0.45	0.69	0.29	0.31	0.00	0.45
665			350.07	350.52		Start of proposed lining							0.69	9	Trap lining	0.016	1.00	1.10	0.0013	0.31	0.45	0.69	0.29	0.31	0.00	0.45
700	350.16	350.37	349.51	349.86	1.63								0.69	9	Trap lining	0.016	1.00	1.10	0.0013	0.31	0.45	0.69	0.29	0.31	0.00	0.45
750	350.08	350.16	349.47		1.12								0.69	9	Trap lining	0.016	1.00	1.10	0.0013	0.31	0.45	0.69	0.29	0.31	0.00	0.45
760			349.78			VRB	VR Bridge	retained with bed protection					0.69	9	Trap lining	0.016	1.00	1.10	0.0013	0.31	0.45	0.69	0.29	0.31	0.00	0.45
790			349.46	349.51			Bifurcator					0.000	0.69	9	Trap lining	0.016	1.00	1.10	0.0013	0.31	0.45	0.69	0.29	0.31	0.00	0.45
790			349.38	349.51		Outlet/R	Bifurcator	double, right with culvert		349.42	65	0.046	0.64	9	Trap lining	0.016	1.00	1.10	0.0013	0.29	0.45	0.69	0.29	0.31	0.00	0.45
790			349.38	349.51			Bifurcator						0.64	9	Trap lining	0.016	1.00	1.10	0.0013	0.29	0.45	0.69	0.29	0.31	0.00	0.45
790			349.38	349.51			Bifurcator		0.15				0.64	9	Trap lining	0.016	1.00	1.10	0.0013	0.28	0.43	0.66	0.28	0.28	0.00	0.43
800	350.00	349.96	349.17	349.74	0.79								0.64	9	Trap lining	0.016	1.00	1.10	0.0013	0.28	0.43	0.66	0.28	0.28	0.00	0.43

Table 10.5 DESIGN SPREADSHEET FOR TORU MINOR

Section groupings: CULATIONS (AB–AE) · CROSS REGULATORS (AF–AI) · DESIGN LEVELS (AJ–AQ) · OUTLETS AND WEIR SUBMERGENCE (AR–BD)

RD (m)	normal w.l. (m)	design bed (m)	froude no	velocity (m/s)	design crest (M)	crest width	disch coeff	tot head (m)	regulated design w.l.	lining freebd (m)	lining top (M)	bank freebd (m)	bank design (M)	w.l. width (m)	depth/top ratio d/t	bank width (m)	flume h or bif throat	flumes B (m)	flume c /double throat	q (m³/s)	flumes	aosm B	aosm y	aosm bif or submgce	aosm q check	outlet head loss	w.l. d/s of outlet	command M check	bif submgce check
0	355.83	355.30	0.38	0.87					355.83	0.15	355.98	0.50	356.33	2.00	0.26	1.00													
20	355.81	355.28	0.38	0.87	355.44	2.00	2.00	0.37	355.81	0.15	355.96	0.50	356.31	2.00	0.26	1.00						high flow modularity:		0.60	OK				
20	355.66	355.13	0.38	0.87					355.66	0.15	355.81	0.50	356.16	2.00	0.26	1.00													
30	355.65	355.05	0.26	0.69					355.65	0.15	#N/A	0.50	356.15	2.55	0.28	1.00													
50	355.64	355.04	0.26	0.69					355.64	0.15	#N/A	0.50	356.14	2.55	0.28	1.00													
100	355.60	355.00	0.26	0.69					355.60	0.15	#N/A	0.50	356.10	2.55	0.28	1.00													
150	355.56	354.96	0.26	0.69					355.56	0.15	#N/A	0.50	356.06	2.55	0.28	1.00													
200	355.52	354.92	0.26	0.69					355.52	0.15	#N/A	0.50	356.02	2.55	0.28	1.00													
250	355.48	354.88	0.26	0.69	354.82 OPEN FLUME (OF)				355.48	0.15	#N/A	0.50	355.98	2.55	0.28	1.00	0.654	0.018	1.66	0.02	OK	width check		too narrow		0.13	355.35	OK	
250	355.48	354.88	0.26	0.69					355.48	0.15	#N/A	0.50	355.98	2.55	0.28	1.00													
250	355.48	354.88	0.26	0.69	354.97 AOSM				355.48	0.15	#N/A	0.50	355.98	2.55	0.28	1.00	0.509	roof block cover…							low command					
250	355.47	354.74	0.26	0.69					355.48	0.15	#N/A	0.50	355.98	2.55	0.28	1.00						0.12	0.12	0.04	too big	0.26	355.22		
260	355.47	354.74	0.26	0.68	354.79	1.32	1.70	0.53	355.32	0.15	#N/A	0.50	355.97	2.51	0.25	1.00						B/y=	1.00	should be between .8 and 1.2					
260	354.22	353.60	0.39	0.91		OK			354.07	0.15	354.15	0.50	354.72	2.24	0.25	1.00													
300	354.00	353.53	0.56	1.19					354.00	0.15	354.06	0.50	354.50	2.03	0.23	1.00													
350	353.91	353.44	0.56	1.19					353.91	0.15	353.97	0.50	354.41	2.03	0.23	1.00													
400	353.82	353.35	0.56	1.19					353.82	0.15	353.95	0.50	354.32	2.03	0.23	1.00													
410	353.80	353.33	0.56	1.19	353.34	1.65	1.70	0.48	353.80	0.15	351.95	0.50	354.30	2.03	0.23	1.00													
410	351.80	351.25	0.41	0.96		OK			351.80	0.15	351.93	0.50	352.30	2.19	0.25	1.00													
430	351.78	351.23	0.41	0.96					351.78	0.15	351.91	0.50	352.28	2.19	0.25	1.00													
450	351.76	351.21	0.41	0.96					351.76	0.15	351.86	0.50	352.26	2.19	0.25	1.00													
500	351.71	351.16	0.41	0.96					351.71	0.15	351.81	0.50	352.21	2.19	0.25	1.00													
550	351.66	351.11	0.41	0.96					351.66	0.15	351.77	0.50	352.16	2.19	0.25	1.00													
590	351.62	351.07	0.41	0.96					351.62	0.15	351.76	0.50	352.12	2.19	0.25	1.00													
600	351.61	351.06	0.41	0.96					351.61	0.15	351.75	0.50	352.11	2.19	0.25	1.00													
610	351.60	351.05	0.41	0.96					351.60	0.15	351.72	0.50	352.10	2.19	0.25	1.00													
635	351.57	351.03	0.41	0.96	Bifurcator			0.10	351.57	0.15	351.72	0.50	352.07	2.19	0.25	1.00	0.000	0.000	wL	0.00	OK	submergence =		#########	OK		351.47	OK	0.00
635	351.57	351.03	0.41	0.96	width splitter walls:	2.10	2.00	0.35	351.57	0.15	351.72	0.50	352.07	2.19	0.25	1.00	0.418	0.000	wR	0.17	OK	submergence =		-0.26	OK		351.47	OK	0.00
635	351.57	351.03	0.41	0.90	351.23				351.57	0.15	351.56	0.50	352.07	2.07	0.23	1.00	1.682	2nd split	wC	0.87	OK	submergence =		0.57	OK			OK	
645	351.42	350.94	0.41	0.90	crest height check:	OK			351.42	0.15	351.55	0.50	351.91	2.07	0.23	1.00	1st split											OK	
655	351.41	350.93	0.41	0.90					351.41	0.15	350.15	0.50	351.90	2.07	0.23	1.00													
655	351.40	350.92	0.41	0.90	350.97	1.43	1.70	0.43	351.40	0.15	350.14	0.50	351.90	2.07	0.23	1.00													
665	350.00	349.56	0.47	0.99		OK			350.00	0.15	350.10	0.50	350.50	2.00	0.22	1.00													
700	349.99	349.54	0.47	0.99					349.99	0.15	350.03	0.50	350.49	2.00	0.22	1.00													
750	349.95	349.50	0.47	0.99					349.95	0.15	350.02	0.50	350.45	2.00	0.22	1.00													
760	349.88	349.43	0.47	0.99					349.88	0.15	349.98	0.50	350.38	2.00	0.22	1.00													
790	349.87	349.42	0.47	0.99					349.87	0.15	349.98	0.50	350.37	2.00	0.22	1.00													
790	349.83	349.38	0.47	0.99	Bifurcator			0.10	349.83	0.15	349.98	0.50	350.33	2.00	0.22	1.00	0.000	0.000	wL	0.00	OK	submergence =		#########	OK		349.75	OK	0.00
790	349.83	349.38	0.47	0.97	width splitter walls:	2.80	2.00	0.25	349.83	0.15	349.83	0.50	350.33	2.00	0.22	1.00	0.944	0.472	wR	0.05	OK	submergence =		-1.21	OK		349.68	OK	0.73
790	349.83	349.38	0.47	0.97	349.58				349.83	0.15	349.82	0.50	350.33	1.96	0.22	1.00	1.856	2nd split	wC	0.69	OK	submergence =		0.39	OK			2nd submerged	
800	349.68	349.25	0.47	0.97	crest height check:	OK		OK	349.68	0.15		0.50	350.18	1.96	0.22	1.00	1st split												
800	349.67	349.23	0.47	0.97					349.67	0.15		0.50	350.17	1.96	0.22	1.00													

The design of these canals was computerised to the extent that all checking only had to be done once, on the design spreadsheet. From there to the final drawings everything was automated and carried out by the same design engineer. The conventional way of designing and drawing by draughtsmen would involve an engineer checking the calculations and drawings at least four or five times, so these designs were completed in less than half the time required for conventional methods, and with far greater accuracy.

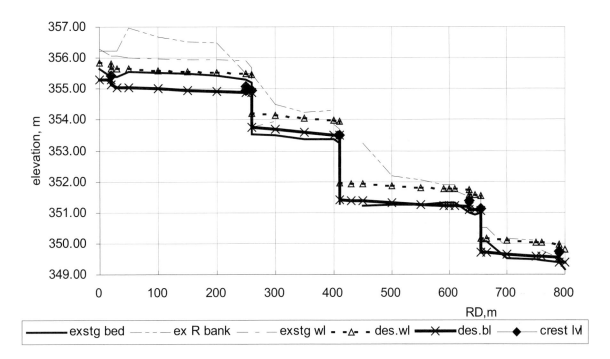

figure 10.5   Toru Minor longitudinal section

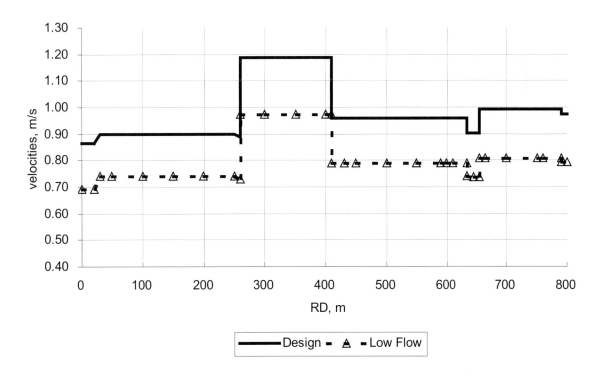

figure 10.6   Toru Minor velocity profile

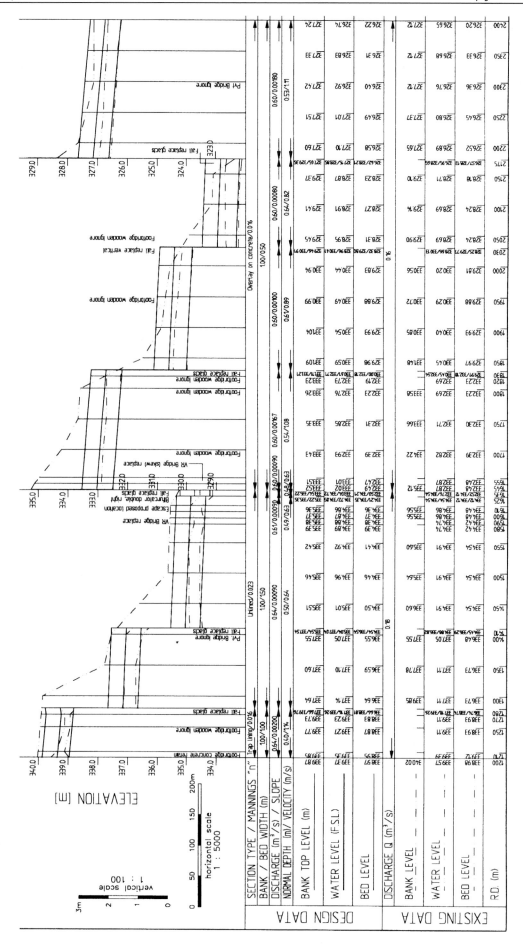

figure 10.7   Longitudinal section for Toru Minor

10.6 Manning - a program for canal design charts

This is a powerful program for the design of canals of various cross-sectional profiles including parabolic, circular, rectangular, trapezoidal and triangular. It produces the design charts shown in figure 10.8. The software is provided on the CD.

The usefulness of charts

It might seem that with the current state of computer art, when the Manning formula can be solved in nanoseconds, design charts are no longer necessary. Before computers became readily accessible the *Biblia Hydraulica* provided complex charts (Ref. 4) and nomograms in order to facilitate calculating the normal depth using only a slide rule. First it was necessary to calculate some intermediate parameters such as the hydraulic mean depth divided by the wetted perimeter, or the section factor divided by the bed width raised to the power of 2.67. From this the chart gave some other figure to be multiplied by the side slope to give the normal depth. Now a computer can give an instantaneous answer. So why bother with charts?

There is an advantage in using charts, which present an overview of the whole range of operation for a particular canal. It is possible to see at a glance what depth and velocity pertain for any given slope or discharge, or the range of discharge for a given depth or slope. It is easy to see where the velocity becomes critical, and how to get a non-silting or non-scouring velocity. The chart indicates the canal's performance in all its relevant facets. Accurate figures for depth, discharge or velocity can be produced if required.

Using the chart

Sample charts are shown in figures 10.8 to 10.13. The charts are plotted with the x-axis as a dynamic logarithmic scale of discharge, from 1 l/s to 1000 cumecs. The y-axis is a dynamic logarithmic scale of the longitudinal slope of the canal, from 0.00001 to 0.1. Strictly the slope refers to the hydraulic gradient, but in practice it can be taken as the canal water surface, and for uniform flow it is also the bed slope. In the case of a pipe or inverted siphon, it is the friction slope.

The lines sloping from bottom left to top right are the depth curves. They are labelled along the bottom axis by the vertical figures which give the normal depth in metres, each line relating to a single depth. Any required range of water depths and depth increment can be keyed in to the on-screen menu. For each depth the program calculates a curve of discharge over the full range of canal slope. For a given depth and slope the discharge is read off along the bottom axis. For a given slope and discharge the depth can be read or interpolated from the depth lines.

The lines sloping from upper left to lower right are the velocity curves. They are labelled at either end with the mean water velocity in m/s. The program calculates these automatically in steps of 0.1 m/sec up to 1.5 m/sec, and thereafter in steps of 0.5 m/sec. Interpolating between them gives the velocity for any combination of slope, depth or discharge. The velocity curves are almost straight for most shapes of canal, the main exception being for circular pipes or any other closed conduit. The heavy dotted line across the top is the critical velocity curve. It is useful to know the range of critical flow, in order to avoid the practical problems of uncontrolled turbulence.

The menu box shows the profile shape selected from trapezoidal, rectangular, triangular, parabolic, circular or rectangular. There is an option for circular or rectangular inverted siphons. For a trapezoidal profile the size and shape is fixed by the bed width b and the side slope m (meaning 1 unit vertical to m units horizontal.) For a rectangular profile m is zero and only the bed width counts. For a triangular profile b is zero and only the side slope counts. A parabolic profile is described by the open top width t and maximum internal depth d, (without confusing this with the water depth, and assuming a first order parabola.) A circular channel is described only by its diameter.

In the special case of a circular or rectangular closed conduit or inverted siphon, the slope on the y axis is not the physical slope of the conduit, but the hydraulic friction slope, i.e. the friction head loss per unit length of the conduit. In the design of an inverted siphon the inlet and outlet losses, which are usually taken as a fraction of the velocity head $v^2/2g$, will need to be added. However, it should be noted that the use of the Manning formula for closed conduits often results in a friction loss that is conservatively high, and should be used with caution or checked in conjunction with another method such as the pipe flow formulae of Hazen-Williams or Colebrook-White.

These charts can be custom designed for individual canals and a specific project. A set of charts can be tailor-made for the use of field engineers in setting out new channels or for water managers to see at a glance in the field the flow conditions in their own canals.

The same procedures can be done using imperial units, by a switch on the menu.

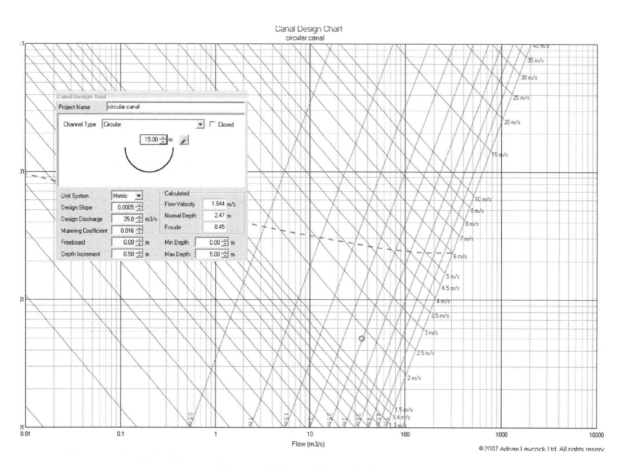

figure 10.8   ***Sample charts***

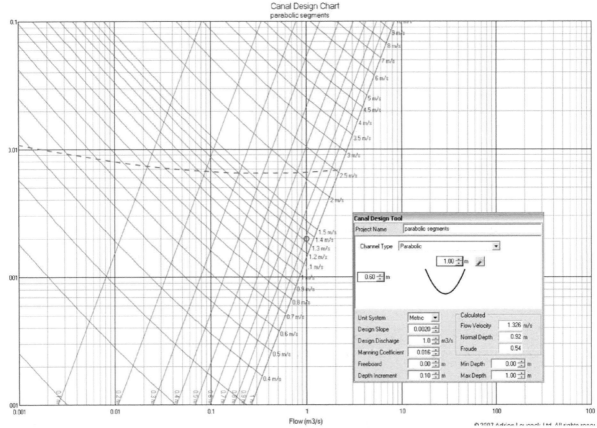

figure 10.9   Sample charts

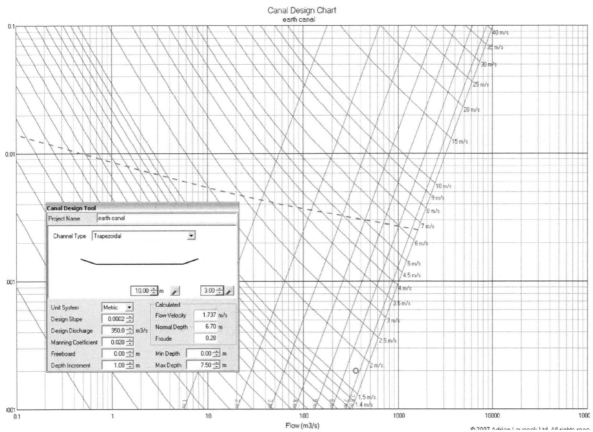

figure 10.10   Sample charts

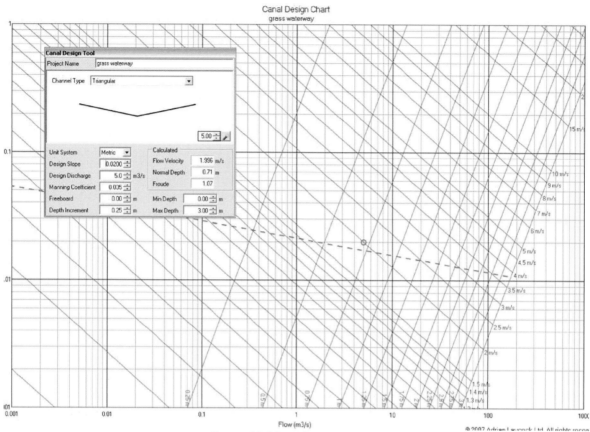

figure 10.11   Sample charts

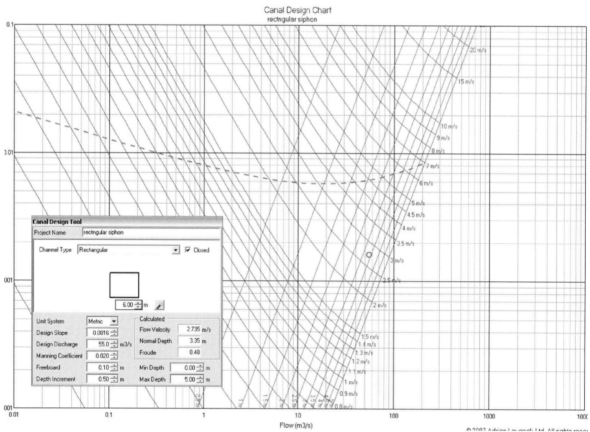

figure 10.12   Sample charts

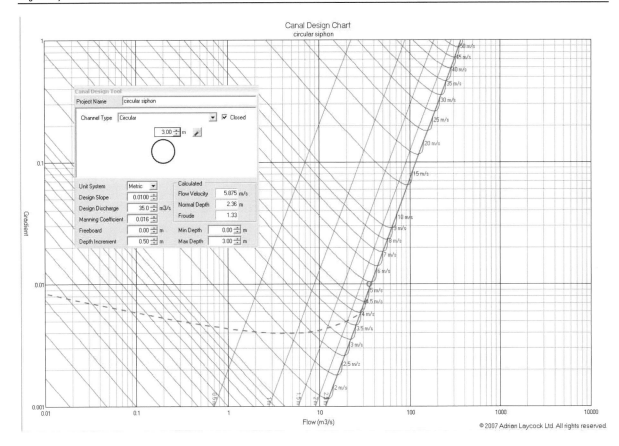

figure 10.13   Sample charts

10.7 Formulae relating to chapter 10

Nomenclature:

v = velocity in m / second

A = cross-sectional area of water in square metres

q = discharge in cubic metres/second = Av

d = normal depth in metres

T = water surface width in metres

P = wetted perimeter in metres

R = hydraulic mean depth or hydraulic radius = A/P

D = hydraulic depth = A/T

n = Manning's roughness coefficient

g = gravitational acceleration = 9.81 m/s^2

s = Hydraulic gradient = longitudinal slope for uniform flow

K_s = Nikuradse roughness coefficient

υ = kinematic viscosity

Open channel flow

Froude number $\boldsymbol{F} = v/(gD)^{0.5}$

Manning

$q = 1/n \; AR^{0.67} \, s^{0.5}$

$v = q/a = 1/n \; R^{0.67} \, s^{0.5}$

(in imperial units, $q = 1.49/n\ AR^{0.67}\ s^{0.5}$ cusecs)

Strickler formula:
This gives an estimate of the roughness for granular soils.
Manning's $n = 0.0342\ d_{50}{}^{1/6}$
where d_{50} is the median size (in feet) of sand or gravel in the canal bed.

Colebrook-White
$v = -(32\ g\ Rs)^{0.5}\ \log[(K_s/14.8R) + (1.255\upsilon/R(32\ g\ Rs)^{0.5})]$
where υ = viscosity = 1.141×10^{-6} at $15°C$
K_s = roughness height, mm

Geometry of common shapes related to hydraulic parameters...
parabola
first order parabola: $y = ax^2$
defined by top width T and depth d
area = $2/3\ Td$

second order parabola: $y = ax^3$
third order parabola: $y = ax^4$

circle
$x^2 + y^2 = a^2$
radius = a
diameter = $2a$
area = πa^2

ellipse
$x^2/a^2 + y^2/b^2 = 1$
area = $\pi\ ab$

special case: when $a = b$, we have a circle.

trapezoid
bed width b, depth d, side slopes m (1 Vertical to m Horizontal.)
area = $d(b + md)$
top width = $b + 2md$
perimeter = $b + 2d(1 + m^2)0.5$

rectangle
bed width b, depth d, side slopes 0
area = db
top width = b
perimeter = $b + 2d$

triangle
bed width 0, depth d, side slopes m
area = md^2
top width = $2md$
perimeter = $2d(1 + m^2)0.5$

References and further reading for chapter 10

(1) Ven Te Chow, 'Open Channel Hydraulics', McGraw-Hill, 1959.

(2) French, R.H., - 'Open-Channel Hydraulics', McGraw-Hill, 1986.

(3) Henderson, F.M., 'Open Channel Flow', Macmillan, 1966.

(4) Hydraulics Research Ltd, Wallingford, 'Charts for the Hydraulic Design of Channels and Pipes', 1990.

(5) Butler,D & Pinkerton,B, - 'Gravity Flow Pipe Design Charts', Thos. Telford, 1987.

(6) International Commission on Irrigation and Drainage, 'Design Practices of Irrigation Canals', 1972.

(7) Justo *et al*, 'A Finite Element Model for Lined Canals on Expansive Collapsing Soil', proc.11th conf.soil mech foundation eng, San Francisco, 1985.

(8) Laycock, A., 'Precast Parabolic Canals, Revelation and Revolution in Pakistan', proc. IWASRI/HR workshop on canal lining and seepage, Lahore, October 1993.

(9) Hydraulics Research Ltd, Wallingford, 'DORC software for design of regime canals', 1992.

(10) Barragan, M., 'Circular Canals', ICID Bulletin, January 1976.

PART 3 – EXPERIENCE

Because irrigation schemes tend to be long-term projects it is rare indeed for an engineer to see a scheme through from planning to design to construction to operation, a process which could take 20 years or more. Perhaps this is another reason why it takes so long for new ideas to gain acceptance. As in most practical professions, an engineer learns a lot from his own mistakes, but because his mistakes in design may not show up on the ground for another decade an irrigation engineer has to rely on learning from the mistakes of others. This section invites you to learn from some of mine, as well as others that I've stumbled across or tripped up on over the years.

There is no substitute for field experience, and the sometimes painful results of it are the best lessons we can learn. The rehabilitation projects are particularly useful for showing up where things went wrong. Sometimes the reasons are not obvious and the wider social environment needs consideration before answers can be found to problems that only become apparent after a scheme is built. This chapter is largely the result of several years of troubleshooting on schemes whose performance ranged from mediocre to disastrous. When the farmers pelt you with stones when you set foot on the scheme then you know there's something seriously wrong. Often the local irrigation staff will just blame the farmers for being ungrateful trouble-makers. But you have to look deeper than that. The fault could lie in the management, the construction, the design or the planning, but rarely with the farmers. Once the problem is recognised it's usually not difficult to put right, and you'll surely do it right the next time. But if you don't take the opportunity to find out what went wrong, it'll just happen again and again.

Even the best designs come to nought if there is no maintenance or if the operation is poor. Or if the construction supervision is inadequate. Or if corruption reigns supreme. Then it's up to the strategists and planners, the politicians and the financiers, to ensure that the entire scheme is properly thought out and goes ahead in a workable environment.

CHAPTER 11 TROUBLESHOOTING - FEEDBACK FROM THE FIELD

Many of the defects shown here are compounded through ignorance, and perpetuated through unchanging habits and inappropriate methods of construction. If we look at the signs of damage and faults we can usually trace the underlying problems of management and social stress. With this in mind, defects can be considered in five groups, namely construction defects, inappropriate design, poor maintenance and management, natural causes, and farmers' interference. Using what we learn from all this we may then be able to avoid these problems next time.

figure 11. 1   The most likely time for a canal breach is during first filling after construction or, as in this case, after extensive remodelling with downstream control gates. (Maira Branch Canal, Pakistan. Photo by C. G. Swayne)

11.1 Canals That Don't Work

Recent bitter experience of malfunctioning irrigation projects in Indonesia, India, Sri Lanka and elsewhere has highlighted a considerable number of physical defects that repeatedly arise, sometimes disrupting the project so completely that total reconstruction of the canal system is necessary.

In south India, 30 small irrigation schemes were constructed under a foreign aid project in the early 1980s. Yet only 3 years after supposed completion, none was working properly, and only six were delivering any water at all! When a journalist from the donor country stopped by to see how well their tax money had been spent, he was dismayed to discover nothing very much.

This sorry state of affairs is not unusual; in fact it is common in many countries. It was only international political repercussions that led to a rehabilitation programme for these particular schemes. The majority of similar schemes rapidly disappear into oblivion.

All defects and damage have a root cause, which may not always be immediately obvious. They can usually offer clues to major problems on the scheme, which may have broken down due to poor management, social problems, or wrong strategies during design and construction. They are often purely the physical manifestations of deeper social problems, and as such are useful indicators for rehabilitation and renewed management strategies.

It did not actually need very much concrete to rehabilitate the schemes in the Indian example. All it took was correct recognition of the problems and a lot of supervision. And inevitably there were a few awkward questions to answer:

- What went wrong?
- Was it avoidable?
- Would we recognise the signs again in time to prevent a repetition?
- Whose fault was it?
- What would we do different next time?

11.2 Spotting the Trouble

A walk around the scheme immediately reveals something wrong. Either water doesn't flow, or it overflows, or it disappears, or the drains are running.

Water doesn't flow

If water is stagnant and doesn't flow anywhere, look for these reasons:

Canal goes uphill. Yes, canals go uphill quite often. Maybe an office-bound official had looked at a map without contours and decided that water had to be delivered to a field that turns out to be at a higher elevation than the canal. A kind of logic, albeit not based on any principles of engineering, dictates that the canal should therefore be oriented uphill. Don't laugh, it happens. The young site supervisor fresh on the job might be afraid to countermand the orders of his superiors, However, misinformed they may be. He might also not be very competent.

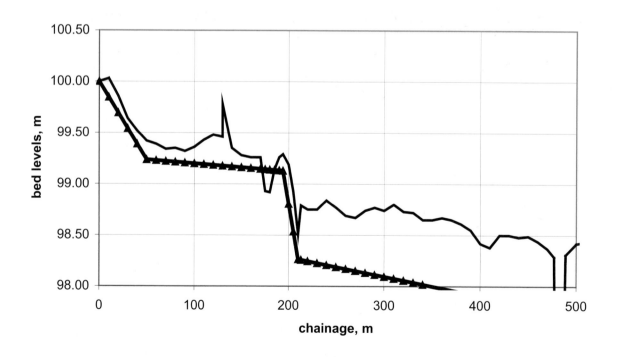

figure 11. 2   *Bogribail Canal long section*

figure 11. 3   ***Bogribail Canal through rock***

Canal blocked with bed-rock. The example in figure 11.2 is a survey of as-built canal bed levels, which quickly revealed why this canal wasn't flowing. The bed level went up and down like a roller coaster track. And a site inspection soon revealed why. As is apparent in figure 11.3, the canal went through a section of solid hard rock. It had been difficult to cut through with the poor quality hand tools available and the contractor had prematurely given up without telling anybody.

Canal blocked by paddy. This type of blockage is common in Indonesia. The canal in figure 11.4 is completely blocked with a fine crop of paddy rice. It has an assured water supply being right in the canal. If the same thing happened in Pakistan there would be a shoot-out with kalashnikovs.

figure 11. 4   ***Paddy in canal***

Water overflows

If water is spilling over the sides of the canal, look out for these:

- Gates wrongly operated. Canal headgates upstream might be open too far. Cross regulators and turnouts further downstream might be closed. Does the gate operator have proper working instructions?

- Wrong gradients. The canal might flow uphill again, only for a short distance. Uneven bed gradients are especially disruptive on small canals, and their construction tolerances have to be tighter than large canals.

- Deliberately or accidentally blocked. It is easy to block a small canal. Search downstream for a blockage, like a farmers' temporary dam, or a dead donkey. A common trick in Pakistan is for farmers to bathe their buffalo in the canal immediately downstream of their watercourse outlet, thereby heading-up the water in the canal and increasing the discharge through the outlet!

Water disappears

If water goes in at the top end of a canal reach but doesn't come out the bottom where it ought to, then look out for the following:

- Diverted by farmers. Some farmers will go to great lengths to get water when they're not supposed to. But don't be too quick to blame the farmers. There's always a logical reason for it.

- Leakage through faulty lining. Leaks at specific locations, rather than steady seepage over the whole length of the canal, may be easy to identify. Look for areas of waterlogging nearby. Likely places are leaking turnout gates, damaged lining or structures, and crab holes.

- Oversized reaches give a slow response time. If the discharge is small compared with the size of the canal, it can take a surprisingly long time for water to travel along a reach, especially if the canal is unlined and in bad condition.

The drains are running

Check the drains first. If they are flowing as heavily as the canals, there could be a management problem. Water is being wasted somewhere, but it's probably easy to stop. Look for deliberate diversions.

> In 1978 I was in Sri Lanka, checking out some water management practices on one of the ancient schemes at Minipe Anicut that had been resurrected by the British. The first thing I noticed was the drains flowing with about as much water as in the canals. The reason was easy enough to spot. I questioned a farmer who was purposely diverting water around his field and into the nearest drain. He explained that he had just put a top dressing of fertiliser on his paddy field and didn't want the water to leach out the fertiliser. He didn't really care that the water going down his drain could have been used by someone else further down the canal system, had he been public-spirited enough to close the minor turnout rather than take the easier recourse of diversion.

On the proportional flow schemes in northern Pakistan there is no effective day-to-day regulation of minor canals, along which the warabandi rotation schedule is used. If a farmer is ready to harvest his crop, the only way to avoid waterlogging at a crucial moment will be to divert incoming water into the drain.

Categories of problem

We can categorise the problems depending on where they emanate from; construction, design, operation, mother nature, society, or that ever-present fallback of corruption.

11.3 Bad Construction

It's easy to point to a broken canal and say it was badly constructed. But it's more informative to find out in what way, and why. Then it might be constructed better next time.

Construction tolerances

Tolerances are a bone of contention on construction sites. The tolerance is the permissible deviation in dimensions or levels between that specified on the design drawings and what is actually constructed. The permissible deviation from canal bed level might be specified as +/- 10 mm, but for an accurate measuring structure the tolerance will need to be much tighter, perhaps down to +/- 1 mm. The tolerance in canal profile or side slope might be much greater, say +/- 50 mm, because this dimension is less critical from the point of view of the canal being able to operate properly. But tolerances cost money. It is little use the designer specifying very close tolerances unless they are really essential for the guaranteed operation of the finished product.

The contractor will be looking for as large a tolerance as he can get away with. Unless the correct tolerance is properly specified, the site supervision staff will have a hard time convincing him that any degree of accuracy is necessary. A fact not often appreciated is that small canals actually need tighter tolerances than large canals. A large canal, 10 m wide and 2 m deep might need a tolerance of +/- 100 mm in the bed level, with which the water flow would not be significantly affected. But the same tolerance applied to a small canal 0.5 m wide and 0.5 m deep, could result in a built-in obstruction to the flow 200 mm high which would restrict the discharge capacity of the canal by 50 per cent. It takes more supervision effort to construct a small canal than it does a large one.

Supervision

Poor supervision, or no supervision at all, is quite usual on small, low-cost projects. It is a common mistake on the part of funding agencies not to spend enough money on supervision.

Inadequate supervision is inevitable when large numbers of small contractors are required to be supervised by a small number of engineers. The paperwork gets out of hand and the engineers don't have any time in the field. This often happens on big projects if it is a government strategy to generate employment for small contractors.

> Once in Indonesia I was visiting some on-going canal lining work that I could see was turning into a disaster. The design was for concrete lining but the mason was skimming a thin layer of mortar over the bare earth and hoping it looked like solid concrete. I asked him if the engineer had inspected it. He said yes, and all his work was approved. 'Where can I meet this engineer?', I persisted. 'Right here,' he said. 'I'm the engineer myself.' So he was being paid two salaries, one from the government and one from the contractor, and the contractor was killing two birds with one stone; no supervision of his own needed and no-one from the client watching over him all the time.

Unqualified staff

If the job is low cost and profit margins are low, the contractor might try to cut costs by not employing adequately experienced staff. A prime example is when a young, inexperienced engineer is put in charge of a project without enough supervision from more senior staff. The young engineer may not have enough force of personality to insist on good workmanship from his own artisans. Worse still, he may never acquire enough familiarity with good workmanship to be able to recognise it. In social cultures in which youth automatically respects and reveres old age (no bad thing in itself), it can be very difficult for a bright young engineer to tell an old but incompetent mason what to do.

Bad workmanship

In some countries there is a vicious circle of ignorance of good workmanship, which is perpetuated by these attitudes. Many engineers may go through their entire career not knowing how to recognise good or bad concrete, all because they never had the right back-up from their seniors when they first went out on site. And now that they themselves are senior engineers, how are they going to explain to their juniors how to recognise good quality ? And just as knowledge is power, the lack of knowledge is to be hidden from view rather than to be publicly admitted and augmented through shared experiences. This is one of several vicious circles in the field of irrigation which is very difficult to break out of.

Obsolete construction techniques

Perhaps ignorance can be forgiven. Perhaps also we can reasonably expect our bright young engineers to know all about concrete vibration, mixing and curing. Many of them don't, though. On those Indian schemes I found a huge difference in knowledge between the few engineers who had worked overseas in the Gulf and been exposed to international standards of construction, and the rest who had spent their careers in the local government departments which still used obsolete design and construction methods. Government engineering departments still use designs of masonry structures dating from the days of the Raj, oblivious of the fact that concrete is nowadays much cheaper and better, not to mention the fact that few modern masons have the skill to emulate their predecessors of 100 years ago.

Difficult access

Many small schemes are in remote areas, far away from major roads and supply centres. Indeed this is one of the attractions of small schemes; they can bring development into remote rural areas without investing in massive infrastructure costs. However,, by their very nature, they are more difficult to construct and supervise than a larger scheme would be. Contractors are likely to cut corners on materials and supervision because of it. And the supervising engineers may not visit the site as often as they should, merely because of the difficulty in getting there.

Poor soil compaction

All earthfill structures need proper compaction. Whether a large earth dam or a canal embankment less than a metre high, if the soil is not compacted properly water will seep through it and rapidly erode a passage with devastating results. This process of internal erosion is known as piping. The real test of an earth canal comes with its first filling. Any zones that are not properly compacted are soon washed out. Structures embedded in the banks can easily wash out because compaction around them is difficult.

The process of compaction rearranges the soil particles to form a dense mass, which binds together for strength and restricts the seepage of water through it. Different types of soil compact at different densities, permeability and strengths. All soils are critically dependent on moisture content to achieve optimum compaction. Too dry, and no amount of compactive effort will work. Too wet, and the soil turns to slurry and the machine gets stuck, or it cracks on drying out. Compaction can be done using rollers of various types, hand or powered rammers, or feet. (The ancient Lankans built all their tanks using elephants and cattle for compaction, driving them back and forth across the earthwork.)

In order to achieve adequate compaction, clay soils need a heavy compactive effort which is often not possible with the limited resources of small contractors. Achieving proper moisture control during compaction is rarely possible, either because water is in short supply or, as in the case of many existing canals in the conjunctive use areas of east Java, the soil is often wet to begin with and takes too long to dry out.

Poor compaction results not only in leakage but in soil movement. It may settle under its own weight. The easy passage of water through the soil can create rapid changes of moisture content. As it dries out it may shrink, and large cracks may form. The montmorillonite clays are famous for this, like the black cotton soils of Africa and India, and the volcanic grumosols of Java. As it gets wet it may expand, and exert tremendous pressures on solid structures or canal lining.

Canal lining that has failed through poor compaction of the underlying soil is easy to spot. Poor compaction is indicated by longitudinal cracks in the lining, caused as the soil in the canal banks settles or consolidates (figures A11.1, A11.2.) It is common to find cracks 10 mm wide in brand new masonry lining. Leakage starts immediately, and the cracks are soon widened by root penetration, soil washout and animal activity.

Consolidation is a related thing. Here the water in the soil is slowly forced out of the pores under the pressure of the soil's own weight. Clay embankments may gradually settle over many years through consolidation. Careful control of moisture content during construction can alleviate future movement through consolidation.

Poor concrete compaction

'All concrete to be vibrated'. So says every international consultant's specification. Compaction is the most important thing in laying concrete for a watertight structure or canal lining. It used to be done in the old days by tamping but now it is done far more effectively by vibration.

Concrete is porous. It is even more porous if it's not vibrated. And even more so if it is a dry mix and not vibrated. Concrete strength and watertightness does not come from high cement content and low water-cement ratios. It comes from vibration. Figure 9.1 shows a rare sight in India. A simple vibrating plate compactor being used in the canal bed. The concrete thickness is only 50 mm, and the strength only a nominal 200 kg/cm^2 from a standard 1:2:4 mix. But that concrete will still be there in 50 years time.

Vibration agitates the concrete mix until most of the air voids are evacuated. The air bubbles erupt on the surface, and when they stop is the time to stop vibrating. Over-vibration will cause the concrete to separate and the 'fat' cement slurry will come to the surface. There are several types of vibrator in common use, including needles or pokers, plate compactors, vibrating tables for precast concrete, and external formwork vibrators. The rotating striker tube described in Chapter 9 does not vibrate but achieves compaction through a pressing action.

In rehabilitating an old canal with a concrete overlay in the bed, vibration is the only means of ensuring that all the holes are filled in properly. Holes in the bed and especially at the junction of bed and sides are the most common source of leakage in a badly lined canal.

There are three things to remember about laying concrete. The first is vibration. The second is vibration. And the third is vibration.

With the advent of self-compacting concrete at the end of the 20th century, some of the potential difficulties of vibration may be alleviated. Moreover, the extra workability imparted by polymeric additives and cement substitution by pulverised fuel ash can also simplify the placement of sloping canal side lining.

Poor concrete curing

The process of curing cement mortar and concrete is often not fully understood by small contractors, nor regrettably by some supervising engineers. If concrete is allowed to dry out too rapidly, incomplete curing leads to a weak finished product. It has a porous and dusty or crumbling surface, which is easily penetrated by roots and animals.

What most people do not realise is that the process of concrete hardening is a chemical reaction which needs water, not just a physical process of drying out. The actual chemical reaction involves complex macro-molecular transformations of calcium hydroxide and silicon dioxide to form a space lattice of calcium carbonate and silica. If there is not enough water in the initial stages of setting, the reaction cannot take place properly. The reaction is also sensitive to temperature, being faster in tropical climates than in temperate ones.

So curing basically means keeping the concrete wet for a few days and not allowing it to dry out in hot sunshine. In canals the best way is to pond water in the canal soon after placing the concrete. It does not matter if the concrete is still soft, it will harden under water. Be aware of the risk of erosion when the water is let in. If it is not possible to cover the concrete with ponded water, cover it with polythene sheet and check regularly that there is moisture beneath the plastic all the time. Do not permit intermittent wetting with buckets of water every morning. In hot weather it dries out in minutes and it can never be kept continuously wet. Hessian or straw used to be recommended as a cover for curing but the latest British Standards reject these as causing cracking through uneven evaporative cooling. Curing compound sprayed on soon after placing is the modern way, and saves a great deal of labour in keeping a large area of concrete damp.

Steam curing of precast concrete provides both moisture and heat at the same time and is usually done at an optimum of around 120°C for just a few hours.

Poor concrete finishing

In order to easily achieve a smooth finish to plasterwork or a concrete surface, an excessively wet mix is sometimes used. This results in a weak surface layer that is liable to shrink on drying, and is evidenced by map-like cracking over the whole surface. In a short time the cracks develop to such an extent that large pieces of plaster peel off under the action of water.

Poor control of concrete mixing

In order to save money it is quite common for contractors to substitute soil for sand, or clay instead of cement during concrete mixing. Even when the intent is less unscrupulous, the inclusion of dirt or organic material is difficult to avoid under the conditions of small contracts. The correct storage of cement and aggregate is also difficult to achieve. The result is variable quality and weak concrete that breaks up rapidly under the action of soil and water.

In India we had to rehabilitate a diversion weir that had been built with a masonry skin and mass concrete infill. Water leaked through this structure so badly that the entire scheme of 500 ha was inoperable. We demolished the upstream half of the weir in order to replace it with properly constructed mass concrete, and in so doing we found the old concrete so weak that you could demolish it with your bare hands. It was full of holes and pockets of soil. Some of the villagers told us that the cement had been sold off in the local market and replaced with soil while the engineer wasn't looking.

Then as we were concreting the foundations again, I caught the engineer in charge doing some underwater concreting. He figured that since there was already water standing in the foundations all you had to do was throw the dry sand, gravel and cement into the standing water and the concrete would mix itself. That's the same as baking a cake by throwing the flour, eggs and sugar into a bowl of water and not stirring it. The end result is indigestible. Anyway, some applied initiative is no bad thing and after I'd bawled him out over that incident he went on to become an excellent engineer.

Inadequate thickness control

In Indonesia, stone masonry canal lining is usually specified to a thickness of 200 mm. In practice the variable size of individual stones makes it difficult to achieve a constant thickness. More serious is the fact that this specification is easily prone to abuse by the contractor. In extreme cases only a thin smear of mortar is applied to bare soil, and the result is difficult to distinguish from the specified product without continuous supervision or destructive testing. The resulting lining is easily damaged by the action of people, animals, water and soil.

Even with good intentions, it is difficult to get an even thickness, especially with a thin lining like plaster or ferrocement. The secret is to prepare the bed or foundation neatly, filling in voids or undulations with rough plaster.

Bad formwork

For any concrete structure with a vertical or sloping face steeper than about 45° you can't have good concrete without good formwork.[1] Make sure that there are no holes in the formwork that the cement slurry can leak out of, and that the formwork is fully supported and doesn't collapse or deform under pressure of vibrating wet concrete. For repetitive work like small canal structures use steel formwork that can be used time and again without damage.

Consider also using permanent formwork. The flumes in the foreground of figure 7.3 were accurately fabricated in steel sheet in a workshop, in one piece, then placed on site and filled with concrete. The flume in the background was formed accurately in precast concrete and backfilled with earth. Either way, critical dimensions can be assured when it would be very difficult to guarantee them with normal in situ formwork.

> I almost threw a carpenter in the river once. He had been sub-contracted as a formwork specialist on building a small pump intake, basically three sides of a box. Being an inaccessible site he had floated the timber planks across the river on a raft. The planks had to be cut accurately and properly jointed at the corners, but he refused to cut it on the grounds that its resale value would be reduced. I ended up cutting it myself. Formwork design and construction is a specialised job and has to be done properly.

11.4 Bad Design

This is where the engineers can really be blamed. If it's not designed right at the outset, no amount of rehabilitation work or juggling with the management later in the project's life will get it going properly.

Obsolescence

Government irrigation departments seem to be generally unadventurous when it comes to designing irrigation schemes. They tend to rely on tried and tested standard designs which might have been originally done when labour and material costs were relatively cheaper. Modern techniques of concrete placing and recent developments in structure design seem to have gone unnoticed. The scene in figure A11.46 is fairly typical of a stone masonry canal which does not work. The water flows not along the canal but from the field on the left, into and out of the porous sides of the canal, and into the field on the right, just as if the canal wasn't there at all.

The concept of 'built-in obsolescence' seems to have crept into irrigation design from the car manufacturing industry, and for similar reasons, although in the case of irrigation it is misguided.

Poor planning

This is another fault that is particularly common in small irrigation schemes. Capital construction cost for the entire scheme is often low. This in turn encourages the belief that engineering costs[2] can also be low. In fact the engineering costs need to be proportionately higher because survey and supervision still needs to be done, whatever the size of the scheme.

---

1 *Formwork is also called 'shuttering' or 'centering' depending where you are.*
2 *'Engineering' here includes survey, planning, feasibility studies, design, contract document preparation, tender evaluation and supervision of construction, all of which is normally carried out by consulting engineers. It can also be done 'in-house' by the government department concerned.*

The canal in figure 11.10 was built in the wrong place and diverted by the farmers to run along their village boundary rather than across fields belonging to the adjacent village. The planning in this case amounted to a line drawn on a map with no recourse to field checking of the actual situation.

But big schemes too can fall foul of inadequate planning procedures, especially long-established ones. In rehabilitating the Upper Swat Scheme in Pakistan, the watercourses are routed by a man sitting in an office of the Irrigation Department and drawing lines on a 100-year-old cadastral survey map with no contours or levels on it. Not surprisingly when we started a more thorough check in the field we found that the farmers were far more aware of the engineering problems than the engineers were. With a century of land boundaries changing, land being levelled and buildings being constructed on it, the old maps bore little resemblance to present-day reality. The farmers' demands for sometimes drastic re-routing of canals were invariably sensible.

Sedimentation

Potential difficulties caused by silt deposition need to be fully appreciated at the design stage. Extract it, eject it, exclude it, or flush it through the canals, or else be prepared to dig it out. Section 7.5 details several strategies.

Oversizing

Oversize canals give a long response time. It is a common mistake of inexperienced engineers to be over-conservative in estimating the roughness coefficient of a canal. The canal ends up larger than it needs to be and response times are increased.

In a recent rehabilitation job in India I was horrified to find that most of the canals had been built in deep cut; in some cases 5 or 10 m deep. There were all kinds of problems with landslips in the rainy season and siltation because of sheer inaccessibility for cleaning. Cattle and buffalo used to fall in and so the farmers were demanding fences and culverts and tunnels. The logical route for the canal would have been a few metres down-slope, following the command level contour, and costing a fraction of the amount that was actually spent. It transpired that at the time of construction the farmers had not permitted the canal to be built on their land because they didn't trust the promises of monetary compensation for land lost. So the route had been altered to traverse forest land that was usually in steep terrain suitable for neither cultivation nor canal building.

Unsuitable local materials

Stone masonry is a very common material in Indonesia and India, widely used in colonial times when labour and materials were cheap. But it is totally unsuitable for modern hydraulic structures besides being unduly expensive. Mud mortar and laterite blocks may be suitable for houses but they are no good at all for hydraulic structures. This may sound obvious, but amazingly enough there are plenty of examples like this around, usually on small schemes where the engineering input has been minimal.

Not enough turnouts

Inadequate provision of outlet structures will result in farmers resorting to their own devices. Canals cannot withstand the abuse shown in figures A11.7 -10.

In east Java the normal turnout design using heavy stone masonry construction renders the structure too expensive. The result is that there are never enough, because the standard tertiary turnout spacing of 400 m is determined by government rules, and not by farmer needs. The original basis for rules such as this one are lost in the mists of time, and nobody ever thought to question them since they are printed in government manuals.

Undershot turnouts

Undershot turnout gates are usually designed so that their sill level coincides with canal bed level. When the gate is closed, the lower seal is always under a head corresponding to the depth of flow in the canal, and it invariably leaks.

It pays to raise the turnout sill above the bed level of the parent canal, to reduce the hydraulic pressure on the lower gate seal when it is closed. Gates made of thin steel plate tend to warp or bend and lose their sealing capability so avoid using steel plate that is thinner than 4 mm. Gate seating grooves tend to get blocked with debris and sediment. All in all, undershot turnout gates are not very satisfactory.

Regulators and division structures

These structures are often adaptations of standard undershot turnouts, with the principles of hydraulic flow ignored in the interests of simplifying construction. This results in square division boxes that are too large and have internal sharp corners where flow velocity is reduced and sediment collects. Head loss is higher than necessary, due to their crude design.

Quaternaries

Here is another Indonesian example, from one of the big conjunctive use schemes in east Java. Quaternary channels are designed to serve up to 20 small farms each, delivering water from the tertiary canal turnouts direct to the fields. Due to the 400 m distance between turnouts, quaternaries are often excessively long, difficult to regulate, and partially duplicate the route of the tertiaries by running alongside them. The design philosophy for this comes from the grouping of offtakes on large surface-water schemes, when it is convenient for the operators to have all the control gates for several minor canals in the same place. In many instances there is a clear case for abstracting water direct from the tertiary. Direct abstraction is only possible provided it is accommodated in the design, but where it is not, damage like that shown in figure A11.7 is the result.

This idea of grouping regulators and turnouts is perhaps a throwback to colonial times when water control was easier to enforce than it is now. Some interesting ideas are thrown up by Lucas Horst (Ref. 2) in examining the relationship between available management skills and the technology level for which the canal system is designed.

Design for short life

Durability of canal lining is not always an important factor in present design philosophies. Early failure of lining is sometimes accepted as inevitable, and even desirable in order that future work in reconstruction may be assured. Sadly, the principles of sound engineering together with the needs of the farmer are not necessarily to the fore.

Every time the canal breaks up, you need a maintenance contractor to rebuild it. If you get a percentage of every maintenance contract price, it makes good sense to design the canal to break up quickly in time for the next round of maintenance contracts, so you can rebuild it all over again.

> A young engineer was designing a concrete-lined canal in a Javanese project. He had designed the lining as 100 mm thick mass concrete, trapezoidal profile with side slopes of 1 on 1.5. The whole canal was about 2 m wide and 0.5m deep. Nothing wrong with that, except for the concrete being twice as thick as it needs to be. It's the most common design of canal section there is, but normally you would lay the concrete lining direct onto a prepared earth sub-grade, compacted and trimmed to shape. But he was actually planning to cast the canal sides on top of formwork sloping at 1 on 1.5, then remove the formwork and backfill with earth beneath the canal sides. Imagine how much compaction he would have got. Zero, and zero support for the concrete which would have cracked under its own weight as soon as the formwork was removed. Not to mention doubling the cost. When I pointed this out he modified his design to reinforced concrete, thereby increasing the cost even more. And then I realised why he was doing it this way. Imminent collapse was assured, as was his rake-off from the next reconstruction contract.

Over-design for high cost

If like the previous example you're a designer with a vested interest, the same end result is achieved by simply designing things bigger and heavier than they need to be. But it's a thinly drawn and uncertain line between what is necessary for reasonable durability and what is patently too much. And over-design in itself doesn't necessarily mean a badly operating scheme.

11.5 Bad Water Management

Bad management covers a multitude of sins. But there are some fundamental causes here, which should be considered.

Lack of training

Many schemes, particularly small ones, suffer because there are no firmly established rules of operation, and incomplete training of Canal Operators. Operating rules and maintenance schedules should be considered an essential part of detailed design.

Political interference

Modern ideals of democracy often conflict with the need for firm control over canal operation. Take the ancient Sri Lankan irrigation schemes. In the days of Parakrama Bahu, there was complete order amongst the irrigators. There was no water stealing, everyone took his turn in receiving water, and there were no arguments. Today, there is continual conflict between farmers and scheme managers, with the local politicians as likely as not siding with the farmers in pressing for unscheduled water issues. There are votes in water. But 1000 years ago there was only one penalty for water theft: *DEATH.*

Pressure from farmers

Put yourself in the position of a typical canal operator. You're young, just out of school. You're in charge of delivering water on a supply schedule to a couple of hundred farmers on a shared canal system. One farmer older than your father demands water out of turn. Another threatens you with a beating if you don't give him water tomorrow. What do you do? It takes a strong and hard man to stand up to that kind of pressure. So you give in, just this once. And so on until before you know it the entire water management schedule has disintegrated. And we're not only talking about developing countries here. This is a real-life example from Montana, USA. I could mention a few more gruesome examples However,. There's more than one Sri Lankan canal operator who paid the price of his own self-esteem with his life.

Improper gate operation

A canal with a lot of manually-operated gates can be a nightmare to control. No matter how many rules might be laid down, spillage like that in figure A11.32 is common. And spillage in one place means a water shortage somewhere else. This is a good advert for automatic gates, or passive automations such as proportional division.

11.6 Natural Causes

Natural causes does not mean that damage is inevitable. It does mean that design of the canal has to take into account the possibility of damage.

Figure 11.5 Inappropriate use of rigid stone masonry in a swelling soil led to this crack soon after construction

Swelling soils

Figure 11.5 shows a brand new stone masonry canal in Java. The crack in the bed is 3 cm wide, another advertisement for precast parabolics!

The fertile volcanic clays of the east Java plains are extremely sensitive to changes in moisture content. On drying they become extremely hard, shrinking to create deep cracks. When wetted they expand by as much as 20 per cent in volume, losing strength to become unstable and unsuitable as free-standing banks or bunds.

Rigid canal linings that depend on the soil for their support are easily damaged by this soil movement, which may take place gradually over a period of months between wet and dry seasons, or rapidly between irrigations and during isolated rainstorms. Soil movement is the biggest single factor in the destruction of stone masonry linings. The flexural strength of masonry is far too weak to resist the horizontal forces created by soil swelling. The roughness of the underside in contact with the soil inhibits any tendency for the soil to slide over it, and the result is the dislodging of individual stones as they move with the surrounding soil.

figure 11. 6   *Lining burst by external hydrostatic pressure from surface-water flooding*

External hydrostatic pressure

This includes pore water pressure developed behind the lining due to leakage through joints and cracks, and natural inflows of groundwater. When the water level in the canal is drawn down quickly, the external pressure can burst the lining as in figure 11.6. Soil particles can be washed out through the joints, resulting in a piping failure such as in figure 11.7.

More pictures are shown in figures A11.12 to A11.15. The cause of these failures was an over-engineered edge to the concrete lining, which should have been cast directly against bare soil. The shuttered edge and loosely backfilled soil against it had allowed penetration of rainwater during a sudden heavy storm.

figure 11. 7   *Lining undermined by piping of soil through concrete joints*

Animal damage

Crabs

The most destructive animals are land crabs, which flourish in the wet conditions of tropical clay soils. They burrow deep into canal banks and create a network of holes and tunnels up to 50 mm in diameter. Not only are these holes a major source of direct leakage, but they contribute to collapse of the canal banks and lining.

A land crab like the one in A11.18 can bore a hole through an earth canal bank in thirty seconds flat. It takes only a little longer to go through the joints in the average Indonesian stone masonry.

Rats

Rodents tend to limit the damage to above the water-line, but they can undermine hard lining from the dry side and generally weaken the banks. Their excavations deposit distinctive tell-tale heaps of soil in the canal (figure A11.20) which then have to be removed.

Invertebrates

Worms and burrowing insects can beaver away just beneath a hard canal lining, unseen and undisturbed by man or predator. So what soon arises beneath an apparently solid brickwork lining is an intense maze of tiny tunnels that serve as a very efficient drain to any water seeping through the lining. Local brickwork is notoriously porous, and lets water both through the mortar joints and through the bricks themselves. Due to the highly effective worm-hole drainage system, the water is carried away as fast as it can get through the lining. The lined canal then leaks more than it did before. There was some research done in India (Ref. 3) which came to the mistaken conclusion that lining was a waste of time. Do it right and it is not a waste of time. But do it wrong and it surely is.

Cattle and water buffalo

In countries like India, the sacred cow is the biggest threat of all to the average small canal. As soon as a canal with a flat bed is constructed, all the local cattle start using it as a highway, and unless it is made of rock-hard concrete, they will break it up in no time (see figure 6.9.) This is one of the most compelling arguments in favour of small parabolic canals. Cattle do not like walking on a curved surface.

Where buffaloes are used as draught animals, it is essential to provide frequent purpose-made wallows in the canal. Otherwise the animals will make their own.

People

The man in figure A11.21 is doing four things at once. He's washing his bicycle, he's saving time by going faster, he's saving further damage to the road on the right, and he's breaking up the canal bed. (But try riding a bike in a parabolic canal!)

Tractors

Vehicles running alongside a canal can destroy it without even making contact. The downward wheel pressure has a horizontal component which exerts a strong sideways force on the canal lining. This is a common cause of failure in the rectangular brick canals of Pakistan.

Even women washing clothes in the canal, a common practice everywhere, can be damaging in the long term. Structures like measuring flumes are especially popular for doing the washing. So build in custom-designed washing facilities like the ones shown in figure A11.23, which is at the head of a Javanese tube-well scheme. And it's no good the engineers just saying they're not allowed to wash there, and fence off the well with barbed wire, as has happened a few times in some other schemes nearby. A tube-well like this is the only source of pure water in the entire village, and it's a valuable resource that needs some lateral thinking to get the most out of it.

Root penetration

Vegetation can destroy a badly-constructed canal in a single season. Roots of weeds and grass, aquatic or not, can exert enormous pressure once they penetrate a small crack. Cracks are progressively widened and under tropical conditions the deterioration of lining can be extremely rapid, especially if its condition is exacerbated by poor construction or maintenance. Stone masonry is particularly susceptible to root penetration as it effectively has a much greater length of joint, and hence of potential lines of weakness, than other types of lining. Good maintenance or, better still, parabolic canals are the answer.

> There's no answer to some roots. In India I was mystified by a canal that took an abrupt detour around an ordinary-looking tree that could have been easily chopped down and used for firewood like most of the other trees in the vicinity. What was more, one of its large roots was protruding across the canal and firmly embedded in the concrete lining, almost blocking the water flow. Then I was shown the little shrine with the statue of the Hindu god Shiva. The tree was sacred.

Landslips

Earth slips are just one of the many things that engineers have to provide against in design. With small canals in rough terrain it is not always easy to predict where the earth will move. But with experience an engineer will come to know how to avoid damage. Slips are most likely during construction, when excavations are left open, and it rains. The pipes in figure A11.27 floated because the deep trench in which they were being laid suffered a landslip at one end, and the trench filled with water.

Thermal Expansion

Temperature fluctuations can have different effects on the two constituents of stone masonry, namely the stones themselves and the mortar in between. Differential rates of expansion lead to cracking of the mortar joints, which is then made progressively worse by root penetration and the action of flowing water and animals.

11.7 Bad Maintenance

The importance of maintenance is often not realised until it's far too late. It is not a popular subject with engineers. But it is quite a challenge to design the canals to be almost maintenance-free.

Siltation

Heavy sedimentation can drastically reduce the canal capacity if the cleaning programme lags behind schedule. The ultimate stage is shown in figure 7.29, with the canal capacity almost reduced to zero. This occurred on the Gezira Scheme in the early 1980s, when there was no money for maintenance of ageing earthmoving machinery. Within a year, canals became unusable and a progressive breakdown in water management ensued.

Periodically heavy silt loads can occur in run-of-river schemes which divert water direct into the canal system from unregulated rivers. Unless it is plastered, stone masonry has a rough surface which, by creating a thick boundary layer, tends to attract a film of deposited silt on its surface. It is not long before plants take root in the silt deposits, and root action soon destroys the stonework.

Incoming seepage flowing in the bottom of an otherwise dry trapezoidal channel may tend to form a meandering thalweg, depositing sediment in miniature sandbanks on alternate sides of the channel bed. These can rapidly encourage vegetation growth, which in turn traps further sediment. Over the course of 2 or 3 months if the canal is not being used for irrigation, the accumulation of sediment in this way can block the canal completely.

figure 11. 8   ***Heavy weed growth in clear water from a large reservoir. Yeleru Left Bank Canal, Andhra Pradesh***

Silt removal must either be done manually or mechanically by digging out, or hydraulically by flushing. However, the latter method is not usually possible in trapezoidal channels, especially if like stone masonry they have a rough surface. Parabolic or semi-circular channels are easier to flush and any sediment or debris is concentrated at the lowest point, making it easier to clean and more likely to be flushed through with the normal discharge of the canal.

Over excavation

In the routine cleaning of earth canals, it is easy to dig out too much silt and inadvertently enlarge the canal profile. This results in lower velocities, a longer response time, and more rapid siltation. Another vicious circle.

Vegetation

The only way to control weeds and roots in very small canals is to get the farmers to clean them out regularly. Unfortunately most farmers don't see it that way. What's the point of cutting weeds that are going to grow again next week? Another advert for parabolics!

Masonry repair

The repair of masonry is often difficult to organise, especially if it is under government control. Farmers may not have the necessary expertise or resources to carry out repairs themselves. Leaks in masonry may be difficult to detect and the amount of work and materials required may be difficult to estimate. It is often more effective and easier to plaster an entire canal rather than plug numerous small leaks in stone masonry.

The expense and difficulty involved in repairing masonry adds considerably to its long-term cost and is another factor against its use.

Closing of holes

The blocking of holes and breaches in unlined canals is tedious and must be done very frequently to have any effect. Crabs can in the space of 1 or 2 days cause enough damage to render an unlined canal unserviceable, so the closing of crab holes must be done constantly by farmers. Once a reach of canal is neglected for any length of time it may take a considerable effort to make it useable again.

Leaking structures

Structures such as turnouts and cross regulators are often built to standard designs as undershot sluice gates, set in stone masonry (figure A11.27.) This type of structure is not only heavy and expensive, but almost always leaks. Steel gates invariably leak around the sides and base, contributing to seepage inflows and leakage losses. The main problem is that the close tolerances required for gate seals can rarely be achieved during construction in the field. Gate seatings are also prone to blockage by soil, stones and other debris.

11.8 Bad Farmers

When schemes don't work properly, it has long been a fault of engineers and irrigation managers to lay all the blame at the farmer's door. I have never subscribed to that view. The average farmer is full of good sense and he never does anything without good reason. His livelihood depends on it, after all. But it's a little naive to think that everything he does is right.

Incomplete understanding of irrigation

Especially on new schemes, farmers are likely to be unused to the irrigation methods and schedules that the designers intended. A more common thing is the farmers being used to only one crop, usually paddy rice, and assuming that there is only one way to irrigate. It is very difficult to persuade an old Indian farmer that groundnuts do not need to be inundated with water all the time, when that's what he's been doing all his life with rice. There is often insurmountable resistance to trying unfamiliar methods like using plastic siphons, or even opening and closing turnout gates, when for generations the only method of spreading water around has been to break a hole in the canal bank and patch it up afterwards. This is not so easy when the canal is lined with concrete.

Water charges encourage wastage

One of the main reasons why water charges don't work with small farmers is that they lead the farmer to think that as he's paying for the water he's entitled to waste it. See chapter 3 for a fuller discussion. Once again, these attitudes apply to the developed world as well. What's the opinion of the average British tax-payer when confronted with a summer hosepipe ban?

Social quarrels and wilful damage

There may always be a reason for it, but bitter quarrels can arise over trivial disputes and end in physical damage that to the outsider seems out of all proportion to the alleged injustice that started it all. I was in Afghanistan rehabilitating small schemes that used scarce water from ancient underground quanats. One farmer was sitting in his field when another crept up behind him and shot him dead. The reason turned out to be a long-standing family feud arising from the time when the dead man's grandfather insulted the other's great uncle during an argument over water 50 years before. The whole village accepted it as a reasonable excuse for a gunfight.

Crops on canal banks

I can sympathise with those farmers in eastern Java trying to raise a family on a tenth of a hectare. Four crops a year is not unusual, and every scrap of land is put to use. So no wonder the canal banks get covered in crops, notably pigeon pea, pepper and legumes for fertiliser and fodder. But perhaps the worst of all is cassava, which has to be dug up when it's harvested. Stone masonry was never intended for the kind of treatment shown in figure 11.9. But the parabolic canals of figure 9.11 are not affected.

figure 11. 9   ***Crops on canal banks damage lining. Cassava, Java***

Theft

If it's moveable, it will be moved. Small precast concrete slabs, structures or canal units should always be made heavy enough to be not easily shifted.

Some people will go to great lengths to steal things. One night in India they dug up and stole a truckload of heavy parabolics from one scheme. We never did discover what they were used for, but if it was to spread the gospel about parabolics them I'm all in favour.

On the Gezira Scheme in the Sudan, there was in the 1980s a well-established secondary industry. The water control gates had been designed with heavy brass thrust nuts on the rising spindle gates. They were just the right size to conceal in a pocket. An important local industry melts down scrap brass and makes new nuts to sell to the irrigation department to replace all the old nuts that mysteriously disappear with alarming regularity. Where did the scrap brass come from.... ?

11.9 Farmer Interference

It is useful to differentiate between wanton vandalism, and deliberate damage that can be traced back to an underlying reason. Such defects are often valuable indicators of social problems or engineering defects in planning or design.

Unofficial turnouts

Like most aspects of interference by farmers, the construction of unofficial turnouts is a symptom either of bad system design or poor water management. Farmers often construct their own turnout structures, either by breaching the canal or by damming it and heading up the water so that it overflows the bank.

The reason for this interference in figure A11.9 is the long distance between turnout structures, typically 400 metres. Due to the high cost of structures as presently designed, a closer spacing is not affordable. Damage to the lining is inevitable, and where the banks are breached continual leakage is normal, hampering water management.

Drainage

Field drainage is rarely considered adequately at the design stage, and farmers are forced to adopt the easiest measures to drain their fields, especially in the case of paddy fields that need to be dried off prior to harvesting. With no field drains provided, it is common practice for farmers to break into the canal and use it as a drain, as in figure A11.7. Stones or slabs from the lining are removed, and although sometimes replaced when not in use they are a source of permanent leakage. With inflexible supply-scheduled systems such as the proportional flow warabandi schemes in Pakistan, there is no means for a farmer to turn off the water when he doesn't need it. So he just turns it into the nearest drain, which may happen to be someone else's watercourse.

Redundant canals

Tertiary canals in poor condition are sometimes by-passed and water conveyed from field to field without using the canal. Paddy is then planted in the canal, as in figure 11.4. This is usually a sign of inadequate maintenance, by which the canals leak so badly that they cannot be used, or by bad design which may set the canals below command level for the surrounding fields. But the example of A11.46 is typical of a stone masonry canal that has rapidly disintegrated and should have been designed in another material. Occasionally, administrative boundaries intervene and the designs of adjacent schemes are not properly co-ordinated, leaving two canals serving the same field.

The unofficial diversion of figure 11.10 is one such indication of a village boundary dispute.

Quaternary canals might be designed to run partly alongside the tertiaries, as in figure 11.11. Farmers are then tempted to incorporate the unlined quaternaries into their paddy fields in order to increase their cropped area. Once one farmer does this, he cuts off the official means of water supply to his neighbours, and all others downstream of him are obliged to follow suit, abstracting directly from the tertiaries.

figure 11. 10   *Diverted watercourse in a boundary dispute*

figure 11. 11   ***Redundant tertiary takes up useful land***

Damaged structures

Wanton damage of control structures is a sign of poor social cohesion or bad water management, in which farmers desperate for water are forced to resort to desperate measures in order to get it. Interestingly enough it was not very common in east Java which is a hotbed of defective schemes in most other respects. This was most likely due to the well-developed social structure that was in existence long before the formal irrigation schemes were built.

The concepts of automation, particularly of downstream control, may be alien to farmers who have been used to a supply schedule for generations. Self-regulating float operated gates are frequently subjected to wanton damage in efforts to by-pass the automatic operation that gives unaccustomed priority to the tail-enders. For that reason the Pehur and Maira Branch gates were enclosed in secure walled structures (figures A11.29-30).

11.10 Corruption

The vicious circle of corruption seems to engulf everyone involved in irrigation development in all the countries that can least afford it. Irrigation schemes seem especially prone to corruption, because they often involve large amounts of capital expenditure. Furthermore their management, maintenance and construction often puts underpaid government employees in positions of relative power which are easily abused. We see it here on two levels, micro and macro, financial and economic.

Corruption can be found at all stages of a scheme's development, from construction to maintenance to operation and water management. Recent literature has referred to corruption in water management as 'rent-seeking behaviour' but why attempt to hide it by dressing it up in flowery language?

Honest John

Almost 40 years ago it took me precisely a week of working in Africa to change from an idealistic, wet-behind-the-ears graduate who thought he was going to change the world, to a wiser and more cynical person who recognised engineering for what it really is. That is, more often than not, merely a means to someone else's political end.

And of course, the longer you go on the more you learn to live with it and you soon realise that corruption in one form or another goes on all the time, in every country in the world including your own. Whether it's a case of the contractor paying off the engineers to accept bad workmanship, or an unjust aberration of geopolitics, there's often no means of fighting it. All you can do is draw up your own ground rules and live by them. Mine are the following:

- Never compromise on engineering quality, safety, or correct technical principles.

- Design for cheapness if and when it is required, but if you have a choice then think laterally and go for the best.

- Never take or give a bribe.

- Accept that 10 or 20 per cent[3] of a contract cost might go under the table, but not at the expense of sacrificing quality, safety or technical principles. Frankly I don't waste any sleep worrying that money is being hived off provided the job gets done properly. If that money ends up in the local economy it's probably no bad thing anyway.

- If you're in charge of supervising a project, insist on controlling the money using the normal engineering procedure of certification of work done. It's the only lever you've got to ensure good quality.

- Be prepared to walk off the job rather than break your own rules.

Perhaps this viewpoint is one of the luxuries generated in a rich society which is not directly transposable to a poor one. Like pollution of the environment, preserving wildlife, destroying the rain-forest. Having wrecked our own countries, who are we to tell others what to do? Perhaps, but I'm a great believer in learning from other peoples mistakes.

Endemic micro-corruption

The most damaging manifestation of corruption is the theft of key materials such as cement in order to finance bribes. This means that the construction quality suffers to the extent that concrete is too weak or too thin to do its job, and the structure falls down. This is common practice in India and Pakistan, where government engineers are obliged to obtain bribe money from contractors to pass on to their superiors in fixed percentages, all the way up the line, according to a set of rules that is so well-known as to be almost officially sanctioned.[4] This would be less of a problem if the small local contractors were allowed to top up their tender prices by the 40 per cent required for bribes. However, the contractors are obliged to work to government schedules of rates, which are invariably out of date and lower than the prevailing market rates for labour and materials. This means that the contractor has only two options if he wants to stay in business, he can cheat on quantities by pretending to have done more work than he actually has, or he can cheat on materials by putting too little cement into the concrete.

All of this we can call micro-corruption. And it is endemic, like malaria, and just as difficult to get rid of. Either way, this form of corruption is crippling the development plans of many countries, and it is compounded by the naiveté of international aid agencies and financing institutions.

---

3 One explanation for the corruption in Pakistan is that under the British colonial administration, junior engineers were encouraged to assist semi-literate contractors in interpreting drawings and preparing tenders, a service for which the engineers were paid a fee by the contractors. When the British left and the junior engineers became senior engineers, the tradition was continued. But the ex-junior engineers were by the time they became senior engineers so accustomed to receiving these perks that went with job that they felt obliged to secure a certain pe centage from the juniors for old times' sake. And as the seniors were promoted higher and higher up the system, so the total pe centage to be got from the contractor increased, until it arrived at the ridiculous level in excess of 40 per cent where it stands at present.

4 Look at it this way; if the engineers suddenly all got a raise in pay of 20 per cent, they would just spend the extra money and boost the local market economy. But if everyone got a raise in pay, prices would all rise in line with it, inflation would sky-rocket, and the economy would be no better off. Ergo, corruption is a good thing???

Some theorise that corruption is only a passing phase as a country changes to a market economy. Who knows. There are plenty of market economies where corruption is rife. Who can truthfully stand up and say there is none in his own country?

> Here's a story from Indonesia in which yet more of my illusions of engineering were shattered. I had developed those beautiful precast parabolic canals to rehabilitate a not-so-old canal system lined with the traditional stone masonry. The parabolics were half the cost, ten times as good hydraulically, and would last 100 years rather than the 1 or 2 years of their predecessors. You would think any self-respecting engineer would be proud of that. Not a bit of it. The parabolics were desperately unpopular with the local engineers, precisely because they were too good. As one of my counterparts said to me as we applied the finishing touches to the parabolic proving trials, 'If we install these canals everywhere, what are we going to do in the future when there are no more broken canals to repair?' He meant that since two-thirds of his income derived from bribes from maintenance contractors, he would be out of a job if the canals lasted too long. I could understand his point. But it goes against all accepted principles of engineering. I never became an engineer in order to build things that fall down. But that's just what some engineers do.

You may not be able to stamp out corruption, but you can at least guarantee good quality of construction that will last for a century, if need be. The way to do that is to employ properly qualified independent supervising engineers, following long-established procedures aimed solely at getting a high-quality end product. It may appear to be more expensive in the short term, but it is a lot cheaper than having to rebuild everything after 5 years.

Macro-corruption

Micro-corruption is an affront to right-minded engineers and an irritant to high-minded sociologists. It can have severe side effects in that the quality of construction may suffer. It can upset the financial balance of a project. But it does not necessarily have to be a drain on the country's economy. Accepting that it goes on is the first step towards limiting its bad side effects. Ensuring good engineering is the next step. Then at least the project gets built properly, and the money that is hived off stays in circulation in the local economy.

But macro-corruption is different. When project money ends up in a Swiss bank account it can be of no value to the project, nor to its local economy, nor to its country. Prime targets for macro-corruption are overseas contracts for the supply of machinery. That 20 per cent of the bulldozer price that goes into a politician's dollar bank account is lost forever to his country. It is an economic loss, whereas micro-corruption is a financial loss.

Caterpillar, a maker of earthmoving machinery, used to boast a proud record of never bribing in order to win contracts. Their machines were and still no doubt are simply the best, and despite being expensive, were economic in the long run because they were reliable. Then came fierce competition from all over the world. From companies selling inferior products, at cheaper capital cost but with hugely marked-up prices for spare parts. And with big bribes to boot. Visit the plant yards in almost any third-world irrigation department, and you will see vast graveyards of heavy machinery that doesn't have Caterpillar written on it. The Swiss bank accounts of some of their ex-politicians are also vast.

We now step into the dirty and unpredictable world of geopolitics. Maybe there is some way of countering macro-corruption, but I don't know what it is.

Violent micro-corruption

In Sri Lanka in about 1980 there was a young engineer, enthusiastic, competent, and eager to do a good job of supervising the construction of an earth dam. The contractor on this small government project had other ideas. The engineer saw that the contractor was doing a bad job of earth filling and compaction, and tried to redress matters by preventing payment to the contractor. The engineer's superiors overruled him However, because they had been adequately bribed. Moreover the contractor resolved to avoid any similar disputes in future and went for the final solution. That young engineer disappeared from the site and was never seen again. His name passed into local legend and his grave is 50 feet down inside the embankment of the dam.

Corruption in water management

Large supply-scheduled irrigation networks are ideal for exploitation if you are put in charge of delivering water to every farmer. It is common for influential farmers to demand more water, often by enlarging their outlet, in return for supplementary bribe payments to management staff. It is also easy for the operating staff to withhold water from farmers who won't or can't pay bribes. This is extortion, and it is becoming increasingly common in some of the large schemes in Pakistan. Some farmers may also recoup their expenditure on bribes by pressuring their neighbours downstream. I know of more than one case in Pakistan where the upstream farmer on a watercourse takes most of the canal water, then sells water from his own pumped tubewell at an inflated price to his neighbours. However,, in Pakistan you need social and political influence to get away with this, otherwise you will very soon end up full of Kalashnikov bullets.

In many schemes in India, Pakistan and Indonesia and no doubt many other countries as well, there are recognised scales of bribes which everyone accepts. Thus the ditch-rider might be paid 100 rupees for opening a gate on time; 75 per cent of this will be passed on to his supervising engineer; 50 per cent of this in turn will be passed on to the area engineer, and so on up the system. In some countries being an irrigation engineer is a top job, for which you may have to bribe several years salary to get the job in the first place.

11.11 The Answers

In short, better engineering. Everyone with any influence in the progress of the project from its inception to its operation and management needs to be aware of the scope of engineering and the ways in which engineers are able or unable to do the job.

Proper supervision

Engineers have received a lot of bad publicity in recent years, being accused of all manner of insensitivities in the design of huge irrigation schemes that ended up not working very well. Perhaps not without reason, civil engineers were not very popular with those more concerned with the social side of irrigation projects, who could see some disastrous end results and huge amounts of money being wasted. By and large, irrigation engineers have taken the hint and I don't know of any that do not derive great satisfaction from doing a good job all round and seeing the end users fully satisfied. In fact the wheel has now gone full circle, and there is often too much emphasis on the sociological side. Love us or hate us, there is really no substitute for a few good engineers on the job.

Good supervision and design might cost as much as the construction itself, but it's worth it. The cost of supervision needs to be built in to the overall cost of the project. On small schemes it may be almost as much as the hardware, but the cost of not doing it can be much more.

Design with the times

Here are some ideas that have been around for a long time but have not yet been absorbed by most irrigation departments.

- Vibrated concrete, properly mixed and cured, is better than stone masonry in every respect.

- Intermediate storage is probably essential but should be designed in from the start.

- Think about how the scheme will be managed before you ever put pencil to drawing board.

- Limiting discharge gates. There's more than one way to design a turnout gate.

- Parabolic canals are easier to build than you might think.

Proper training

This is a neglected sphere although things are beginning to change with aid agencies pouring more money into this kind of low-risk software rather than high-risk, high profile hardware made of concrete.

Management training is a basic necessity. But don't send someone to Harvard Business school when she needs to know about farmers' attitudes to government officers, recognising water wastage, and operating sluice gates. There are some excellent computer models around now that simulate scheme operation and allow management staff to learn from their mistakes before they even get out on the job. Learning from your mistakes is the best way there is.

Farmers are normally trained through the extension services. Over-stretched extension services often find farmers stubbornly refusing to accept new ideas. It's not helped by many aid agencies' equally stubborn persistence with inappropriate extension philosophies such as the T & V system, in which very little contact between farmers and trainers seems to come about.

Women and youth. India is one country now realising the potential of the formerly neglected three-quarters of its population which is both more receptive to new ideas and surprisingly influential behind the scenes, when it comes to persuading farmers about the benefits of new ideas. In South India's Women and Youth Training Project the women extension staff showed a lot more initiative than the men, asked more incisive questions, and were much more willing to try out new ideas such as siphons and get their hands dirty in the field. Most of their male colleagues considered it beneath their dignity to walk around in the field or wield a hoe or siphon. Or perhaps, never having done either, they just didn't want to demonstrate their ignorance.

Training of trainers. One well-trained trainer can train a hundred farmers. But a badly trained one can do untold damage.

Supplier Training. If you buy pumps, or computers, or tractors, make sure the supplier provides training for the operators as part of the contract.

What engineering is

Civil Engineers like to see themselves both as guardians and manipulators of nature and its resources. Their Institution's[5] charter, established nearly 200 years ago, grandly talks about 'directing the great sources of power in nature for the use and convenience of man'. A bit bumptious perhaps, in today's climate of awareness, ecology, social side-effects and things green. But nevertheless, there are few engineers who do not derive satisfaction from seeing a job well done, that is to say a project which fulfils the needs of the people it is intended to serve.

---

5 That is, the British Institution of Civil Engineers, founded in 1818.

When engineers talk about 'The Engineer', they usually mean one of their number who happens to be in the unenviable position of refereeing the project and taking responsibility for its successful completion. Quite often 'The Engineer' is also a consultant firm, in between the contractor and the client. That is how consultants really began. But both the contractor and the client will also have their own engineers. It takes one to know one.

In brief, 'Engineering', from a consultant's viewpoint, sets about the realisation of a project in this way:

The Planning Stage

Planning, survey, field investigation, feasibility studies. This stage, which may take several years for a big project, is designed to examine all relevant options and forms that a project may take, and to set the project off on the right course. A good planning engineer has to be broad-minded enough to understand other viewpoints and visionary enough to relate all facets of a project to the whole.

The Design Stage

Outline design, cost estimates, detailed design. To ensure that the project will perform its functions in the intended way, and endure over its intended life-span, whilst complying with any other imposed constraints such as cost, aesthetics, or avoidance of bad ecological side effects. A good design engineer has to be both technically proficient and up-to-date with modern methods and technology.

The Contract Stage

Tender documents, specification, conditions of contract, bills of quantity, tender evaluation, selection of contractors. These procedures are intended to safeguard all parties in ensuring fair and satisfactory construction. A good contracts engineer needs a mind like an encyclopaedia and an eye for the unexpected detail.

The Construction Stage

Site supervision and administration of payments, quality control, project management. This critical stage enforces the contractual procedures laid down previously. Any weakness here can lead to calamitous breakdown of the project. There are conflicting interests between those parties who want to make money and those who want to see the project completed as cheaply as possible. The engineer falls in between. A good construction engineer has to be tough, fair, and single-minded enough to see the project through. Perhaps it is this necessary single-mindedness in the end that gives engineers a reputation for inflexibility.

This, then, is the essence of an engineer's job. Not many people know that.

References and further reading for chapter 11

(1) British Standard 8000, 'Workmanship on Building Sites', 1990.

(2) Horst, L., 'Interactions between Technical Infrastructure and Management', ODI/IIMI Irrigation Management Network Paper 90/3b, 1990.

(3) 'Canal Lining Poses Questions', World Water, April 1989.

(4) Goldsmith and Makin, 'Canal Lining: From the Laboratory to the field and back again', Proc. Intl. Conf. on Irrigation, Southampton University, Sept 1989.

(5) Diamant, R.M.E., 'Chemistry for Engineers', Pitman, 1962.

CHAPTER 12   COSTS AND ECONOMICS

Since this is not a textbook on economic theory it does not include even a précis version of material that can be found in numerous excellent books on that subject. However, it is important to draw attention to some contentious issues involving cost, which are often ignored or approached in the wrong way. Excessive capital cost is a common excuse for bad engineering, but cheapness is not everything. Often a low capital cost belies the vast expense and economic loss incurred later when the scheme does not work. Here are a few examples which would have benefited from some lateral thinking at the planning and economic evaluation stage.

12.1 Small Scale Irrigation Projects - the Hidden Costs

This section is derived from rehabilitation work on small irrigation projects in India. The projects were typical of countless others which had failed from the outset through poor engineering. The cost of engineering failure is assessed through its various manifestations of lost crop production, increased maintenance, rehabilitation, design errors, social conflict, lost revenue and slack supervision. It is argued that in order to prevent such failures it is necessary to spend a larger-than-normal percentage of the construction costs on engineering design, supervision and planning. In the cases examined here, fault could be attributed to the project instigators, social scientists who had little appreciation of the value or necessity of engineering procedures, and consequently did not bring in suitably qualified engineers until it was too late. Although it is fashionable to blame the engineers for disasters like these, it is all too easy to do so because only engineers leave tangible evidence behind them. It seems that engineers, for too long *'les betes noirs'* of development projects, are misunderstood by other professions more now than at any time in the past. The case for engineering, its aims and procedures are therefore restated here in a bid to foster closer and more beneficial links with everyone else concerned with irrigation projects.

Cheap schemes - a myth exploded

During the 1980s small-scale irrigation schemes became a popular feature of many third world development projects. Prompted by numerous well-documented failures of large schemes, governments and aid agencies began to regard small-scale schemes as economically less risky, less disruptive on local people and offering a short implementation time and hence more rapid returns on investment. Remote rural regions could be developed without huge financial commitments to new infrastructure. Farmer participation, on-farm water management, were the watchwords. Get the farmers interested in their scheme, and success would be assured.

On paper these are valid arguments. Coupled with relatively cheap construction costs due to the avoidance of large and expensive dams, pumping installations, barrages or infrastructure, small scale schemes (here meaning between 20 and 500 ha in size) appeared cheap indeed.

Often, since the design of small schemes is usually straightforward, the use of local unskilled labour is promoted for construction, thus reducing the costs even further, and generating work for the rural unemployed. Local construction methods such as the use of stone masonry for hydraulic structures and canals are adopted on the premise that they are already understood by local artisans.

But because the schemes are relatively simple, no-one pays much attention to planning, design or supervision. And too often the construction methods used are entirely unsuitable for the job. The end result is a cheap scheme. But low construction cost. low survey cost, low planning cost, and low supervision cost, leads to low quality, low success rate, and a high failure rate.

And so, when the canals start to disintegrate after 1 or 2 years, the scheme does not appear so cheap after all. In order to keep it going more and more money has to be spent each year on rehabilitation and maintenance. In some countries this is seen as a good thing by the engineers, because it keeps them in work. And very often their schemes are deliberately designed wrongly and constructed badly, in order to hasten the disintegration process. But there are several levels of engineering, and it is unfortunate that engineers in general should be damned by the deeds of the few.

12.2 Counting the Cost of Failure - the cost of bad quality

The cost of a cheap scheme that does not work can be enormous. An example of four lift irrigation schemes in southern India is given here. All were constructed in the early 1980s, under a foreign aid programme.

None of them worked properly, and by 1986 most of them were completely out of use due to physical disintegration of the system or incorrect initial construction. All were rehabilitated fully by 1990, but only at considerable expense. Their projected financial cash flows from initial construction, through rehabilitation and for a future period of 25 years, showed a continuing decline over the first decade followed by a long and slow recovery period. The worst schemes have overall financial internal rates of return close to zero, and none are financially viable.

The cost can be counted in terms of crop production foregone, in social conflict arising from water disputes, in maintenance effort required (which may be so great as to be beyond the capability of the responsible agency), in lost revenue as the farmers refuse to pay water taxes, and in the results of poor design.

Lost crop production

The schemes as planned were to have trebled crop production, not only by allowing three crops per year to be grown instead of one, but also by guaranteeing water in time to ensure that correct planting dates could be met, thereby making optimum use of rainfall and other climatic conditions for each crop. Climatic conditions in this part of coastal Karnataka are characterised by a short but intensive Kharif (wet season, having 3000 mm of rain between June and September), scattered rainfall during November, followed by a 6 month dry season in Rabi and summer, which starts cool and dry but becomes increasingly hot and humid by April-May.

To take full advantage of these conditions, high yielding rice is grown between June and October, conditions being too wet for any other field crop. This is followed by groundnuts or rice, depending on soil conditions, from October to January, and pulses or vegetables from February to April. With irrigation, rice nurseries can be sown from mid-May and three crops are possible.

A cropping pattern such as this is dependent on timely planting. If irrigation is unreliable, farmers often delay planting until the rains are well established in mid-July. This means that rice harvesting is delayed until November, and due to the likelihood of rain at that time many farmers prefer to plant a long duration local variety that is then harvested in December or January. By this time it is too late to utilise any residual moisture for a pulse crop, so nobody bothers.

So for irrigation to be effectively utilised, the farmer must be able to rely on the irrigation system 100 per cent. If there is any risk of crop failure due to breakdown of the irrigation system, the farmer will choose the option that his father knew, a single long duration rice crop, of a local variety that will tolerate some dry spells.

None of these schemes initially instilled any confidence amongst the farmers. The net farm-gate incremental value of crop production foregone was estimated at USD625 / ha / year .

Maintenance effort

The proposed extent of government involvement in maintenance was restricted to running the electric pumps and minor concreting or welding jobs on the main canals and structures. Routine desilting and cleaning of minor canals was to have been the responsibility of farmers.

However, the durability of all works was so poor that in order to keep the canal system fully operational it would have been necessary to reconstruct everything after 2 years! Canal lining disintegrated almost as soon as it was laid due to poor quality concrete, materials and methods of construction. Pumps were already second-hand when they were installed, and none were adequately protected against fluctuations in power supply. Motor burn-outs were frequent as a result.

Canal levels were often incorrect. Hard soil or rock encountered during construction often resulted in canals running uphill. Careless setting-out had left many canals without adequate carrying capacity. Minor structures such as drops or turnouts leaked so badly as to be rendered useless. Pipe structures such as aqueducts and inverted siphons were jointed so badly that in some cases leakage losses were 100 per cent.

The government's maintenance cost was supposed to be met through a water charge of around USD 10 per hectare per crop. However,, no farmer would contemplate payment unless the water supply was reliable. So because the supply was unreliable from the start, no money or maintenance effort was ever contributed by the farmers and little or no maintenance was ever done by the government. This vicious circle can only be broken by means of a complete rehabilitation, or else a change in funding policy.

It is therefore misleading to consider the cost of maintenance if its quality is no better than that of the original construction. Money could have been poured into faulty maintenance in the same way as it had been wasted in faulty construction. The fact that it was not was due more to a complete lack of caring rather than any awareness of the futility of spending money on perpetuating a faulty irrigation system. In all these schemes it can be said that the maintenance requirement was so great as to be beyond the capability of the government agency responsible. In the event a full rehabilitation was carried out. The cost amounted to USD1400 / ha .

Social conflict

The social problems arising from a badly executed scheme can be long-lasting and divisive to the community. Any imbalance or inequity of water distribution along the system leads to conflict. Whatever the water management philosophy in use, if water is in short supply the head-end farmers commandeer it. On these schemes, the tail-end problem began in the upper reaches of the canal system.

Tail-end farmers deprived of water and jealous of their neighbours deliberately damaged turnout gates and structures. Regulators and measuring flumes, perceived as obstructions to water flow, were destroyed. Herds of buffalo were made to rampage over the fields of head-end farmers. Ancient family feuds were re-awakened.

Many farmers had not been compensated for their land taken over for useless irrigation works. Government staff and consultants alike were pelted with stones. On larger schemes which served more than one village, old animosities and rivalries resurfaced, and physical violence was not uncommon.

How do we assess the cost of all this? Such communities rely on mutual co-operation for their existence. When levels of co-operation diminish, it must cost something. What is the cost of repairing damaged fields, of mounting guard to keep out your neighbours' buffalo, of not having your neighbour assist with harvesting, of having your water stolen? Any attempt to cost all this must be very subjective. So, a subjective estimate would be say USD50 / ha / year.

Lost Revenue

Water charges in this situation only work if the scheme works well, but otherwise encourage wastage and abuse. These schemes were intended to collect around USD 30 / ha / year from grateful farmers, to be spent on power costs and maintenance. When farmers' confidence in the schemes evaporated as quickly as the trickle of water in the canals, so did any chance of revenue. <u>Cost to the Government = USD 30 / ha / year.</u>

The cost of poor design

Poor Design on these schemes led to excessive power cost because pump delivery chambers were wrongly sited at too high an elevation. They also incorporated unnecessarily large chambers which created excessive head loss through turbulence and free air discharge.

Unduly expensive canals were routed through steeply undulating terrain when alternative cheaper routes were possible. Canals were invariably two or three times as large as they needed to be on hydraulic capacity considerations alone. Much time and money was wasted in building things that were not necessary.

Canal lining materials were inappropriate. Laterite blocks or mud mortar were specified for watertight lining, but neither of these is either durable or watertight. Stone masonry or stone slabs were specified on some schemes, but these are expensive and easily destroyed by root action and swelling soils. In all cases the philosophy of design had been to utilise local materials with which the semi-skilled local artisans were already familiar. These materials are adequate for building small houses or garden walls, but not for hydraulic structures. Estimated cost = USD510 / ha .

The cost of bad supervision

Bad supervision can mean either inadequate staffing levels or incompetent supervisors, or both. In this example there was no engineering supervision at all, and the construction contractor was allowed to do whatever he wanted. In other instances supervisors themselves might have vested interests in poor construction. Either way, the result is deficient work in time-honoured fashion such as concrete with too little cement in it, reinforcing steel left out, plastering only millimetres thick instead of centimetres, no earth compaction, no concrete vibration or curing, and a host of others. The cost of rectifying all this was USD1400 / ha .

The real cost

If all these costs are combined and assuming a 10 per cent discount factor we get an overall present value of about USD10,000 per Hectare, excluding the initial cost:

	USD/Ha
maintenance	1,400
bad supervision	1,400
poor design	510
social	500
crops foregone	6,250
lost revenue	(30 but assume in crops foregone)
initial construction cost	(2,045 but leave out of this)
total	**10,060**

12.3 The Cost of Good Supervision

All of the above problems could have been avoided with good construction and establishment supervision. However,, good supervision is not, on the face of it, cheap.

Supervision cost

The average construction costs of these schemes (adjusted to 1990 prices) were as follows:

	USD/Ha
Original construction cost	2,125
Rehabilitation construction cost	786
Total	**2,911**

But this is not all. The supervision costs were significant:

Supervision and administration before rehabilitation	1,580
Supervision and administration during rehabilitation	2,055
Total	**3,635**

So the supervision cost was more than the construction cost alone, and in the rehabilitation stage it was more than double the capital cost. This included expatriate advisor time, not only for construction supervision but also for planning and design, and a continuing but futile presence of agricultural staff during the first decade when the scheme was not working. But even a more pragmatic analysis, based on the actual experience of rehabilitation supervision, suggested that in order to have constructed the schemes properly in the first place it would have been necessary to spend more than 50 per cent of the construction cost on supervision, planning and design (i.e. on 'Engineering'.)

Supervision cost in perspective

This would put a real anticipated cost for these schemes at:

	USD/Ha
Construction	2,800
Supervision	1,400
total	**4,200**

which is not cheap, but realistic nevertheless.

Compare this with the actual cost incurred without supervision, of USD10,060 /ha. What these figures suggest is that it is necessary to spend another USD1,400 or 14 per cent on supervision in order to save USD6,000 or 60 per cent on later rectification, not to mention all the political and public embarrassments that often go with failure.

Does the supervision cost have to be so high? And is there really no way of having a cheap scheme that works?

Firstly, the cost of good supervision only looks high because it is relative to a low construction cost. Herein lies one of the hidden costs of small schemes. By nature, schemes of this size do not involve large quantities of concrete or other construction inputs, but they still require the same supervision effort as would larger projects costing several times as much. Furthermore, the rehabilitation work often entails extensive survey, redesign and site checking for only a small amount of construction work.

Secondly, good supervision may cost money, but the civil engineering profession has over the past 2 centuries developed its own codes of practice, ethics and procedures designed to forestall all of these problems. Employing reputable engineers is the only way of guaranteeing success.

The schemes mentioned here stood little chance of success, because they were administered by social scientists intent on social engineering, and who did not understand the value or roles of the civil engineer.

12.4 Spending Money on Simplicity – the Cost of Automation

In Chapter 4, various philosophies and methods of automation were looked at. The advantages of automation are clear, less management effort, for more reliable water supplies. Some forms of passive automation can be designed-in to a scheme for no cost at all, provided that management and operation of the scheme is considered from the outset. The self-regulating gates on the Pehur High Level Canal cost some 20 per cent more than conventional radials, but this amounted to less than 5 per cent of the total project cost, and their effect on the project economics was positive, since they required far less management cost, would reduce wastage of water, would remove the tail-end problem, and hence bring more land into useful production. The social benefits are always difficult to quantify in monetary terms, but they always have a value, in reducing social tensions, and in freeing up peoples' time. Time previously spent on manual labour can consequently be expended more profitably in education, commerce and social interaction of many kinds.

12.5 Spending Money on Posterity – the Cost of Parabolics

Precast parabolic canals might at first sight appear more expensive than brickwork, but if they last considerably longer then they are effectively cheaper. In Indonesia the ubiquitous stone masonry was twice the capital cost of alternative precast parabolics and had one-twentieth the life expectancy, but nevertheless it was still used for all the wrong reasons. In Pakistan, precast parabolics were cheaper than conventional brick watercourses, and the irrigation authorities' resistance to their use was based on a web of excuses for not wishing to relinquish the lucrative kickbacks from repeated maintenance contracts. In the end it was the farmers who brought market pressure to bear by insisting on a superior product for a cheaper price.

12.6 Spending Money on Flexibility – the Cost of Pipelines

The arguments for flexibility presented in Chapter 3 ought to be convincing enough, but too often the question of construction cost is brought to the fore at an early stage of planning, and would-be pipeline schemes get delayed for half a century for want of a better understanding of economic principles. Flexibility is an essential thing to have, but what is the cost of it?

Pipelines are often more expensive to construct than an equivalent open canal. They are not suited to situations with very flat topography which are short of available head, and they cannot economically be used as an alternative to canals carrying more than about 1 cumec. They can often be economically used at levels 3 and 4 in the distribution system, in conjunction with higher capacity open canals.

Extra flexibility can be got by increasing the pipe diameter, thereby reducing operating head losses and increasing intermediate storage. Doubling the capacity of a pipeline increases pipe costs by 10 per cent to 15 per cent. Pipe costs often are about 1/5 to 1/4 total project costs, giving an increase of perhaps 2 per cent to 5 per cent in net annual costs attributable to the farmer as water charges.

I would like to end this book with a few words by Professor John Merriam, who died in February 2007 at the age of 96 and after a lifetime of promoting flexible irrigation through the use of low-pressure pipelines:

'A system that is initially too small cannot be economically enlarged later. Future labour cost, crop values, and water value, will all increase with time. ***Don't limit the future by what is built now.***'

References and further reading for Chapter 12

1) Gittinger, J. Price - 'Economic Analysis of Agricultural Projects' EDI World Bank, 1972

INDEX